Hazardous Area Classification in
PETROLEUM AND CHEMICAL PLANTS
A Guide to Mitigating Risk

Hazardous Area Classification in
PETROLEUM AND CHEMICAL PLANTS

A Guide to Mitigating Risk

Alireza Bahadori

CRC Press
Taylor & Francis Group
Boca Raton London New York

CRC Press is an imprint of the
Taylor & Francis Group, an **informa** business

CRC Press
Taylor & Francis Group
6000 Broken Sound Parkway NW, Suite 300
Boca Raton, FL 33487-2742

First issued in paperback 2017

© 2014 by Taylor & Francis Group, LLC
CRC Press is an imprint of Taylor & Francis Group, an Informa business

No claim to original U.S. Government works

Version Date: 20131021

ISBN 13: 978-1-4822-0645-6 (hbk)
ISBN 13: 978-1-138-07464-4 (pbk)

Visit the Taylor & Francis Web site at
http://www.taylorandfrancis.com

and the CRC Press Web site at
http://www.crcpress.com

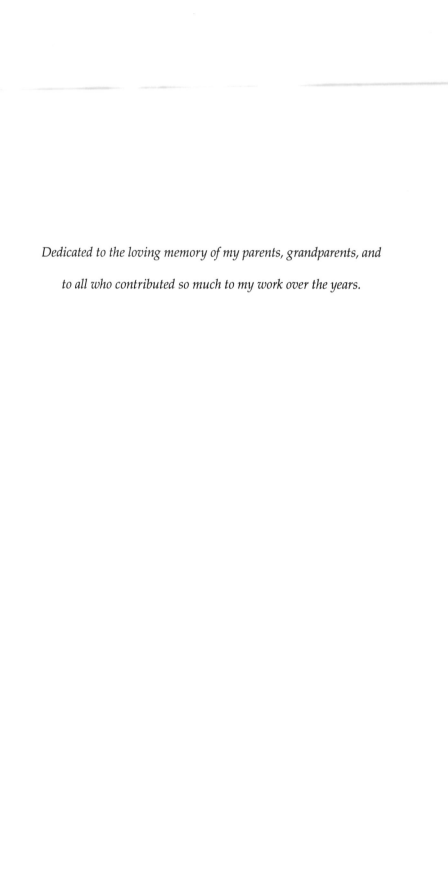

Dedicated to the loving memory of my parents, grandparents, and

to all who contributed so much to my work over the years.

Contents

Preface

The increasing use of a wide variety of chemicals, many of which introduce problems other than flammability, led to the need for a simple hazard identification system. The purpose of such a system would be to safeguard the lives of those individuals who may be concerned with fires occurring in an industrial plant or storage location.

In plants where flammable gases, vapors, liquids, or dusts are present, a flammable atmosphere may be formed if they are released. The flammable atmosphere may also exist inside plant equipment if air or oxygen is present together with a flammable material.

Hazardous locations can also be described as those locations where electrical equipment might be installed and which, by their nature, might present a condition that could become explosive if the elements for ignition are present. Unfortunately, flammable substances are not always avoidable (e.g., methane and coal dust in mines). Therefore, it is of great importance that a user of electrical equipment, such as push buttons and pilot lights, be aware of the environment in which these products will be installed. The user's understanding of the hazard will help ensure that the electrical equipment is properly selected, installed, and operated to provide a safe operating system.

The primary step in recognition of danger is the classification of a plant into zones in which the probability of the existence of a flammable atmosphere is broadly assessed. This procedure is known as area classification.

This book provides simple, readily recognizable, and easily understood markings that will give, at a glance, a general idea of the inherent hazards of any material and the order of severity of these hazards as they relate to fire prevention, exposure, and control in chemical- and petroleum-processing industries.

Its objectives are to provide an appropriate alerting signal and on-the-spot information to safeguard the lives of both public and fire-fighting personnel during fire emergencies. It will also assist in planning for effective fire-fighting operations and may be used by plant design engineers and plant protection and safety personnel.

This book also covers the general features and rules and minimum requirements for extinguishing fires in the related industries. Fire hazards, occupancy hazards, extinguishing methods, hazard identifications and classification, and so forth, are also discussed.

By using this book, an area classification map could be sketched for each plant. It is intended to be applied in oil industries, where there may be a risk due to the presence of flammable gas or vapor mixed with air under normal atmospheric conditions. Therefore, this book covers the following areas:

- Petroleum refineries
- Petroleum and gas pipeline transportation facilities
- Natural gas liquid-processing plants
- Drilling rigs, production facilities on land, and marine fixed or mobile platforms
- Chemical process areas

Area classification methods provide a succinct description of the hazardous material that may be present and the probability that it is present, so that the appropriate equipment may be selected and safe installation practices may be followed. It is intended that each room, section, or area of a facility should be considered individually in determining its classification. Actually determining the classification of a specific location requires a thorough understanding of the particular site.

An exhaustive study of the site must be undertaken before a decision can be made as to what class, zone, and group is to be assigned. The local inspection authority has the responsibility for defining a class, zone, and group classification for specific areas.

This book provides a brief overview of the essential aspects of explosion protection. Ultimately, safety in a potentially explosive atmosphere is a team effort. Manufacturers have a responsibility to ensure only safe equipment is placed on the market. Installers must follow the instructions provided and use the equipment only for its intended purpose.

Finally, the user has a duty to inspect and maintain the equipment in a safe working order. The directives and national and international standards provide a basis for a safer future.

Last but not least I would like to thank the CRC Press editorial and production team, Allison Shatkin and Kate Gallo, for their editorial assistance.

Author

Alireza Bahadori, PhD, is a research staff member in the School of Environment, Science & Engineering, at Southern Cross University, Lismore, NSW, Australia. He received his PhD from Curtin University, Western Australia. For the better part of 20 years, Dr. Bahadori had held various process engineering positions and was involved in many large-scale projects at NIOC, Petroleum Development Oman (PDO), and Clough AMEC PTY LTD.

He is the author of over 200 articles and six books. His books have been accepted/published by prestigious publishers such as John Wiley & Sons, Springer, Taylor & Francis, and Elsevier. Dr. Bahadori is the recipient of highly competitive and prestigious Australian Government's Endeavour International Postgraduate Research Award as part of his research in the oil and gas area. He also received a top-up award from State Government of Western Australia through Western Australia Energy Research Alliance (WA:ERA) in 2009. Dr. Bahadori serves as a member of the editorial board for a number of journals such as *Journal of Sustainable Energy Engineering*, which is published by Wiley-Scrivener.

1

Classification of Fires and Fire Hazard Properties

1.1 Introduction

A major safety concern in industrial plants is the occurrence of fires and explosions. Hazardous locations are areas where flammable liquids, gases, or vapors or combustible dusts exist in sufficient quantities to produce an explosion or fire. In hazardous locations, specially designed equipment and special installation techniques must be used to protect against the explosive and flammable potential of these substances.

Flammable materials are substances that can ignite easily and burn rapidly. Flammable substances can be divided into three subgroups:

- Flammable gas
- Flammable liquids
- Flammable solids

They can be common materials that are at most work sites in gas, liquid, and solid forms. Some examples of flammable materials include the following:

1. *Gases.* Flammable gases are usually gases with a lower explosive limit (LEL) of less than 13% in air or have a flammable range in air of at least 12%. For example, butane is a flammable gas because its LEL in air is 20%. Carbon monoxide has a LEL of 13% and an upper explosive limit (UEL) of 74% in air, and is flammable over a range of 61%. Typical flammable gases are natural gas, propane, butane, methane, acetylene, carbon monoxide, and hydrogen sulfide.

2. *Liquids.* Flammable liquids have a flashpoint below 37.8°C (100°F). Typical flammable liquids are gasoline, many solvents such as acetone, alcohols, and toluene, paints and paint thinners, adhesives, degreasers, cleaners, waxes, and polishes.

3. *Solids.* Some types of coal, pyrophoric metals (metals that burn in contact with air or water, such as sodium and potassium), solid wastes that are soaked with flammable liquids (rags, paper, spill cleanup products), gunpowder, matches, dusts, and ignitable fibers are typical flammable solids.

Table 1.1 provides more information about flammable substances.

A *combustible liquid* is a liquid having a flash point at or above 38°C. Combustible liquids are subdivided as follows:

- Class IA liquids shall include those having flash points at or above 38°C and below 60°C.

- Class IIA liquids shall include those having flash points at or above 60°C and below 93°C.

- Class IIIB liquids shall include those having flash points at or above 93°C.

TABLE 1.1

Flammable Substances Categories

Flammable Substance	Examples	Description
Flammable gas	Hydrogen, etc.	Often compounds of hydrogen and carbon that require very little to react with atmospheric oxygen.
Flammable liquids/ vapors	Hydrocarbons such as ether, acetone, lighter fluids, etc.	Even at room temperature, sufficient quantities of these hydrocarbons can evaporate to form a potentially explosive atmosphere at their surface. Other liquids require higher temperature for this to occur. The flash point of a flammable liquid is the lowest temperature at which a sufficient quantity of vapor will arise to permit ignition under laboratory conditions. This is an important factor in the classification of hazardous areas. Flammable liquids with a high flash point are less hazardous than liquids with a low flash point.
Flammable solids	Dust, fibers, and flyings	The cumulative nature of the dust hazard is the most significant difference between a gas/vapor hazard and the dust hazard. A dust cloud will settle on nearby surfaces if it is not ignited. Unless removed, layers of dust can build up and will serve as fuel for subsequent ignition. The typical dust explosion starts with the ignition of a small dust cloud, resulting in relatively small damages. Pressure waves of the small initial explosion are the most damaging part of the dust explosions. These pressure waves release dust layers from surrounding vertical or horizontal surfaces to produce a larger cloud that is ignited by the burning particles of the initial cloud. In this way, the small initial explosion can produce a much larger explosion. In some cases a series of explosions occur, each stronger than the previous.

A *flammable liquid* is a liquid having a flash point below 38°C and having a vapor pressure not exceeding 275.79 kPa (absolute) (2.068 mm Hg) at 38°C and shall be known as a Class I liquid.

Class I liquids shall be subdivided as follows:

- Class IA liquids shall include those having flash points below 23°C and having a boiling point below 38°C.
- Class IB liquids shall include those having flash points below 23°C and having a boiling point at or above 38°C.
- Class IC liquids shall include those having flash points at or above 23°C and below 38°C.

1.1.1 National Fire Protection Association System for Classification of Hazards

The National Fire Protection Association (NFPA) diamond is a symbol used to identify the hazards associated with a given chemical to rescue workers. Frequently this symbol is found on the sides of buildings where chemicals are stored and on chemical containers. Figure 1.1 shows various hazards

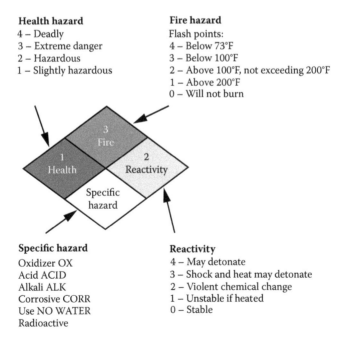

Health hazard
4 – Deadly
3 – Extreme danger
2 – Hazardous
1 – Slightly hazardous

Fire hazard
Flash points:
4 – Below 73°F
3 – Below 100°F
2 – Above 100°F, not exceeding 200°F
1 – Above 200°F
0 – Will not burn

Specific hazard
Oxidizer OX
Acid ACID
Alkali ALK
Corrosive CORR
Use NO WATER
Radioactive

Reactivity
4 – May detonate
3 – Shock and heat may detonate
2 – Violent chemical change
1 – Unstable if heated
0 – Stable

FIGURE 1.1
Various hazards symbolized by the diamond and the numerical code based on NFPA Fire Hazard ratings.

TABLE 1.2

NFPA Fire Hazard Ratings of Some Common Laboratory Chemicals

Chemical Name	NFPA Rating	Flash Point (°C)	Boiling Point (°C)	Ignition Temperature (°C)
Acetaldehyde	4	−37.8	21.1	175
Acetic acid (glacial)	2	39	118	463
Acetone	3	−18	56.07	465
Acetonitrile	3	6	82	524
Carbon disulfide	3	−30	46.1	90
Cyclohexane	3	−20	81.7	245
Diethylamine	3	−23	57	312
Diethyl ether	4	−45	35	160
Dimethyl sulfoxide	1	95	189	215
Ethyl alcohol	3	12.8	78.3	365
Heptane	3	−3.9	98.3	204
Hexane	3	−21.7	68.9	225
Hydrogen	4	−	−252	500
Isopropyl alcohol	3	11.7	82.8	3
Methyl alcohol	3	11.1	64.9	385
Methyl ethyl ketone	3	−6.1	80	515
Pentane	4	−40	36.1	260
Styrene	3	32.2	146.1	490
Tetrahydrofuran	3	−14	66	321
Toluene	3	4.4	110	480
p-Xylene	3	27.2	138.3	530

symbolized by the diamond and the numerical code that indicates the severity of the hazard. Class IA and IB flammable liquids have an NFPA fire hazard rating of 4. Class IC flammable liquids are rated as 3. Combustible liquids are rated as 1 or 2.

NFPA fire hazard ratings of some common laboratory chemicals are listed in Table 1.2.

1.2 Classification of Fires

Fire is the rapid oxidation of any combustible material. It is a chemical reaction involving fuel, heat, and oxygen. These three elements, commonly referred to as the *fire triangle*, in the right proportions will always produce a fire. Remove any one side of the triangle and the fire will be extinguished.

It should be noted that every flammable gas or vapor has specific lower and upper flammability limits. If the substance or concentration in the oxidizer is either below a specific value (lower flammability limit) or above a specific value (upper flammability limit), ignition might occur; however, a flame will not propagate. This can be exploited by diluting the flammable substances with air or preventing the ingress of air/oxygen. The latter option is ruled out in environments where people work regularly and is feasible only in a chemical plant where there are no human beings (Tommasini and Pons 2011).

If a flammable gas or vapor cloud is released and ignited, all the material may be consumed in one explosion. If the flammable gas or vapor cloud is not ignited, convection and diffusion will eventually disperse the flammable cloud, the immediate danger passes, and the particular fuel source is lost (Kantawong 2012).

The most common types of reaction are between flammable gases, vapors, or dust with oxygen contained in the surrounding air.

As a rule, three basic requirements must be met for an explosion to take place in atmospheric air:

1. A *flammable substance* needs to be present in sufficient quantity to produce an ignitable or explosive mixture.
2. An *oxidizer* must be present in sufficient quantity in combination with the flammable substance to produce an explosive mixture. The most common is air (O_2).
3. A *source of ignition*—a spark or high heat—must be present.

Figure 1.2 shows the fire triangle. It is commonly used as a model to understand how a fire starts and how it can be prevented. The presence of these three elements makes up the sides of the ignition triangle. If any one of the three elements is missing, an explosion will not occur. All three elements must exist simultaneously for an explosion to occur.

Scholars have also introduced a fourth element in the equation, known as the uninhibited chain reaction, thereby giving the fire chemical reaction an additional side. This is referred to as the fire tetrahedron.

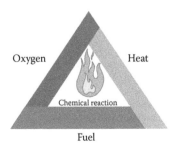

FIGURE 1.2
Fire triangle.

1.2.1 Oxidizer

The oxidizer referred to in all common hazardous location standards and explosion-proof equipment is air at normal atmospheric conditions. The oxygen in the air is only enough for the combustion of a certain quantity of flammable material. Air must be present in sufficient volume to propagate a flame before the air–fuel mixture becomes a hazard.

When the amount of available atmospheric oxygen is more or less in equilibrium with the quantity of flammable material, the effect of an explosion—both temperature and pressure—is most violent. If the quantity of flammable material is too small, combustion will spread with difficulty or cease altogether. The same applies if the quantity of flammable material is too great for the available oxygen.

The presence of an oxygen-enriched atmosphere or a pressurized enclosure alters the conditions for ignition and dictates the use of special means for prevention and containment of explosions. No means of explosion protection considered safe for atmospheric mixtures should be used in either oxygen-enriched or pressurized situations without careful study (Schröder and Molnarne 2005).

1.2.2 Ignition Source

The amount of energy required to cause ignition is dependent upon these factors:

- The concentration of the hazardous substance within its specific flammability limits
- The explosive characteristics of the particular hazardous substance
- The volume of the location in which the hazardous substance is present

Further classification of sources of ignition in industrial electrical equipment is detailed in Table 1.3.

TABLE 1.3

Classification of Ignition Source

Ignition Sources (Industrial Electrical Equipment)	Examples
Hot surfaces	Surfaces heated by coils, resistors, lamps, brakes, or hot bearings. Hot surface ignition can occur at the autoignition temperature (AIT) or spontaneous ignition temperature at which a hazardous substance will spontaneously ignite without further energy.
Electrical sparks	Occur when circuits are broken or static discharge takes place. In low-voltage circuits, arcs are often created through the making and breaking of electrical contacts.
Friction and impact sparks	When casings or enclosures are struck.

1.2.3 Designation of Classes of Fires

The classification of fire depends mainly upon the fuel involved. There are five classes of fire. The following designations shall be used for the purpose of classifying fires of different natures in order to simplify spoken and written reference to them.

1.2.3.1 Class A

Class A fires are fuelled by ordinary combustible materials, such as wood, cloth, paper, and many plastics, and many other combustible materials are some examples. This type of fire burns with an ember, leaves an ash, and is best extinguished by removing the heat side of the triangle. Extinguishers suitable for Class A fires should be identified by a triangle containing the letter A (Figure 1.3); if color-coded, the triangle will be green. These are fires involving solid materials, usually of organic nature in which combustion normally takes place with the formation of glowing embers.

1.2.3.2 Class B

Class B fires involve liquids or liquefiable solids that are flammable or combustible, including gases or greases. These fires are fuelled by flammable liquids, combustible liquids, petroleum greases, tars, oils, oil-based paints, solvents, lacquers, alcohols, and flammable gases. This type of fire burns on the surface of the fuels and is best extinguished by a blanketing or smothering action. A fire of this type is fast-spreading and capable of engulfing a large area in a very short time. Extinguishers suitable for Class B fires should be identified by a square containing the letter B (Figure 1.4). If color-coded, the square is red.

FIGURE 1.3
Fire class A.

FIGURE 1.4
Fire class B.

1.2.3.3 Class C

Class C fires occur in energized electrical equipment, where the electrical nonconductivity of the extinguishing media is of importance. Blanketing or smothering this type of fire with a nonconducting extinguishing agent is of prime importance. Water or solutions containing water should never be used on a Class C fire. Extinguishers suitable for Class C fires should be identified by a circle containing the letter C (Figure 1.5); if color-coded, the circle is blue.

In brief, these are fires involving gases or materials in contact with energized power such as motors and switch gears. When electrical equipment is de-energized the fire may continue to burn as a Class A, B, or D fire.

1.2.3.4 Class D

Class D fires involve combustible metals, such as magnesium, titanium, zirconium, sodium, lithium, and potassium. Generally the extinguishing agent is referred to as *dry powder*. These extinguishers should be identified by a star containing the letter D (Figure 1.6); if color-coded, the star is yellow.

FIGURE 1.5
Fire class C.

FIGURE 1.6
Fire class D.

FIGURE 1.7
Fire Class K.

1.2.3.5 *Class K*

Class K fires are fires in cooking appliances that involve combustible cooking media such as vegetable or animal oils and fats. The extinguishing agent is referred to as *wet chemical*. These extinguishers should be identified by the letter K (Figure 1.7).

1.3 Fire Hazard Properties

A single fire-hazard property such as flash point or ignition temperature should not be used to describe or appraise the hazard or fire risk of a material, product, assembly, or system under actual fire conditions. The subject fire-hazard properties have been determined under controlled laboratory conditions and may properly be employed to measure or describe the response of materials, products, assemblies, or systems under these conditions (Shen et al. 2009).

Properties measured under such conditions may be used as elements of a fire risk assessment only when such assessment takes into account all of the factors that are pertinent to the evaluation of the fire hazard of a given situation. Properties of the flammable materials are generally for materials in the pure form and may be different if there are impurities or where there are mixtures of materials.

1.3.1 Flash Point

Flash point of the liquid is the minimum temperature at which it gives off sufficient vapor to form an ignitable mixture with the air near the surface of the liquid within the vessel used. An "ignitable mixture" is a mixture within the flammable range (between upper and lower limits) that is capable of the propagation of flame away from the source of ignition when ignited (Hernandez et al. 1995).

Propagation of flame is the spread of flame from the source of ignition through a flammable mixture. A gas or vapor mixed with air in proportions below the lower limit of flammability may burn at the source of ignition, that is, in the zone immediately surrounding the source of ignition, without propagating (spreading) away from the source of ignition. However, if the mixture is within the flammable range, the flame will spread through it when a source of ignition is supplied. The use of the term *flame propagation* is therefore convenient to distinguish between combustion, which takes place only at the source of ignition and travels (propagates) through the mixture.

Some evaporation takes place below the flash point but not in sufficient quantities to form an ignitable mixture. This term applies mostly to flammable and combustible liquids, although there are certain solids, such as camphor and naphthalene, that slowly evaporate or volatilize at ordinary room temperature, or liquids such as benzene that freeze at relatively high temperatures (5.5°C) and therefore have flash points while in the solid state.

The test apparatus used for the measurement of flash point is normally one of two types, of which there are several variants. These are called generally open cup and closed cup flash point testers. For most liquids the flash point determined by the closed cup method is slightly lower (in the region of 5%–10% when measured in degrees Celsius) than that determined by the open cup method.

Figures 1.8 and 1.9 illustrate a closed cup flash point tester and a Cleveland open cup flash point tester, respectively.

FIGURE 1.8
A closed cup flash point tester. (Reprinted with permission from Utherm Group.)

FIGURE 1.9
Cleveland open cup flash point tester. (Reprinted with permission from Clarkson Laboratory and Supply Inc.)

1.3.2 Ignition Temperature

Ignition temperature of a substance, whether solid, liquid, or gaseous, is the minimum temperature required to initiate or cause self-sustained combustion independently of the heating or heated element.

Ignition temperatures observed under one set of conditions may be changed substantially by a change of conditions. For this reason, ignition temperatures should be looked upon only as approximations.

Some of the variables known to affect ignition temperatures are percentage composition of the vapor or gas–air mixture, shape and size of the space where the ignition occurs, rate and duration of heating, kind and temperature of the ignition source, catalytic or other effect of materials that may be present, and oxygen concentration (Berkowitz et al. 2004).

As there are many differences in ignition temperature test methods, such as size and shape of containers, method of heating, and ignition source, it is not surprising that ignition temperatures are affected by the test method.

Ignition temperature of a substance, whether solid, liquid, or gaseous, is the minimum temperature required to initiate or cause self-sustained combustion in the absence of any source of ignition.

Ignition temperatures observed under one set of conditions may be changed substantially by a change of conditions. For this reason, variables known to affect ignition temperatures are percentage composition of the vapor or gas–air mixture, shape and size of the space in which the ignition occurs, rate and duration of heating, and type and reactivity of other materials present in the space in which the ignition occurs.

As there are many differences in ignition temperature test methods, such as size and shape of ignition chambers, composition of ignition chambers, method of heating, rate of heating, residence time, and method of flame detection, it is not surprising that reported ignition temperatures are affected by the test methods employed (Heinemann and Hoef 2008).

1.3.2.1 Flammable (Explosive) Limits

All combustible gases and vapors are characterized by flammable limits between which the gas or vapor mixed with air is capable of sustaining the propagation of flame.

The limits are called the lower flammable limit (LFL), LEL, upper flammable limit (UFL), and UEL, and are usually expressed as percentages of the material mixed with air by volume. Figure 1.10 shows the LEL and UEL for gasoline.

Where flammability is presented for materials with flash points above 40°C, the determinations have been made at an elevated temperature sufficient to give the quoted concentration.

In popular terms, a mixture below the LFL is too lean to burn or explode and a mixture above the UFL is too rich to burn or explode.

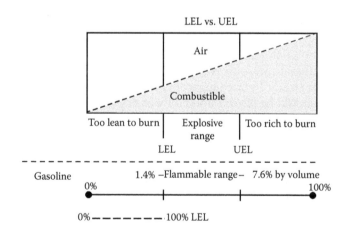

FIGURE 1.10
LEL and upper UEL for gasoline.

Research has shown that the limits of flammability are not a fundamental combustion parameter but are dependent on many variables that include, in part, the surface-to-volume ratio of reaction chamber, the flow direction, and velocity. In some experiments it appears that with laminar flow, the upper limit increases with increasing velocity, reaches a maximum value independent of the tube diameter, and then decreases with turbulent flow. In contrast, the lower limit has been unaffected by flow rate.

ASTM (American Society for Testing and Materials) E681 is an existing standard method for determination of flammability limits. Gases and vapors may form flammable mixtures in atmospheres other than air or oxygen, as for example, hydrogen in chlorine.

1.3.2.2 Flammable (Explosive) Range

The range of flammable vapor or gas air mixture between the UFL and LFL is known as the flammability range, also often referred to as the *explosive range*. For example, the lower limit of flammability of butane at ordinary ambient temperatures is approximately 1.6% vapor in air by volume, while the upper limit of flammability is about 8.4%. All concentrations by volume of butane vapor in air falling between 1.6% and 8.4% are in the flammable or explosive range.

No attempt is made to differentiate between the terms "flammable" and "explosive" as applied to the lower and upper limits of flammability.

Gas–air mixtures outside this range are nonflammable under normal atmospheric condition. Concentration above the UFL in free atmospheric conditions cannot be controlled and further dilution with air will produce mixtures within the flammable range (Dennis 1998).

1.3.3 Specific Gravity

The specific gravity of a substance is the ratio of the weight of the substance to the weight of the same volume of another substance. (Temperature affects the volume of liquids, and temperature and pressure affect the volume of gases. It is therefore necessary to make corrections for effects of temperature and pressure when making accurate specific gravity determinations.)

Specific gravity, as commonly used and as used in Table 1.3, refers to the ratio of the weight of a substance to the weight of an equal volume of water.

The figures given in Table 1.3 for specific gravity are rounded off to the nearest tenth. For materials having a specific gravity from 0.95 to 1.0, the information is given as 1.0. For those materials with a specific gravity of 1.0 to 1.05, the information is given as 1.0+.

In a few cases, such as the fuel oils, where percentage composition of the substance varies, specific gravity information is given as greater than 1 (>1) or less than 1 (<1).

1.3.4 Relative Vapor Density

The relative vapor density of a material is the mass of a given volume of the material in its gaseous or vapor form compared with the mass of an equal volume of dry air at the same temperature and pressure. It is often calculated as the ratio of the relative molecular mass of the material to the average relative molecular mass of air (the value of the latter being approximately 29).

1.3.5 Boiling Point

The boiling point of a liquid is the temperature of a liquid at which the vapor pressure of the liquid equals the atmospheric pressure. Therefore, the lower the boiling point is, the more volatile and generally the more hazardous the flammable liquid becomes.

Where an accurate boiling point is unavailable for the material in question or for mixtures that do not have a constant boiling point, for purposes of this classification the 10% point of a distillation performed in accordance with ASTM D86 (Standard Method of Test for Distillation of Petroleum Products) may be used as the boiling point of the liquid.

1.3.6 Melting Point

Melting point is the temperature at which a solid of a pure substance changes to a liquid.

1.3.7 Boil-Over

Boil-over is an event in the burning of certain oils in an open top tank when, after a long period of quiescent burning, there is a sudden increase in fire intensity associated with expulsion of burning oil from the tank. Boil-over occurs when the residues from surface burning become more dense than the unburned oil and sink below the surface to form a hot layer, which progresses downward much faster than the regression of the liquid surface. When this hot layer, called a *heat wave*, reaches water or water in oil emulsion in the bottom of the tank, the water is first superheated and subsequently boils almost explosively, overflowing the tank. Oils subject to boil-over consist of components having a wide range of boiling points, including both light ends and viscous residues. These characteristics are present in most crude oils and can be produced in synthetic mixtures.

It should be noted that a boil-over is an entirely different phenomenon from a slop-over or froth-over. Slop-over involves a minor frothing that occurs when water is sprayed onto the hot surface of a burning oil. Froth-over is not associated with a fire but results when water is present or enters a tank containing hot viscous oil. Upon mixing, the sudden conversion of water to steam causes a portion of the tank contents to overflow.

1.3.8 Water Solubility of Flammable Liquid

Information of the degree to which a flammable liquid is soluble in water is useful in determining effective extinguishing agents and methods. Alcohol-resistant type foam, for example, is usually recommended for water-soluble flammable liquids. Water-soluble flammable liquids may also be extinguished by dilution, although this method is not commonly used because of the amount of water required to make most liquids nonflammable

TABLE 1.4

Properties of Various Chemical Components

Product	Flash Point (°C)	AIT (°C)	Lower Explosive Limits in Air% Vol	Upper Explosive Limits in Air% Vol	Miscible with Water?	Vapor Density (Air = 1)
Products with a Flash Point of 61°C or Less						
Acetone	−18	465	2.6	13	Yes	2
Acetonitrile	6	524	4.4	16	No	1.4
Benzene	−11	498	1.3	7.9	No	2.7
Butyl acetate	27	370	1.7	7.6	No	4
Cyclohexane	−20	245	1.3	8.4	No	2.9
Cyclopentane	−37	380	1.1	8.7	No	2.4
Cyclopentene	−30	395	1.2	9	No	2.3
Dichloropropane	−45	180	1.7	48	No	2.5
Diethyl ether	4	557	3.1	14.5	No	No data
Ethanol	13	363	3.3	19	No	1.6
Ethoxy propanol	40	255	1.3	12	Yes	No data
Ethoxy propyl acetate	54	325	1	9.9	Yes	No data
Ethyl acetate	−4	426	2.2	11	No	3
Isoamyl acetate	25	360	1	7.5	No	4.5
Isoamylene	−45	290	1	9	No	2.4
Isobutanol	25	415	1.6	10.9	No	2.6
Isobutyl acetate	17	421	2	15	No	4
Isododecane	46	430	0.5	5	No	5.9
Isooctane	−14	420	0.7	5.5	No	3.9
Isopropanol	14	399	2	12	Yes	2.1
Isopropyl acetate	5	460	1.8	7.8	No	3.5
Metaxylene	25	527	1.1	7	No	3.7
Methanol	12	464	7.3	36	Yes	1.1
Methoxy propanol	35	290	1.6	13.8	Yes	3.1
Methoxy propyl acetate	50	340	1.2	10.6	No	4.6
Methyl ethyl ketone	−9	404	1.8	11.5	No	2.5

(continued)

TABLE 1.4 (Continued)

Properties of Various Chemical Components

Product	Flash Point (°C)	AIT (°C)	Lower Explosive Limits in Air% Vol	Upper Explosive Limits in Air% Vol	Miscible with Water?	Vapor Density (Air = 1)
Monopropylene glycol	9	370	2.6	12.5	No	2.6
n-Butanol	37	343	1.4	11.2	No	2.6
n-Propanol	–15	360	2	12	No	2.1
Paraxylene	24	528	1.1	7	No	3.7
Solvent blends	Values depend on specific blend	Values depend on specific blend	Values depend on specific blend	Values depend on specific blend	No	Values depend on specific blend
Tertiary amyl alcohol	20.5	435	1.2	9	No	No data
Tertiary butyl alcohol	11	480	2.4	8	No	2.5
Toluene	4	480	1.3	7.8	No	3.1
Triisobutylene	42	355	0.5	6.5	No	No data
Vinyl acetate monomer	–8	402	2.6	13.4	No	3
White spirit	43.5	215	1.1	8	No	4.9
Xylene	32	463	1.1	7	No	3.7
Products with a Flash Point above 61°C						
2-Ethylhexanol	77	270	0.9	9.7	No	4.5
Butyl diglycol acetate	116	299	0.8	10.7	No	5.6
Butyl diglycol ether	114	228	0.7	9.9	Yes	5.6
Butyl glycol acetate	84	340	0.9	8.5	No	5.4
Butyl glycol ether	68	245	1.1	12.7	Yes	4.1
Diethanolamine	138	662	1.6	9.8	Yes	3.7
Diethylene glyco	143	380	1.8	12.2	Yes	3.7
Ethoxy propoxy propanol	86.5	190	0.8	9.6	Yes	No data
Methoxy propoxy propanol	79	270	1.4	10.4	Yes	5.1
Monoethylene glycol	116	432	3.2	33	No	No data
Monoethanolamine	93	420	5.5	17	Yes	2.1
Odorless kerosene	75	250	0.9	8	No	6.3
Triethanolamine	179	324	1.3	8.5	Yes	5.1
Triethylene glycol	165.5	371	0.2	9.2	Yes	5.2

Note: AIT = autoignition temperature.

and there may be danger of frothing in this method if the burning liquid is heated to over 100°C. Table 1.4 shows various property values for several chemical components.

1.4 Classification of Hazards

This section denotes selection of extinguishing agents for the specific class(es) of occupancy hazards to be protected and is classified as follows.

1.4.1 Light (Low) Hazard

Light hazard occupancies are locations where the total amount of Class A combustible materials, including furnishings, decorations, and contents, are of minor quantity. This may include some buildings or rooms occupied as offices, classrooms, assembly halls, and so forth. This classification anticipates that the majority of content items are either noncombustible or so arranged that a fire is not likely to spread rapidly. Small amounts of Class B flammables are included, provided that they are kept in closed containers and safely stored.

1.4.2 Ordinary (Moderate) Hazard

Ordinary hazard occupancies are locations where the total amount of Class A combustibles and Class B flammables are present in greater amounts than expected under light (low) hazard occupancies. These occupancies could consist of offices, classrooms, mercantile shops and allied storage, light manufacturing halls, research operations centers, auto showrooms, parking garages, workshops or support service areas of light (low) hazard occupancies, and warehouses containing Class I or Class II commodities.

1.4.3 Extra (High) Hazard

Extra hazard occupancies are locations where the total amount of Class A combustibles and Class B flammables present in storage, production, use, and/or finished product is over and above those expected and classed as ordinary (moderate) hazards. These occupancies could consist of wood-working, vehicle repair, aircraft and boat servicing, individual product display showrooms, product convention center displays, and storage and manufacturing processes such as painting, dipping, coating, and including flammable liquid handling refineries, petrochemical gas treating plants, and stations. Also included is warehousing of or in-process storage of other than Class I and Class II commodities.

1.5 Classification of Occupancy Hazards

1.5.1 General

This section classifies occupancy hazards in relation to their quantity and combustibility of contents by surveying and allocating proper occupancy hazard classification numbers to different occupancies where adequate and reliable water supply systems do not exist.

Responsible authorities upon completing the survey specified below shall determine the occupancy hazard classification number as specified hereunder.

Responsible authorities shall perform an onsite survey of all buildings, including type of construction, occupancies, and exposure, within the applicable jurisdiction to obtain the information needed to compute the total water supplies required. At the time of the onsite survey, a record shall be prepared of available water supplies. This information is to be utilized for prefire planning purposes.

Residential areas may be surveyed as an area to determine square meters or cubic meters of each structure and distance to structural exposure hazards, but without a survey of contents (Zhang and Zheng 2012).

These surveys may be combined with fire prevention or prefire planning inspections.

Occupancy hazard classification numbers shall not be assigned to any structure not surveyed as specified in the above sections.

An occupancy hazard classification number shall not be assigned to any building when such building is protected by an automatic sprinkler system.

Storage of products potentially hazardous from the standpoint of increased fire volume or of an explosive nature exists at many rural locations, and such products may be in sufficient quantities to increase the occupancy hazard classification number of the building.

The water supply requirements described in this section shall be performance-oriented, and the authority having jurisdiction shall specify how these water supplies are made available.

The minimum water supply required under this work shall be determined by obtaining the following information:

- Step 1: Classification of occupancy hazard
- Step 2: Classification of construction
- Step 3: Structure dimensions
- Step 4: Exposures, if any

1.5.2 Occupancy Hazard Classification Number

The occupancy hazard classification number is a series of numbers from 3 through 7 that are mathematical factors used in a formula to determine total water supply requirements only.

The occupancies listed in each section are only examples of types of occupancies for the particular classification, and these lists of examples shall not be interpreted as being comprehensive. Similar occupancies shall be assigned the same occupancy hazard classification number.

Where more than one occupancy is present in a structure, the occupancy hazard classification number for the most hazardous occupancy shall be used for the entire structure.

1.5.3 Occupancy Hazard Classification Number 3

Occupancy Hazard Classification Number 3 is considered a severe hazard occupancy where quantity and combustibility of contents are very high. Fire in these occupancies can be expected to develop very rapidly and have high rates of heat release.

When an exposing structure is of Occupancy Hazard Classification Number 3, it is considered an exposure hazard if within a distance of 15.2 m regardless of size.

Occupancy Hazard Classification Number 3 examples in oil, gas, and petrochemical industries include

- Aircraft hangars
- Flour mills
- Chemical works and plants
- Petrochemical works and plants
- Explosive storage
- Oil terminals
- Grain elevators and warehouses
- Lumberyards
- Gas compressor stations
- Oil refineries
- Gas refineries
- Plastic manufacturing and storage
- Crude oil production units
- Crude oil booster stations
- Solvent extracting
- Varnish and paint storage

1.5.4 Occupancy Hazard Classification Number 4

Occupancy Hazard Classification Number 4 is considered a high hazard occupancy, where quantity and combustibility of contents are high. Fires in these occupancies can be expected to develop rapidly and have high rates of heat release.

When an exposing structure is of Occupancy Hazard Classification Number 4, it is considered an exposure hazard if within a distance of 15.2 m regardless of size.

Occupancy Hazard Classification Number 4 examples include

- Building materials
- Department stores
- Exhibition halls, auditoriums, and theaters
- Feed stores (without processing)
- Mercantiles
- Piers and wharves
- Repair garage
- Rubber products storage
- Warehouses such as
  Paper
  Furniture
  Paint
  Department store
  General storage
- Woodworking shops

1.5.5 Occupancy Hazard Classification Number 5

Occupancy Hazard Classification Number 5 is considered a moderate hazard occupancy, where quantity and combustibility of contents are moderate and stockpiles of combustibles do not exceed 3.7 m in height. Fires in these occupancies can be expected to develop quickly and have moderately high rates of heat release.

Occupancy Hazard Classification Number 5 examples include

- Libraries (with large stockroom areas)
- Machine shops
- Metalworking shops
- Pharmaceutical stores
- Printing and publishing shops
- Restaurants
- Unoccupied buildings

1.5.6 Occupancy Hazard Classification Number 6

Occupancy Hazard Classification Number 6 is considered a low hazard occupancy, where quantity and combustibility of contents are moderate and stockpiles of combustibles do not exceed 2.44 m in height. Fires in these occupancies can be expected to develop at a moderate rate and have moderate rates of heat release.

Occupancy Hazard Classification Number 6 examples include

- Automobile parking garages
- Bakeries
- Barber or beauty shops
- Boiler houses

1.5.7 Occupancy Hazard Classification Number 7

Occupancy Hazard Classification Number 7 is considered a light hazard occupancy, where quantity and combustibility of contents are low. Fires in these occupancies can be expected to develop at a relatively low rate and have relatively low rates of heat release.

Occupancy Hazard Classification Number 7 examples include

- Apartments
- Colleges and universities
- Dormitories
- Dwellings
- Fire stations
- Hospitals
- Hotels and guest houses
- Libraries (except large stockroom areas)
- Offices (including data processing)
- Guard houses
- Schools

1.6 Classification of Construction

Classifying types of construction by surveying and allocating a proper number to different types of structures and a series of numbers from 0.5 to 1.5

that are mathematical factors are used in a formula to determine total water supply requirement.

According to available standards, the slowest burning or lowest hazard type of construction, fire resistive, takes a construction classification number of 0.50. The fastest burning or highest hazard type construction, wood frame, takes a construction class number of 1.50.

1.6.1 General

Responsible authorities upon completing the survey should determine the construction classification number from the sections of this work.

For the purpose of this work, each building surveyed shall be classified according to type of construction and shall be assigned a construction classification number. However, no dwelling shall be assigned a construction classification number higher than 1.0. All dwellings shall be assigned a Construction Classification No. 1 or lower when construction is noncombustible or fire-resistive.

Construction classification numbers shall not be assigned to any structure not surveyed as specified in the above.

Where more than one type of construction is present in a structure, the higher construction classification number shall be used for the entire structure.

When a building is located within 15.2 m of the surveyed building and is 9.3 m² or greater in total area, the building is treated as an exposure with the water requirement calculated by the standard method multiplied by 1.5.

1.6.2 Construction Classification Number

1.6.2.1 Fire-Resistive Construction, Construction Classification Number 0.5

Fire-Resistive Construction, Construction Classification Number 0.5 is a building constructed of noncombustible materials (reinforced concrete, brick, stone, etc. and having any metal members properly fireproofed) with major structural members designed to withstand collapse and prevent the spread of fire.

1.6.2.2 Noncombustible Construction, Construction Classification Number 0.75

Noncombustible Construction, Construction Classification Number 0.75 is a building having all structural members (including walls, floors, and roofs) of noncombustible materials and not qualifying as fire-resistive construction.

1.6.2.3 Ordinary Construction, Construction Classification Number 1.0

Ordinary Construction, Construction Classification Number 1.0 is any structure having exterior walls or masonry or other noncombustible material in which the other structural members are wholly or partly made of wood or other combustible material.

1.6.2.4 Wood Frame Construction, Construction Classification Number 1.50

Wood Frame Construction, Construction Classification Number 1.50 is any structure in which the structural members are wholly or partly made of wood or other combustible material and the construction does not qualify as ordinary construction.

When a dwelling is classified as wood frame construction (i.e., having structural members wholly or partly made of wood or other combustible material), assign a Construction Classification Number of 1.0.

1.6.3 Structure Dimensions

The cubic footage of each residence, including attached garages, covered porches, and so forth, must be determined. This encompasses everything under the horizontal roofline (length × width × height).

1.6.3.1 Calculating Minimum Water Supplies

After completing the structure survey and determining the construction classification number and the occupancy hazard classification number, the authority having jurisdiction should compute the required minimum water supply.

A structure shall be considered an exposure hazard if it is 100 ft² (9.29 m²) or larger in area and is within 50 ft (15.24 m) of another structure. However, if a structure, regardless of size, is of Occupancy Hazard Classification Number 3 or 4, it shall be considered an exposure hazard if within 50 ft (15.24 m) of another structure.

1.6.3.2 Structures without Exposure Hazards

For structures with no exposure hazards, the minimum water supply, in gallons, shall be determined by the total cubic footage of the structure, including any attached structures, divided by the occupancy hazard classification number as determined and multiplied by the construction classification number as determined.

$$\begin{matrix} \text{Minimum} \\ \text{water} \\ \text{supply} \end{matrix} = \cfrac{\begin{matrix} \text{Total volume} \\ \text{of structure} \end{matrix}}{\begin{matrix} \text{Occupancy hazard} \\ \text{classification number} \end{matrix}} \times \begin{matrix} \text{Construction} \\ \text{classification} \\ \text{number} \end{matrix}$$

The minimum water supply required for any structure without exposure hazards shall not be less than 2000 gal (7570 L), for residential purposes.

1.6.3.3 Structures with Exposure Hazards

For structures with unattached structural exposure hazards, the minimum water supply, in gallons, shall be determined by the cubic footage of the structure, divided by the occupancy hazard classification number as determined, multiplied by the construction classification number, and multiplied by 1.5.

$$\begin{matrix} \text{Minimum} \\ \text{water} \\ \text{supply} \end{matrix} = \cfrac{\begin{matrix} \text{Total volume} \\ \text{of structure} \end{matrix}}{\begin{matrix} \text{Occupancy hazard} \\ \text{classification number} \end{matrix}} \times \begin{matrix} \text{Construction} \\ \text{classification} \\ \text{number} \end{matrix} \times 1.5$$

The minimum water supply required for any structure with exposure hazards should not be less than 3000 gal (11,355 L), for residential purposes.

1.6.3.4 Structures with Automatic Sprinkler Protection

The authority having jurisdiction shall be permitted to waive the water supply requirement for residential occupancies (except for the minimum water supply of 2000 gal) by this work when a structure is protected by an automatic sprinkler system that fully meets the requirements of NFPA 13, Standard for the Installation of Sprinkler Systems; NFPA 13D, Standard for the Installation of Sprinkler Systems in One- and Two-Family Dwellings and Manufactured Homes; or NFPA 13R, Standard for the Installation of Sprinkler Systems in Residential Occupancies up to and Including Four Stories in Height.

If a sprinkler system protecting a building does not fully meet the requirements of NFPA 13, NFPA 13D, or NFPA 13R, a water supply shall be provided in accordance with the work.

1.7 Fire Protection in Buildings

This section specifies the minimum requirement for construction, protection, and occupancy features necessary to minimize danger to life from fire, smoke, fumes, or panic. The section identifies the minimum criteria for the design of egress facilities so as to permit prompt escape of occupants from buildings, or where desirable, into safe areas within the building. When in fixed locations and occupied as buildings, vehicles, vessels, or other mobile structures shall be treated as buildings (Mostia 1997).

This section does not attempt to address those general fire prevention or building construction features that are normally a function of fire prevention and building codes. The prevention of accidental personal injuries during the course of normal occupancy of buildings, personal injuries incurred by an individual's own negligence, and the preservation of property from loss by fire have not been considered as the basis for any of the provisions in this section.

This section applies to both new construction and existing buildings. In various sections, there are specific provisions for existing buildings that differ from those for new construction.

Where two or more classes of occupancy occur in the same building or structure, and are so intermingled that separate safeguards are impracticable, means of egress facilities, construction, protection, and other safeguards shall comply with the most restrictive requirements of the occupancies involved (Ramachandran 1995).

1.7.1 Classification of Occupancy

A building or structure shall be classified as detailed below.

1.7.1.1 Assembly

Assembly occupancies include, but are not limited to, all buildings or portions of buildings used for gathering together 50 or more persons for such purposes as conference, entertainment, eating, amusement, exhibition, and training.

Assembly occupancies include

- Assembly halls
- Mosques
- Auditoriums
- Bowling lanes
- Conference rooms
- Exhibition halls
- Libraries
- Movie theaters
- Recreation piers and sports
- Restaurants
- Theaters
- Training centers

Occupancy of any room or space for assembly purposes by less than 50 persons in a building of other occupancy and incidental to such other occupancy should be classed as part of the other occupancy and subject to the provisions applicable thereto.

1.7.1.1.1 Subclassification of Assembly Occupancies

Each assembly occupancy shall be subclassified according to its occupant load as follows:

- Class A: occupant load greater than 1000 persons
- Class B: occupant load greater than 300 but not greater than 1000 persons
- Class C: occupant load of 50 or more but not greater than 300 persons

1.7.1.1.2 Educational

Educational occupancies include all buildings or portions of buildings used for educational purposes by six or more persons for 4 or more hours per day or more than 12 hours per week. Educational occupancies include

- Academic nursery schools
- Kindergarten schools
- Educational occupancies also include daycare facilities of any occupant load

In cases where instruction is incidental to some other occupancy, the section of this chapter governing such other occupancy shall apply (Beck et al. 1992).

1.7.1.1.3 Health Care

Health care occupancies are those used for purposes such as medical or other treatment or care of persons suffering from physical or mental illness, disease, or infirmity, and for the care of infants, convalescents, or infirm aged persons. Health care occupancies provide sleeping facilities for four or more occupants and are occupied by persons who are mostly incapable of self-preservation because of age, physical, or mental disability, or because of security measures not under the occupants' control.

Health care occupancies include

- Hospitals
- Nursing homes
- Limited care facilities

Health care occupancies also include ambulatory health care centers.

1.7.1.1.4 *Residential*

Residential occupancies are those occupancies in which sleeping accommodations are provided for normal residential purposes and include all buildings designed to provide sleeping accommodations.

Residential occupancies consist of the following groups:

- Dormitories
- Apartments
- Guest houses
- Board and care facilities
- One- and two-family dwellings

1.7.1.1.5 *Mercantile*

Mercantile occupancies include stores, markets, and other rooms, buildings, or structures for the display and sale of merchandise. Mercantile occupancies include

- Department stores
- Drugstores
- Supermarkets

Minor merchandising operations in buildings predominantly of other occupancies, such as a newsstand in an office building, shall be subject to the exit requirements of the predominant occupancy (Yu et al. 2013).

1.7.1.1.6 *Business*

Business occupancies are those used for the transaction of business (other than that covered under "Mercantile"), for the keeping of accounts and records and similar purposes. Business occupancies include

- Dentists' offices
- Doctors' offices
- General offices
- Laboratories for basic or applied research not including hazardous chemicals
- Outpatient clinics, ambulatory

Minor office occupancy incidental to operations in another occupancy shall be considered as a part of the predominating occupancy and shall be subject to the provisions of this work applying to the predominating occupancy (Chu and Sun 2008).

1.7.1.1.7 Industrial

Industrial occupancies include factories making products of all kinds and properties devoted to operations such as processing, assembling, mixing, packaging, finishing or decorating, and repairing. Industrial occupancies include

- Dry cleaning plants
- Laundries
- Factories of all kinds
- Power plants
- Gas plants
- Pumping stations
- Laboratories
- Refineries
- Hazardous chemicals
- Smokehouses
- Production units
- Gas compressor and injection plants
- Oil booster stations
- Oil loading terminals
- Distribution depots
- Gas compressor stations
- Liquefied petroleum gas (LPG) bottling plants

1.8 Hazard of Contents

The hazard of contents, for the purpose of this book, shall be the relative danger of the start and spread of fire, the danger of smoke or gases generated, and the danger of explosion or other occurrence potentially endangering the lives and safety of the occupants of the building or structure (Megri 2009).

Hazard of contents shall be determined by the relevant authorities on the basis of the character of the contents and the processes or operations conducted in the building or structure. Where different degrees of hazard of contents exist in different parts of a building or structure, the most hazardous shall govern the classification for the purpose of this work (Ellicott 2006).

1.8.1 Classification of Hazard of Contents

The hazard of contents of any building or structure shall be classified as low, ordinary, or high in accordance with the clauses detailed below.

1.8.1.1 Low Hazard

Low-hazard contents shall be classified as those of such low combustibility that no self-propagating fire therein can occur, such as metal furniture in office buildings.

1.8.1.2 Ordinary Hazard

Ordinary-hazard contents shall be classified as those that are likely to burn with moderate rapidity or to give off a considerable volume of smoke, such as tire.

1.8.1.3 High Hazard

High-hazard contents shall be classified as those that are likely to burn with extreme rapidity or from which explosions are to be expected, such as gasoline (petrol) or petroleum gas.

1.8.2 Classification of Storage

1.8.2.1 Commodity Classification

- A Class I commodity is defined as essentially noncombustible products on combustible pallets, in ordinary corrugated cartons with or without single-thickness dividers, or in ordinary paper wrappings with or without pallets.
- A Class II commodity is defined as Class I products in slatted wooden crates, solid wooden boxes, multiple-thickness paperboard cartons, or equivalent combustible packaging material with or without pallets.
- A Class III commodity is defined as wood, paper, natural fiber cloth, or Group C plastics or products thereof, with or without pallets. Products may contain a limited amount of Group A or B plastics. Metal bicycles with plastic handles, pedals, seats, and tires are examples of a commodity with a limited amount of plastic.
- A Class IV commodity is defined as Class I, II, or III products containing an appreciable amount of Group A plastics in ordinary corrugated cartons and Class I, II, and III products in corrugated cartons with Group A plastic packing, with or without pallets. Group

B plastics and free-flowing Group A plastics are also included in this class. An example of packing material is a metal typewriter in a foamed plastic cocoon in an ordinary corrugated carton.

1.8.3 Classification of Plastics, Elastomers, and Rubber

The following categories are based on unmodified plastic materials. The use of fire or flame-retarding modifiers or the physical form of the material may change the classification.

1.8.3.1 Group A

- Acrylonitrile butadiene styrene (ABS) copolymer
- Acrylic (polymethyl methacrylate)
- Acetal (polyformaldehyde)
- Butyl rubber
- Ethylene-propylene rubber (EPDM)
- Fiberglass-reinforced polyester (FRP)
- Natural rubber (if expanded)
- Nitrile rubber (acrylonitrile-butadiene rubber)
- Thermoplastic polyester (polyethylene terephthalate [PET])
- Polybutadiene
- Polycarbonate
- Polyester elastomer
- Polyethylene
- Polypropylene
- Polystyrene
- Polyurethane
- Polyvinyl chloride (PVC) (highly plasticized, e.g., coated fabric, unsupported film)
- Styrene acrylonitrile (SAN)
- Styrene-butadiene rubber (SBR)

1.8.3.2 Group B

- Cellulosics (cellulose acetate, cellulose acetate butyrate, ethyl cellulose)
- Chloroprene rubber
- Fluoroplastics (ethylene chlorotrifluoroethylene [ECTFE] copolymer)
- Ethylene tetrafluoroethylene (ETFE) copolymer

- Fluorinated ethylene propylene (FEP) copolymer
- Natural rubber (not expanded)
- Nylon (nylon 6, nylon 6/6)
- Silicone rubber

1.8.3.3 Group C

- Fluoroplastics (polychlorotrifluoroethylene [PCTFE])
- Polytetrafluoroethylene (PTFE)
- Melamine (melamine formaldehyde)
- Phenolic
- PVC (rigid or lightly plasticized, e.g., pipe, pipe fittings)
- Polyvinylidene chloride (PVDC)
- PVF (polyvinyl fluoride)
- PVDF (polyvinylidene fluoride)
- Urea (urea formaldehyde)

1.9 Ramps

A ramp, either interior or exterior, shall be used as a component in means of egress where it conforms to the general requirements of Section 5.1 of NFPA-101 and to the special requirements of this subsection.

1.9.1 Classification

A ramp shall be designated as Class A or Class B in accordance with the following (see also Table 1.5):

- Existing Class B ramps with slopes of 10 to 17 cm in 1 m are permitted subject to the approval of the authority concerned.
- All existing Class A ramps and new ramps not exceeding slope of 1 in 15 need not be provided with landings.

TABLE 1.5

Criteria for Ramp Classification

Parameter	Class A	Class B
Minimum width	112 cm	76 cm
Maximum slope	1 in 10	1 in 8
Maximum height between landings	3.7 m	3.7 m

1.10 Fire Resistance

1.10.1 External Walls

1.10.1.1 General

Every external wall of a building other than that of the warehouse class shall comply with the requirements as to noncombustibility and fire resistance specified as appropriate thereto in Column 2 of Table 1.6 according to the distance of the wall from the nearest boundary of the premises.

1.10.1.2 Large Storage Buildings

Every external wall of a building of the warehouse class intended to be used wholly or predominantly for storage shall, if the capacity of the building exceeds 7000 m^3, or if its height exceeds 22 m, be noncombustible throughout and have a fire resistance of three hours.

1.11 Basic Methods of Firefighting

1.11.1 Smothering

Reducing the concentration of oxygen, the vapor phase of the fuel, or both, in the air to the point where combustion stops will smother fires. Carbon dioxide is a good extinguishing agent but has disadvantage when used in a confined room or space. It will cause suffocation and endangers the lives of those people who are in the room.

TABLE 1.6

Requirements to Noncombustibility and Fire Resistance

Distance of Wall in Meters from Nearest Boundary of Premises		Appropriate Requirements to Noncombustibility and Fire Resistance
Not less than	Less than	
1	–	Noncombustible throughout
1	1.5	Noncombustible externally
1.5	3	No special requirement
3	–	–
6 (or a distance equivalent to half of the height of the building, whichever is greater)	12	Noncombustible externally and to have resistance to fire of 1 hour
12 (or a distance the height of the building)	–	Noncombustible externally

Inhibiting the chemical reaction of fuel and oxygen is caused by vaporizing liquid such as bromochlorodifluoromethane (BCF) (or accepted replacement). These agents are specifically used for Class C fires.

1.11.2 Cooling

The quenching and cooling effects of water or of solutions containing large percentages of water are of first importance from a cooling point of view in extinguishing fires involving Class A fires. Where total extinguishment is mandatory, a follow-up with water or other Class A agent is essential.

1.11.3 Starving

Isolation of the sources supplying the fire (i.e., materials supporting combustion) should be isolated by removing or transferring unburnt materials to a safe place.

1.12 Principles of Fire Protection

The principles of fire protection to be followed in a design will be a formulation of a statement of requirements with regard to local site conditions such as the degree of potential fire hazard involved, backup available from the local fire service, and other site characteristics.

The location of a fire station shall be sufficiently remote from installations to ensure that its operation is not impaired by incidents on them.

Equipment and its operating pressures within the works shall be compatible with that of outside fire brigades who may be called in the event of a major fire. This is of particular importance in relation to adapters, hoses, and hose couplings.

Mutual aid arrangements will, where possible, be formulated with other operating companies in the immediate area as part of an overall emergency plan.

Such aid should extend to mobile and portable fire equipment with or without manpower, integration of piped water supplies and fire pumping capability, foam stocks, and use of specialized and selective items of safety equipment and mobile plants.

1.13 Extinguishing Methods

The extinguishing methods that are recommended for flammable liquid fires are in most instances suitable for use on different fires. Carbon dioxide, dry chemical, foam, and vaporizing liquid type extinguishing agents have all been found suitable for use on flammable liquid fires of moderate size, such as fires in dip tanks or small flammable liquid spills. The following comments pertain to other extinguishing methods that have been found suitable for control or extinguishment of certain types of flammable liquid fires.

Water spray (fog) is particularly effective on fires in flammable liquids and volatile solids having flash points above 38°C, but frothing may occur with liquids having flash points above 100°C.

Automatic sprinklers are similar to water spray in extinguishing effectiveness. Their principal value, however, is in absorbing heat and keeping surroundings cool until a flammable liquid fire burns out or is extinguished by other means. Overflow drains on open tanks prevent sprinkler discharge from causing burning liquid to overflow and spread the fire. Sprinklers have a good record of fire control in garages, paint and oil rooms, and storage areas where flammable liquids are largely in closed containers and water from sprinklers keeps the containers cool.

1.13.1 Selecting an Extinguishing Method

The selection of the extinguishing method should be made with caution as there are factors to be considered in any individual problem of extinguishment that may affect the choice of extinguishing agent and the method of application. Flowing fires, such as might be caused by a leaking overhead pipe with the liquid on the ground also burning, are always difficult to extinguish.

The amount and rate and method of application of the extinguishing material in relation to the size and type of fire anticipated must be carefully considered and may call for special engineering judgment, particularly in large-scale applications. The use of standard approved equipment is also of major importance.

The chemical and physical properties of a flammable substance will also affect the choice of an extinguishing method. Ordinary type foam, for example, would not be suitable on a fire involving a water-soluble flammable liquid. These special properties affecting extinguishment were taken into consideration when preparing the statements in the Extinguishing Method. The following paragraphs describe the properties of the flammable liquid, volatile solid, or gas that are responsible for the statements in this column.

1.13.2 Numerically Designated Extinguishing Methods

1. *Water may be ineffective* fighting fires with low flash points. This precautionary wording is used for materials having a flash point below 38°C. Obviously, the lower the flash point is, the less effective water

will be. However, water can be used on low flash point liquids when applied in the form of a spray to absorb much of the heat and to keep exposed material from being damaged by the fire.

Much of the effectiveness of using water spray, particularly from hose lines, will depend on the method of application. With proper nozzles, even gasoline-spill fires of some types have been extinguished when coordinated hose lines were used to sweep the flames off the surface of the liquid. Water also has been used to extinguish fires in water-soluble flammable liquids by cooling, diluting, and mixing the flammable liquid with water. In the distilling industry, spray streams from hose lines have been used effectively to achieve control and extinguishment.

The inclusion of phrase "water may be ineffective" is to indicate that although water can be used to cool and protect exposed material, water may not extinguish the fire unless used under favorable conditions by experienced firefighters trained in fighting all types of flammable liquid fires.

2. *Water or foam may cause frothing* when applied on flammable liquids having flash points above 100°C or the boiling point of water. This remark is included only as a precaution and does not indicate that water or foam should not or could not be used in fighting fires in such liquids. The frothing may be quite violent and could endanger the life of the firefighter, particularly when solid streams are directed into the hot burning liquid. On the other hand, water spray carefully applied has frequently been used with success in extinguishing such fires by causing the frothing to occur only on the surface, and this foaming action blankets and extinguishes the fire. To be considered in this regard are not only those liquids with a flash point above 100°C but also any viscous liquids. For example, certain asphalts have a small amount of low flash point solvent added for fluidity purposes, but because of the viscosity the frothing action may occur (Lakhapate 1998).

3. *Water may be used to blanket fire* and accomplish extinguishment when the flammable liquid has a specific gravity of 1.1 or heavier, and is not water-soluble. Here again, however, the method of applying water is significant since the water must be applied gently to the surface of the liquid.

4. *Water may be ineffective except as a blanket*—this statement is included in the Extinguishing Method column as a warning that because the liquid has a flash point below 38°C, water may be ineffective except when applied gently to the surface to blanket and extinguish the fire. This statement applies only to those liquids that are not soluble in water and are heavier than water.

5. *Alcohol foam* is recommended for all water-soluble or polar-flammable liquids except for those that are only "very slightly" soluble. Certain judgment factors are again introduced since, for some liquids, ordinary foam might be used successfully to extinguish fires for liquids only slightly soluble in water, particularly if regular foam was applied at increased rates over those normally recommended. Conversely, some flammable liquids, such as certain higher molecular weight alcohols and amines, will destroy alcohol foam even when applied at very high rates. Foam should not be used on water-reactive flammable liquids. Recently developed alcohol foams have been listed by independent testing laboratories for use on both polar- and nonpolar-flammable liquids.

In the oil industries, protein foam (regular) is generally used on the most flammable liquids, paints, and so forth.

Fluoroprotein foam is a protein foam incorporating specially selected fluorinated surfactants that is also used to increase foam fluidity and fire knockdown. This type of foam can be used in both premix and concentrated applications only.

6. *Stopping the flow of gas* rather than extinguishing the fire is usually the best procedure to follow when escaping gas is burning. It may be dangerous to extinguish the flame and allow the gas to continue to flow, as an explosive mixture may be formed with air, which, if ignited, may cause far greater damage than if the original fire had been allowed to burn. Extinguishing the flame by carbon dioxide or dry chemical may be desirable where it is necessary to permit immediate access to valves to shut off the supply. In many cases, however, it is preferable to allow the flame to continue, keeping the surroundings cool with water spray to prevent ignition of other combustible materials.

1.14 Suggested Hazard Identification

The suggested hazardous identification system identifies the hazards of a material in terms of three categories; namely, health, flammability, and reactivity, and indicates the order of severity in each of these categories by five divisions ranging from four (4), indicating a severe hazard, to zero (0), indicating no special hazard.

While this system is basically simple in application, the hazard evaluation that is required for the precise use of the signals in a specific location must be made by experienced, technically competent persons. Their judgment must be based on factors encompassing knowledge of the inherent hazards of different materials, including the extent of change in behavior to be anticipated under conditions of fire.

1.14.1 Health Hazards

A health hazard is any property of a material that either directly or indirectly can cause injury or incapacitation, either temporary or permanent, from exposure by contact, inhalation, or ingestion.

This section deals with the capacity of a material to cause personal injury from contact with or absorption into the body.

Only hazards arising out of an inherent property of the material or a property of the products of combustion of the material will be considered. Injury resulting from the heat of a fire or force of an explosion is not included.

In general, a health hazard in firefighting or other emergency conditions is that of a single exposure that may vary from a few seconds up to an hour. The physical exertion demanded in firefighting or other emergency conditions may be expected to intensify the effects of any exposure.

There are two sources of health hazards. One arises out of the inherent properties of the material. The other arises out of the toxic products of combustion or decomposition of the material. The hazard degree shall be assigned on the basis of the greater hazard that could exist under fire or other emergency condition. The common hazards from the burning of ordinary combustible materials are not included.

The degree of hazard shall indicate to firefighting personnel one of the following: that they can work safely only with specialized protective equipment, that they can work safely with suitable respiratory protective equipment, or that they can work safely in the area with ordinary clothing.

1.14.1.1 Degrees of Hazards

Degrees of hazards are ranked according to the probable severity of hazard to personnel in the following sections.

1.14.1.1.1 Degree of Hazard No. 4

The Degree of Hazard No. 4 comprises materials that on very short exposure could cause death or major residual injury even though prompt medical treatment was given, including those which are too dangerous to be approached without specialized protective equipment. This degree should include

- Materials that can penetrate ordinary rubber protective clothing, such as hydrogen chloride (HCL)
- Materials that under normal conditions or under fire conditions give off gases that are extremely hazardous (i.e., toxic or corrosive)

through inhalation or through contact with or absorption through the skin (e.g., carbon tetrachloride [CTC] that gives off phosgene, which is highly toxic when exposed to heat)

1.14.1.1.2 Degree of Hazard No. 3

The Degree of Hazard No. 3 comprises materials that on short exposure could cause serious temporary or residual injury even though prompt medical treatment was given, including those requiring protection from all bodily contact. This degree should include

- Materials giving off highly toxic combustion products, such as hydrogen cyanide (HCN)
- Materials corrosive to living tissue or toxic by skin absorption, such as mercury salts and acetic anhydride

1.14.1.1.3 Degree of Hazard No. 2

The Degree of Hazard No. 2 comprises materials that on intense or continued exposure could cause temporary incapacitation or possible residual injury unless prompt medical treatment is given, including those requiring use of respiratory protective equipment with independent air supply. This degree should include

- Materials giving off toxic combustion products, such as ammonium perchlorate (NH_4ClO_4)
- Materials giving off highly irritating combustion products, such as acrolein ($CH_2:CHCHO$)
- Materials that either under normal conditions or under fire conditions give off toxic vapors lacking waning properties, such as acetonitrile (CH_3CN).

1.14.1.1.4 Degree of Hazard No. 1

The Degree of Hazard No. 1 comprises materials that on exposure would cause irritation but only minor residual injury even if no treatment is given, including those which require use of an approved canister type gas mask. This degree should include

- Materials that under fire conditions would give off irritating combustion products, such as acetic anhydride (($CH_3CO)_2O$)
- Materials that on the skin could cause irritation without destruction of tissue, such as calcium hypochlorite ($Ca(ClO)_2$) (bleaching powder)

1.14.1.1.5 Degree of Hazard No. 0

The Degree of Hazard No. 0 comprises materials that on exposure under fire conditions would offer no hazard beyond that of ordinary combustible material.

1.14.2 Flammability Hazards

This section deals with the degree of susceptibility of materials to burning. Many materials that will burn under one set of conditions will not burn under others. The form or condition of the materials, as well as their inherent properties, affects the hazard.

1.14.2.1 Degrees of Hazards

The degrees of hazard are ranked according to the susceptibility of materials to burning in the following sections.

1.14.2.1.1 Degree of Hazard No. 4

The Degree of Hazard No. 4 comprises materials that will rapidly or completely vaporize at atmospheric pressure and normal ambient temperature or are readily dispersed in air and will burn readily. This degree should include

- Gases
- Cryogenic materials
- Any liquid or gaseous material that is a liquid while under pressure and has a flash point below 23°C and a boiling point below 38°C (Class IA flammable liquids)
- Materials that on account of their physical form or environmental conditions can form explosive mixtures with air and are readily dispersed in air, such as dusts of combustible solids and mists of flammable or combustible liquid droplets

1.14.2.1.2 Degree of Hazard No. 3

The Degree of Hazard No. 3 comprises liquids and solids that can be ignited under almost all ambient temperature conditions. Materials in this degree produce hazards atmospheres with air under almost all ambient temperatures, or, though unaffected by ambient temperatures are readily ignited under almost all conditions. This degree should include

- Liquids having a flash point below 23°C and a boiling point at or above 38°C and those liquids having a flash point at or above 23°C and below 38°C (Class IB and class IC flammable liquids)
- Solid materials in the form of coarse dusts that may burn rapidly but generally do not form explosive atmospheres with air
- Solid materials in a fibrous or shredded form, such as cotton, sisal, and hemp, that may burn rapidly and create flash fire hazards

- Materials that burn with extreme rapidity and usually are the cause of self-contained oxygen (e.g., dry nitrocellulose and many organic peroxides)
- Materials that ignite spontaneously when exposed to air

1.14.2.1.3 Degree of Hazard No. 2

The Degree of Hazard No. 2 comprises materials that must be moderately heated or exposed to relatively high ambient temperatures before ignition can occur. Materials in this degree would not under normal conditions form hazardous atmospheres with air, but under high ambient temperatures or under moderate heating may release vapor in sufficient quantities to produce hazardous atmospheres with air (Li 2011). This degree should include

- Liquids having a flash point above 38°C but not exceeding 93°C
- Solids and semisolids that readily give off flammable vapors

1.14.2.1.4 Degree of Hazard No. 1

The Degree of Hazard No. 1 comprises materials that must be preheated before ignition can occur. Materials in this degree require considerable preheating under all ambient temperature conditions before ignition and combustion can occur. This degree should include

- Materials that will burn in air when exposed to a temperature of 815°C for a period of 5 minutes or less, such as oily woollen material and phenol (carbolic acid) 1319°F C715.
- Liquids, solids, and semisolids having a flash point above 93°C. This degree includes most ordinary combustible materials (e.g., fuel oil no. 2).

1.14.2.1.5 Degree of Hazard No. 0

The Degree of Hazard No. 0 comprises materials that will not burn. This degree should include any material that will not burn in air when exposed to a temperature of 815°C for a period of 5 minutes.

1.14.3 Reactivity (Instability) Hazards

- *Reactive materials* are those that can enter into a chemical reaction with other stable or unstable materials. For purposes of this chapter, the other material to be considered is water and only if its reaction releases energy. Reactions with common materials other than water may release energy violently. Such reactions should be considered in individual cases, but are beyond the scope of this identification system.

- *Unstable materials* are those that in the pure state or as commercially produced will vigorously polymerize, decompose or condense, or become self-reactive and undergo other violent chemical changes.
- *Stable materials* are those that normally have the capacity to resist changes in their chemical composition, despite exposure to air, water, and heat as encountered in fire emergencies.

This section deals with the degree of susceptibility of materials to release energy. Some materials are capable of rapid release of energy by themselves, by self-reaction or polymerization, or can undergo violent eruptive or explosive reaction when combined with other materials.

The violence of reaction or decomposition of materials may be increased by heat or pressure, by mixture with certain other materials to form fuel-oxidizer combinations, or by contact with incompatible substances, sensitizing contaminants, or catalysts.

Because of the wide variations of accidental combinations possible in fire or other emergencies, these extraneous hazard factors (except for the effect of water) cannot be applied in a general numerical scaling of hazards. Such extraneous factors must be considered individually in order to establish appropriate safety factors such as separation or segregation. Such individual consideration is particularly important where significant amounts of materials such as aluminum powder or magnesium powder are to be stored or handled.

The degree of hazard should indicate to firefighting and emergency personnel that the area should be evacuated, that the fire may be fought from a protected location, that caution must be used in approaching the fire and applying extinguishing agents, or that the fire may be fought using standard materials such as chlorine, ammonia, H_2S gas, or LPG.

1.14.3.1 Degrees of Hazards

The degrees of hazards are ranked according to ease, rate, and quantity of energy release in the following sections.

1.14.3.1.1 Degree of Hazard No. 4

The Degree of Hazard No. 4 comprises materials that in themselves are readily capable of detonation or of explosive decomposition or explosive reaction at normal temperatures and pressures. This degree should include materials that are sensitive to mechanical or localized thermal shock at normal temperatures and pressures, such as black powder or gelignite.

1.14.3.1.2 Degree of Hazard No. 3

The Degree of Hazard No. 3 comprises materials that in themselves are capable of detonation or of explosive decomposition or explosive reaction but which require a strong initiating source or must be heated under confinement before initiation. This degree should include materials that are sensitive to thermal or mechanical shock at elevated temperatures and pressures or that react explosively with water without requiring heat or confinement.

1.14.3.1.3 Degree of Hazard No. 2

The Degree of Hazard No. 2 comprises materials that in themselves are normally unstable and readily undergo violent chemical change but do not detonate. This degree should include materials that can undergo chemical change with rapid release of energy at normal temperatures and pressures or can undergo violent chemical change at elevated temperatures and pressures. It should also include those materials that may react violently with water or that may form potentially explosive mixtures with water.

1.14.3.1.4 Degree of Hazard No. 1

The Degree of Hazard No. 1 comprises materials that in themselves are normally stable but can become unstable at elevated temperatures and pressures or may react with water with some release of energy but not violently.

1.14.3.1.5 Degree of Hazard No. 0

The Degree of Hazard No. 0 comprises materials that in themselves are normally stable, even under fire exposure conditions, and are not reactive with water.

1.14.4 Special Hazards

This section deals with other properties of the material that may cause special problems or require special firefighting techniques.

1.14.4.1 Symbols

Materials that demonstrate unusual reactivity with water should be identified with the letter W with a horizontal line through the center.

Materials that possess oxidizing properties should be identified by the letters OX.

Materials possessing radioactivity hazards should be identified by the standard radioactivity symbol.

1.14.4.1.1 Shape and Proportions of a Symbol

The basic symbol for signifying ionizing radiation or radioactive materials should be designed and proportioned as illustrated in Figure 1.11.

For hazardous chemical data, flammability, reactivity and extinguishing methods, reference should be made to page 86 of Table NFPA Manual, Volume 10.

FIGURE 1.11
The basic symbol for signifying ionizing radiation or radioactive materials.

1.15 Hazard and Operability Studies

Essentially, the hazard and operability (HAZOP) examination procedure systematically questions every part of a process or operation to discover qualitatively how deviations from normal operation can occur and whether further protective measures, altered operating procedures, or design changes are required.

The examination procedure uses a full description of the process that will almost invariably include a piping and instrumentation diagram (P&ID) or equivalent and systematically questions every part of it to discover how deviations from the intention of the design can occur, and determine whether these deviations can give rise to hazards.

The questioning is sequentially focused around a number of guide words that are derived from method study techniques. The guide words ensure that the questions posed to test the integrity of each part of the design will explore every conceivable way in which operation could deviate from the design intention.

Some of the causes may be so unlikely that the derived consequences will be rejected as not being meaningful. Some of the consequences may be trivial and need no further consideration. However, there may be some deviations with causes that are conceivable and consequences that arc potentially serious. The potential problems are then noted for remedial action. The immediate solution to a problem may not be obvious and could need further consideration either by a team member or perhaps a specialist. All decisions taken must be recorded.

Secretarial software may be used to assist in recording the HAZOP, but it should not be considered as a replacement for an experienced chairperson and secretary. The main advantage of this technique is its systematic thoroughness in failure case identification. The method may be used at the design stage, when plant alterations or extensions are to be made, or applied to an existing facility.

1.15.1 Sequence of Examination

Figure 1.12 illustrates the logical sequence of steps in conducting a HAZOP. The main elements under consideration are

- Intention
- Deviation
- Causes
- Consequences

  Hazards

  Operating difficulties
- Safeguards
- Corrective action

Typically, a member of the team would outline the purpose of a chosen line in the process and how it is expected to operate. Various guide words, such as "more," are selected in turn. Consideration will then be given to what could cause the deviation.

Following this, the results of a deviation, such as the creation of a hazardous situation or operational difficulty, are considered. When the considered events are credible and the effects significant, existing safeguards should be evaluated and a decision then taken as to what additional measures could be required to eliminate the identified cause. A more detailed reliability analysis such as risk or consequence quantification may be required to determine whether the frequency or outcome of an event is high enough to justify major design changes.

1.15.2 Details of Study Procedure

The study of each section of plant generally follows the following pattern:

1. The process designer very briefly outlines the broad purpose of the section of design under study and displays the P&ID (or equivalent) where it can be readily seen by all team members.
2. Any general questions about the scope and intent of the design are discussed.
3. The first pipeline or relevant part for study is selected, usually one in which a major material flow enters that section of the plant. The pipeline is highlighted on the P&ID with dotted lines using a transparent, pale-colored felt pen.
4. The process designer explains in detail its purpose, design features, operating conditions, fittings, instrumentation and protective systems, and so forth, and details of the vessels immediately upstream or downstream of it.

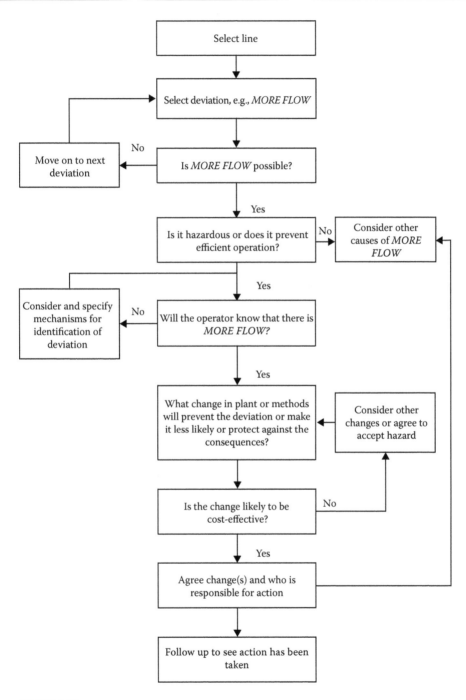

FIGURE 1.12
The logical sequence of steps in conducting a HAZOP.

5. Any general questions about the pipeline or relevant part are then discussed.

6. A detailed line-by-line study commences at this point. The discussion leader takes the group through the guide words chosen as relevant. Each guide word or prompt, such as "high flow," identifies a deviation from normal operating conditions.

This list is used to prompt discussion of the possible causes and effects of flow at an undesirably high rate. If, in the opinion of the study team, the combination of the consequences and the likelihood of occurrence are sufficient to warrant action, then the combination is regarded as a problem and noted in the minutes as such.

If the existing safeguards are deemed to be sufficient then no further action is required. For major risk areas the need for action may be assessed quantitatively using such techniques as hazard analysis (HAZAN) or reliability analysis. For less critical risks the assessment is usually based on experience and judgment. The person responsible for defining the corrective action is also nominated.

The list continues as follows:

7. It should always be remembered that the main aim of the meeting is to find problems needing solutions rather than finding the actual solution. The group should not be tied down by trying to resolve a problem; it is better to proceed with the study, deferring consideration of the unsolved problems to a later date.

8. When the guide word requires no more consideration, the chairperson refers the team to the next guide word.

9. Discussion of each guide word is confined to the section or pipeline marked, the vessels at each end and any equipment, such as pumps or heat exchangers, in between. Any changes agreed at the meeting are noted in the minutes, and where appropriate, marked on the P&ID or layout with red pen.

10. When all guide words have been covered, the line is fully highlighted to show that it has been completed, and the next line is chosen.

11. When all the lines in a plant subsection have been reviewed, additional guide words are used for review (overview) of the P&ID as a whole.

1.15.3 HAZOP Effectiveness

The effectiveness of a HAZOP will depend on

- The accuracy of information (including P&IDs) available to the team; information should be complete and up-to-date.

- The skills and insights of the team members.
- How well the team is able to use the systematic method as an aid to identifying deviations.
- Maintaining a sense of proportion in assessing the seriousness of a hazard and the expenditure of resources in reducing its likelihood.
- The competence of the chairperson in ensuring the study team rigorously follows sound procedures.

Key elements of a HAZOP are

- The HAZOP team
- The full description of the process
- Relevant guide words
- Conditions conducive to brainstorming
- Recording of meeting
- Follow-up plan

1.15.4 The HAZOP Team

The group carrying out the HAZOP will typically consist of a team of approximately five to eight people. Team members should possess a range of relevant skills to ensure all aspects of the plant and its operations are covered; engineering disciplines, management, and plant operating staff should be represented. This will help prevent possible events from being overlooked through lack of expertise and awareness. HAZOP is normally carried out by a team of people whose roles are detailed (with alternative names from other sources) in Table 1.7.

It is essential that the chairperson is experienced in HAZOP techniques. This will ensure that the team follows the procedure without diverging or taking short cuts.

Where the HAZOP is required as a condition of development consent, the name of the chairperson is typically required to be submitted to the director of urban affairs and planning or the director's nominee for approval prior to commencement of the HAZOP.

Apart from the chairperson, it is important that the study team be highly familiar with the information contained in the P&ID of the plant or alternative description of the process being considered. For an existing plant, the group should include experienced operational and maintenance staff.

A HAZOP team assigned to consider a new chemical plant could comprise

- *Chairperson*, an independent person who has a sound knowledge and experience of HAZOP techniques. Some understanding of the proposed plant design would also be beneficial.

TABLE 1.7

HAZOP Team and Their Roles

Name	Alternative	Role
Study leader	Chairman	Someone experienced in HAZOP but not directly involved in the design, to ensure that the method is followed carefully
Recorder	Secretary or scribe	To ensure that problems are documented and recommendations passed on
Designer	(or representative of the team which has designed the process)	To explain any design details or provide further information
User	(or representative of those who will use it)	To consider it in use and question its operability and the effect of deviations
Specialist	(or specialists)	Someone with relevant technical knowledge
Maintainer	(if appropriate)	Someone concerned with maintenance of the process

- *Design engineer,* the project design engineer, usually mechanical, who has been involved in the design and will be concerned with project cost.
- *Process engineer,* usually the chemical engineer responsible for the process flow, diagram, and development of the P&IDs.
- *Electrical engineer,* usually the engineer who was responsible for design of the electrical systems in the plant.
- *Instrument engineer,* the instrument engineer who designed and selected the control systems for the plant.
- *Operations manager,* preferably the person who will be in charge of the plant when it moves to the commissioning and operating stages.

A team with a narrower range of skills is unlikely to be able to satisfactorily conduct a HAZOP of this nature. Other skills may be needed. For example, if the plant uses a new chemical process, a research chemist might be required. Including an experienced supervisor or operator on the team is also often appropriate, especially one from an existing or similar plant already in operation.

At least one member of the team must have sufficient authority to make decisions affecting design or operation of the facility, including those decisions that involve substantial additional costs.

1.15.5 Full Description of Process

A full description of the process is needed to guide the HAZOP team. In the case of conventional chemical plants, detailed P&IDs should be available for the plant under consideration. At least one member of the HAZOP team should be familiar with these diagrams and all instrumentation represented on them. If the plant is too complex or large it may be split into a smaller number of units to be analyzed at separate HAZOP meetings.

In addition to P&IDs, models (either physical or computer-generated) of the plant or photographs of similar existing plants may also be utilized. Both greatly assist in visualizing potential incidents, especially those caused by human error. If the HAZOP is to be carried out on an existing plant, or the proposal is for a new plant but a similar plant is already operating, inspection of this facility before commencing the HAZOP by the team would be highly beneficial.

In conducting a HAZOP on an existing or proposed plant where a similar one has been in operation elsewhere, past incidents could also be considered during the course of the HAZOP.

Key information that may be required during the HAZOP should also be readily available. This could include

- Layout drawings
- Hazardous-area drawings
- Safety data sheets
- Relevant codes or standards
- Plant operating manual (for an existing plant)
- Outline operating procedures (for a new plant)

When carrying out a HAZOP on a facility for which traditional P&IDs are not appropriate, it may be more suitable to use alternative visualization and diagrammatic decision as to which medium will be used well before the HAZOP commences.

In batch processes, additional complexities are introduced into the technique because of the time-dependent nature of batch operations. It is strongly recommended that the references be consulted for guidance and to have a chairperson experienced in batch HAZOPs.

1.15.6 Relevant Guide Words

A set of guide words is chosen as relevant to the operation to be studied and then systematically applied to all parts of that operation. This may entail application of the guide words to each process line within a P&ID, or by following each stage of an operation from start to finish. The key feature is to select appropriate parameters that apply to the design intention. These are general words such as "flow," "temperature," "pressure," and "composition." In the above example, it can be seen that variations in these parameters could constitute deviations from the design Intention. In order to identify deviations, the study leader applies (systematically, in order) a set of guide words

TABLE 1.8

Some Guide Words for HAZOP Studies

Guide Word	Meaning
No or not	Complete negation of the design intent
More	Quantitative increase
Less	Quantitative decrease
As well as	Qualitative modification/increase
Part of	Qualitative modification/decrease
Reverse	Logical opposite of the design intent
Other than	Complete substitution
Early	Relative to the clock time
Late	Relative to the clock time
Before	Relating to order or sequence
After	Relating to order or sequence

to each parameter for each section of the process. The current standard guide words are listed in Table 1.8.

Note that the last four guide words in Table 1.8 are applied to batch or sequential operations. These are therefore combined, e.g., "no flow," "more temperature," and if the combination is meaningful, it is a potential deviation. In this case "less composition" would suggest less than 96% sulfuric acid, whereas "other than composition" would suggest something else such as oil. Table 1.9 gives an overview of commonly used guide word–parameter pairs and common interpretations of them.

Commissioning should be included for new plant and modifications. Human response time and the possibility that the operator or supervisor may take inappropriate action should also be considered in this analysis.

1.15.7 Conditions Conducive to Brainstorming

The HAZOP should be carried out under conditions conducive to brainstorming. The team should conduct the HAZOP in an area that is free from interruptions and includes facilities for displaying diagrams and so forth. White boards or other recording media should also be available. The minutes should be recorded competently and clearly during the HAZOP, preferably not by the chairperson.

1.15.8 Recording of Meeting

There are two possible approaches to record keeping. One is to record only key findings ("reporting by exception"). The other is to record all issues. Experience has shown that reporting by exception can be adopted in most cases since it minimizes the secretarial load and focuses on the issues that

TABLE 1.9

Overview of Commonly Used Guide Word–Parameter Pairs and Common Interpretations of Them

Parameter/Guide Word	More	Less	None	Reverse	As well as	Part of	Other than
Flow	High flow	Low flow	No flow	Reverse flow	Deviating concentration	Contamination	Deviating material
Pressure	High pressure	Low pressure	Vacuum		Delta-p		Explosion
Temperature	High temperature	Low temperature					
Level	High level	Low level	No level		Different level		
Time	Too long/too late	Too short/too soon	Sequence step skipped	Backwards	Missing actions	Extra actions	Wrong time
Agitation	Fast mixing	Slow mixing	No mixing				
Reaction	Fast reaction/runaway	Slow reaction	No reaction				Unwanted reaction
Start-up/Shut-down	Too fast	Too slow			Actions missed		Wrong recipe
Draining/Venting	Too long	Too short	None		Deviating pressure	Wrong timing	
Inertizing	High pressure	Low pressure	None / Failure			Contamination	Wrong material
Utility failure (instrument air, power)							
DCS failure			Failure				
Maintenance			None				
Vibrations	Too low	Too high	None				Wrong frequency

need attention. It is important, however, that the recording of safeguards is retained, even when no further action is required. This record helps ensure that safeguards are not removed through ignorance subsequent to the HAZOP.

It is generally acknowledged that the process becomes tedious over an extended period and that sessions should he kept to half a day if possible when the HAZOP is likely to extend over several days. It is also important to ensure the maximum participation in the study by each team member. Continuous attendance at the sessions should be given the high priority it deserves.

Care should be exercised to provide physical surroundings conducive to such participation. The number of records generated from the HAZOP may be very large. If this is the case, only those records for which possible incidents could occur or where it is not obvious from the identified hazards that such incidents cannot occur need be included with the report. A comprehensive set of all records generated by the HAZOP should be kept, however, for the company's own use and for the use of the relevant department if requested.

1.15.9 Computer HAZOP

The use of electrical, electronic, or programmable electronic (E/E/PE) systems in safety-related applications is steadily growing. This applies to computer-based instrumentation, control, and safety-related functional applications in modern chemical plants and related industrial situations. Difficulties arising due to the malfunction of such systems are also increasing, particularly as experience with such systems flags new types of problems that were not encountered in older plant designs. The interface with modern electronic control/protective systems remains a potential weakness in the overall reliability of these systems.

The E/E/PE systems relating to the operations function of the plant may be considered as being regularly tested "on the run." However, the same may not be true for the safety-related systems that are called upon to perform as intended infrequently in the event of a failure or dangerous situation.

Dangerous situations can arise due to

- The requirements of the functional safety system (hardware/software) being inadequately specified at design stage
- Modifications to software/hardware not being adequately considered
- Common-cause failures
- Human error
- Random hardware faults
- Extreme variations in surrounding conditions (e.g., electromagnetic, temperature, vibration)

- Extreme variations in supply systems (e.g., low or high supply voltage, loss of air pressure for emergency shutdown, voltage spikes on resumption after a power outage)

The hazard analysis determines whether functional safety is necessary to ensure adequate protection. Functional safety is part of the overall safety that depends on a system or equipment operating correctly in response to its inputs. For example, an overpressure protection system using a pressure sensor to initiate the opening of a relief device before dangerous high pressures are reached is an instance of functional safety.

Two types of requirements are necessary to achieve functional safety:

1. Safety function requirements (what the function does)
2. Safety integrity requirements (the likelihood of a safety function being performed satisfactorily)

The safety function requirements are derived from the hazard analysis and the safety integrity requirements are derived from the risk assessment. The HAZOP or computer HAZOP should review the safety-related systems that must operate satisfactorily to achieve a safe outcome in the event of an incident/situation with potential to result in a dangerous failure.

The aim should be to ensure that the safety integrity of the safety function is sufficient to ensure that no one is exposed to an unacceptable risk associated with the hazardous event.

The importance of E/E/PE systems has tended to increase in recent years, particularly with computer control and software logic interlocks. If the computer and instrumentation system is sufficiently complex for the facility, it may be useful to consider this system in a separate HAZOP (sometimes referred to as a CHAZOP; the letter "C" indicating computer-based in terms of both control and protective) or as a discrete component of a more general HAZOP.

Modern plants will almost invariably include/E/PE systems. These typically have a different spectrum of failure modes than those encountered in a conventional HAZOP. The flexibility of E/E/PE systems that offer the capability to control several complex operations can also provide possibilities for making more errors than with conventional control systems. The likelihood of common mode failures increases with such systems; for example, the failure of a single input/output (I/O) card may result in the loss of several control and information channels. A CHAZOP will highlight such issues and lead to corrective solutions such as employing two independent systems or hardwiring key control circuits.

A discrete study of the control systems and safety-related systems can be particularly valuable where the instrumentation has been designed and installed as a package unit by a contractor, as well as allowing the rest of the team to gain an understanding of the system. Treating this part as a discrete component of the HAZOP also allows the operator/computer interaction to be examined. However, plant management should not forget that the overall plant HAZOP will not be complete until the E/E/PE systems have been reviewed by CHAZOP or equivalent technique.

These aspects can be reviewed by other disciplined techniques along the lines of HAZOP. Clearly, for such techniques to be suitable for a particular system, they need to be adapted and refined appropriately.

1.15.10 Failure Modes and Effects Analysis

Failure modes and effects analysis (FMEA) uses a similar "what if" approach to a HAZOP but has as its objective the identification of the effects of all the failure modes of each piece of equipment or its instrumentation. As a result, FMEA identifies single failure modes that can play a significant part in an accident. It is not effective, however, at identifying combinations of equipment failures that lead to accidents.

Human operators are not usually considered specifically in FMEA, even though the effects of operational errors are usually included in the equipment failure mode. FMEA is similar in methodology to a HAZOP but with a different approach. Whereas the HAZOP evaluates the impact of a deviation in the operating conditions to a level outside the design range such as "more flow" or "low temperature fmea" uses a systematic approach to evaluate the impact of a single equipment failure or human error, in turn, on the system or plant.

In FMEA, the reason or cause for the equipment failure is not specifically considered.

This is different from a HAZOP in which the cause/s for the deviation have to be assumed or agreed by judgment and experience, since it is the cause that the HAZOP initially addresses. The FMEA methodology assumes that if a failure can occur, it must be investigated and the consequences evaluated to verify if the failure can be tolerated on safety grounds or if the remaining serviceable equipment is capable of controlling the process safely.

As for HAZOP, to be effective, the FMEA needs a strong, well-led team with wide cumulative experience. The initial briefing by the leader and the contributions expected from each member are similar to that in a HAZOP.

The results of the analysis are recorded as in a HAZOP. The recording should be in the same format for the whole plant in order to facilitate reviews of the analysis and maintenance of records.

In carrying out the FMEA, the process flow diagrams and the P&IDs are first studied to obtain a clear understanding of the plant operation. Where a part of a process is being analyzed, it may be necessary, in addition, to include the failure modes of equipment immediately outside the analysis area and the consequence of the failure on the plant/process section being analyzed.

There is every reason to believe that, with appropriately modified guide words, the HAZOP technique can be applied to situations that are not strictly process ones.

Even if a strictly disciplined technique were not employed, a searching study of materials handling and warehousing and even of mining operations would benefit from the group study approach.

References

ASTM D681. Test Method for Concentration Limits of Flammability of Chemical.

ASTM D86. Standard Method of Test for Distillation of Petroleum Products.

ASTM E681-04. Standard Test Method for Concentration Limits of Flammability of Chemicals (Vapors and Gases).

Beck, V., Thomas, I., Ramsay, G.C., MacLennan, H., Lacey, R., Johnson, P. and Eaton, C. 1992. Risk assessment and the design of fire safety systems in buildings, ASTM Special Technical Publication 1150, pp. 209–223.

Berkowitz, Z., Horton, D.K. and Kaye, W.E. 2004. Hazardous substances releases causing fatalities and/or people transported to hospitals: Rural/agricultural vs. other areas. *Prehospital and Disaster Medicine* 19 (3), 213–220.

Chu, G. and Sun, J. 2008. Decision analysis on fire safety design based on evaluating building fire risk to life. *Safety Science* 46 (7), 1125–1136.

Dennis, N. 1998. Hazardous area classifications and system documentation. *Australian Journal of Instrumentation and Control* 13 (3), 18–20.

Ellicott, G. 2006. Shouldn't buildings critical to the community have extra levels of fire protection? *Building Engineer* 81 (8), 14–15.

Heinemann, J. and Hoef, J.V. 2008. Grounding and bonding practices for hazardous areas. *Consulting-Specifying Engineer* 44 (1), 49–52.

Hernandez, J.E., Bradley, B.A., Crooke, R.W., Faulkner, E.B., Lewis, W.M., Mai, V.Q., Messick, K.A. and Miles, J.K. 1995. One company's guideline for hazardous area classification, Record of Conference Papers—Annual Petroleum and Chemical Industry Conference 1995, pp. 243–265; Proceedings of the 1995 IEEE 42nd Petroleum and Chemical Industry Conference, Denver, CO, September 11–13, 1995, Code 44037.

Kantawong, S. 2012. Hazardous signs and fire exit signs classification using appropriate shape coding algorithm and BPN, JCSSE 2012—9th International Joint Conference on Computer Science and Software Engineering, article no. 6261919, pp. 23–27.

Lakhapate, P.J. 1998. Hazardous area classification. *Chemical Engineering World* 33 (3), 67–70.

Li, W. 2011. Research on countermeasures and methods of disposing incidents of hazardous chemicals reacting with water. *Procedia Engineering* 26, 2278–2286.

Megri, A.C. 2009. Integration of different fire protection/life safety elements into the building design process. *Practice Periodical on Structural Design and Construction* 14 (4), 181–189.

Mostia, W. 1997. New options in hazardous area classification, *Control* 10 (1), 4.

NFC (NFPA) (National Fire Codes) NFC 101, NFC 220, NFC 231.

Ramachandran, G. 1995. Probability-based building design for fire safety: Part 1. *Fire Technology* 31 (3), 265–275.

Schröder, V. and Molnarne, M. 2005. Flammability of gas mixtures: Part 1: Fire potential. *Journal of Hazardous Materials* 121 (1–3), 37–44.

Shen, T.-S., Tseng, W.-W., Hsu, W.-S., Chen, L.-T. and Huang, Y.-H. 2009. A study on the classification and design of fire protection systems for road tunnels in Taiwan. *Journal of Applied Fire Science* 19 (3), 231–246.

Tommasini, R. and Pons, E. 2011. Classification of hazardous areas produced by maintenance interventions on distribution networks and in presence of open surface of flammable liquid. 58th Annual IEEE Petroleum and Chemical Industry Conference, Toronto (Canada), September 19–21, 2011, pp. 1–10.

Yu, C.C., Chen, T.C., Lin, C.S. and Wang, S.C. 2013. Numerical simulation of the performance-based of the building fire protection safety evaluation. *Key Engineering Materials* 531–532, 668–672.

Zhang, H.-D. and Zheng, X.-P. 2012. Characteristics of hazardous chemical accidents in China: A statistical investigation. *Journal of Loss Prevention in the Process Industries* 25 (4), 686–693.

Further Readings

Andow, P. 1991. Guidance on HAZOP procedures for computer-controlled plants. UK Health and Safety Executive Contract Research Report No. 26.

ASTM (American Society for Testing and Materials).

ASTM D5. Test Method for Penetration of Bituminous Materials.

Beck, V.R. 1987. A cost-effective, decision-making model for building fire safety and protection. *Fire Safety Journal* 12 (2), 121–138.

BSI (British Standard Institution). BS 5588 Pt. 2.

Bullock, B.C. The development application of quantitative risk criteria for chemical processes. Fifth Chemical Process Hazard Symposium, I. Chem E., Manchester, April 1974.

Dicken, A.N.A. 1974. *The Quantitative Assessment of Chlorine Emission Hazards,* Chlorine Bicentennial Symposium, Chlorine Institute, New York.

Farmer, R.R. 1971. I. Chem E. Symposium Series No. 34, Major Loss Prevention in the Process Industries, p. 82.

2

Hazardous Area Classification

2.1 Introduction

Gases, vapors, mists, and dusts can all form explosive atmospheres with air. Hazardous area classification is used to identify places where, because of the potential for an explosive atmosphere, special precautions over sources of ignition are needed to prevent fires and explosions.

There are a great variety of hazardous area classifications applications, especially in the chemical and petrochemical industries, that require explosion protected equipment. As a result, there have been principles and technologies developed to allow electrical instrumentation and control devices to be used even in environments where there is a danger of explosion. However, focus on explosion-protected electrical equipment is not limited to utilization and processing of oil and natural gas. It has expanded into new fields such as waste disposal, landfills, and the utilization of biogas.

2.1.1 Groups and Categories of Apparatus

Apparatus are divided into equipment groups and categories detailed below.

Equipment group I applies to equipment intended for use in underground parts of mines and in those parts of surface installations of such mines liable to be endangered by firedamp and/or combustible dust. Equipment group I is further subdivided into categories M1 and M2.

Equipment group II applies to equipment intended for use in other places liable to be endangered by explosive atmospheres. Equipment group II is subdivided into categories 1, 2, and 3. The equipment selection is shown in Table 2.1.

2.1.2 CE Mark

The CE marking (or formerly EC mark) is a mandatory conformity marking for products sold in the European Economic Area (EEA) since 1993. The CE marking is the manufacturer's declaration that the product meets the requirements of the applicable EC directives.

TABLE 2.1

Equipment Selection

	Category of Equipment	Level of Protection	Performance of Protection	Conditions of Operation
Apparatus group I (mines)	M1	Very high	Two independent means of protection or safe even when two faults occur independently of each other.	Equipment remains energized and functioning when explosive atmosphere present.
	M2	High	Suitable for normal operation and severe operating conditions. If applicable, also suitable for frequently occurring disturbances or for faults, which are normally taken into account.	Equipment de-energized when explosive atmosphere is recognized.
Apparatus group II (surface)	1	Very high	Two independent means of protection or safe even when two faults occur independently of each other.	Equipment remains energized and functioning in zones 0, 1, 2 (gas) or 20, 21, 22 (dust D)
	2	High	Suitable for normal operation and frequently occurring disturbances or equipment where faults are normally taken into account.	Equipment remains energized and functioning in zones 1, 2 (gas) or 21, 22 (dust)
	3	Normal	Suitable for normal operation	Equipment remains energized and functioning in zones 2 (gas) or 22 (dust)

The CE marking symbolizes the conformity of the product with the applicable community requirements imposed on the manufacturer. The CE marking affixed to products is a declaration that the product conforms to all applicable Community provisions and the appropriate conformity assessment procedures have been completed. Goods for hazardous areas carrying the CE mark may be placed on the market and put into service.

Through the application of the conformity assessment procedures, manufacturers can issue an EC Declaration of Conformity, stating compliance with the relevant directive(s) and apply the CE mark on their equipment.

Components are not CE marked. The manufacturer or their authorized representative will issue a written attestation that declares the conformity of the components with the provisions of this directive that apply to them, and states their characteristics and how they must be incorporated into equipment or protective systems to assist compliance with the essential requirements applicable to finished equipment or protective systems.

2.1.3 Marking

All equipment and protective systems must be marked legibly and indelibly with the following minimum information:

- Name and address of the manufacturer
- CE marking, if involved the identification number or the notified body
- Designation of series or type
- Serial number, if any
- Year of construction
- The community mark T
- The equipment group and the category
- For group II, the letter G (concerning explosive atmospheres caused by gases, vapors, or mists) and/or the letter D (concerning explosive atmospheres caused by dust)

Furthermore, where necessary, they must also be marked with all information essential to their safe use. This additional marking is typically a requirement from harmonized standards that are used to demonstrate the compliance to the Essential Health and Safety Requirements. Figure 2.1 shows a sample of marking.

FIGURE 2.1
A sample of marking.

2.2 The Basic Physical Principles and Definitions

An explosion is defined as a sudden reaction involving rapid physical or chemical decay accompanied by an increase in temperature or pressure or both. This typically exists in chemical plants, refineries, paint shops, cleaning facilities, mills, flour silos, tanks, and loading facilities for flammable gases, liquids, and solids.

It is a fact that gases, vapors, and mists escape during the production, processing, transportation, and storage of flammable substances in the chemical and petrochemical industries, as well as in the production of mineral oil and natural gas, in mining, and in many other sectors. During many processes, especially in food industries, combustible dusts are also created.

These flammable gases, vapors, mists, and dusts form an explosive atmosphere with the oxygen in the air. If this atmosphere is ignited, explosions take place that can result in severe harm to human life and property. To avoid the danger of explosions, protective specifications in the form of laws, regulations, and standards have been developed in most countries that are aimed at ensuring that a high level of safety is observed. Due to the growing international economic link, extensive progress has been made in harmonizing regulations for explosion protection.

Reduction of hazards is not absolute. There is no absolute safety. Removing one of the elements from the ignition triangle can provide explosion protection and preclude unwanted, uncontrolled, and often disastrous explosions. wIf one of the three elements of the ignition triangle is missing, ignition will not occur. Since flammable substance and oxidizers cannot be frequently eliminated with certainty, inhibiting ignition of a potentially explosive atmosphere can eliminate danger at the source.

The objective of selecting an electrical apparatus and the means of installation is to reduce the hazard of the electrical apparatus to an acceptable level. An acceptable level might be defined as selecting protective measures and installation means to ensure that the probability of an explosion is not significantly greater due to the presence of electrical apparatus than it would have been had there been no electrical apparatus present.

The most certain method of preventing an explosion is to locate electrical equipment outside of hazardous (classified) areas whenever possible. In situations where this is not practical, installation techniques and enclosures are available that meet the requirements for locating electrical equipment in such areas. These methods of reducing hazards are based on the elimination of one or more of the elements of the ignition triangle discussed earlier.

2.2.1 Preventing Explosive Atmospheres (Primary Explosion Protection)

The term primary explosion protection refers to all precautions that prevent a hazardous explosive atmosphere from being created. Figure 2.2

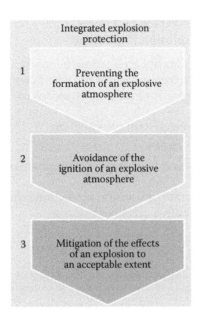

FIGURE 2.2
Basic principles of explosion protection.

shows three steps for integrated explosion protection. This can be achieved by

- Avoiding flammable substances (replacement technologies)
- Inerting (addition of nitrogen, carbon dioxide, etc.)
- Limitation of the concentration by means of natural or technical ventilation
- Avoiding ignition of explosive atmospheres

If the danger of explosion cannot be completely or only partly avoided by measures of preventing the formation of a hazardous explosive atmosphere, then measures must be taken that avoid the ignition of the explosive atmosphere.

The required safety level of these measures depends on the possible danger potential in the installation location. The hazardous areas are therefore divided into zones, according to the probability of an explosive atmosphere being formed, or hazardous locations, and classified into classes and divisions.

For locations classified in this way, requirements must be met concerning the apparatus approved for use in these locations. In addition, it is also necessary to prove that these requirements have been met.

2.2.2 Mitigation of the Explosion Effects
(Constructive Explosion Protection)

If hazardous explosive atmospheres cannot be safely avoided and their ignition cannot be excluded, then measures must be taken that limit the effect of
explosions to a safe degree, for example, by means of

- Explosion-pressure-resistant construction
- Explosion relief devices
- Explosion suppression by means of extinguishers

The principle of integrated explosion protection requires following explosion protection measures in a certain sequence.

2.2.3 Ignition Sources

The following ignition sources are examples that can cause an explosion in
the right circumstances:

- Hot surfaces
- Mechanically generated sparks
- Electrical equipment
- Equalizing current
- Static electricity
- Lightning
- Electromagnetic fields
- Optical radiation
- Ionizing radiation
- Ultrasonics
- Open flames
- Hot gas

Chemical reactions or biological processes that occur spontaneously at certain oxygen levels or temperatures

- Intense electromagnetic radiation
- Adiabatic compression and shock waves
- Static electricity
- Sparks or arcs from electrical equipment or wiring

2.2.4 Characteristics of Gases and Vapors

Different safety-relevant properties of substances can be obtained by laboratory experiments. Basically all gasses and vapors require oxygen to make them flammable. Too much or too little oxygen and the mixture will not ignite. The properties of the mixture gives information about a substance's burning behavior. The details have been discussed in Chapter 1 and some relevant properties are explained.

2.2.4.1 Flash Point

Flash points are normally associated with liquids but a few materials give off vapors when still in the solid state. The flash point of a flammable substance is the minimum temperature at which the material gives off vapors in a quantity such that it is capable of forming an ignitable vapor/air mixture. The combustible gas or vapor will ignite momentarily on application of an effective ignition source. Figure 2.3 shows the flash point for some chemical components and that the lower the flash point, the more dangerous and easier it is to inflame the liquid.

2.2.4.2 Ignition Temperature

The ignition temperature of a flammable substance is the minimum temperature at which the material will ignite and sustain combustion. This is also known as AIT (see Table 1.3).

2.2.4.3 Explosion Limits

If the concentration of a sufficiently dispersed flammable substance in air exceeds a minimum value (LEL), an explosion is possible. No explosion occurs if the concentration exceeds a maximum value (UEL).

Explosion limits change under conditions other than atmospheric. For example, the range of concentrations between the explosion limits widens generally as the pressure and temperature of the mixture increase. An explosive atmosphere can form above a flammable liquid only if the temperature of the liquid exceeds a minimum value. Figure 2.4 shows a sketch of explosion limits.

If the concentration is too high (rich mixture) or too low (lean mixture), no explosion occurs. Instead, there is just a steady-state combustion reaction or none at all. It is only in the range between the LEL and UEL that the mixture reacts explosively when ignited. The explosion limits depend on the ambient pressure and the proportion of oxygen in the air. Table 2.2 shows explosion limits of selected gases and vapors. Figure 2.5 shows the LEL scale for several chemical components and that the lower the LEL, the more dangerous the substance is, as ignitable concentrations can form more easily.

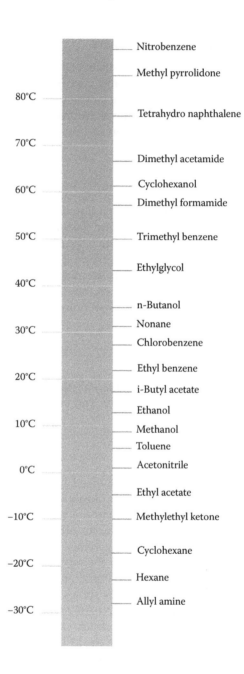

FIGURE 2.3
Flash point scale.

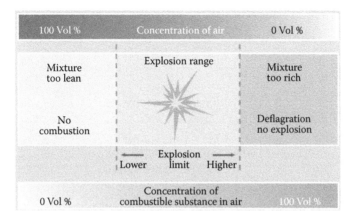

FIGURE 2.4
Explosion limits.

TABLE 2.2

Explosion Limits of Selected Gases and Vapors

Substance Designation	Lower Explosion Limit (Vol.%)	Upper Explosion Limit (Vol.%)
Acetylene	2.3	100 (Self-decomposing!)
Ethylene	2.4	32.6
Gasoline	~0.6	~8
Benzol	1.2	8
Heating oil/diesel	~0.6	~6.5
Methane	4.4	17
Propane	1.7	10.8
Carbon disulfide	0.6	60
Hydrogen	4	77

2.2.4.4 Maximum Oxygen Concentration

The maximum oxygen concentration in a mixture of a flammable substance, air, and an inert gas in which an explosion will not occur, is determined under specified test conditions. The safest way of concentration limiting is the removal of flammable gases and vapors from the process—but this is not normally practical. Where flammable gases and vapors are used, it is normal for gas detection systems to be used to limit concentrations. When the process is closed and the level of flammable gases or vapors is allowed to exceed the LEL level, this is acceptable as long as the oxygen concentration is kept low enough to control the risk of explosion (inertization).

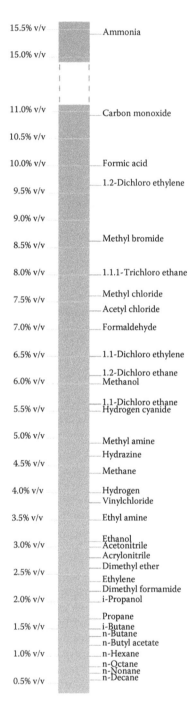

FIGURE 2.5
LEL scale for various chemical components.

2.2.4.5 *Vapor Density*

Vapor density of a gas is given relative to that of air. Many gases are lighter than air. Any vapor release will rise and dilute rapidly. When indoors, these gases will collect in the roof space. Where gases are heavier than air they will fall to the lowest point and fill sumps, trenches, or hollows in the ground. These gases can remain there long after the release has been stopped and continue to pose a danger.

Relative vapor density (RVD) is defined as the mass of a gas or vapor compared to air, which has an arbitrary value of 1. If the value of the RVD of a gas is less than 1 then the gas is lighter than air and hence will rise. The lighter the gas, the faster it rises. If the value of the RVD is greater than 1 then the gas is heavier than air and will sink.

The simple formula below can be used to calculate the RVD of a gas:

$$RVD = (Mw\ of\ gas)/(Mw\ of\ air) = (Mw\ of\ gas)/29$$

2.2.5 Maximum Experimental Safe Gap and Minimum Igniting Current Ratio

2.2.5.1 *Grouping and Classification of Areas Based on the Chemicals Used*

For the purposes of testing and approval of the electrical equipment suitable for the zone area and to help in area classification of the location itself, air mixtures that have not been oxygen enriched have been classified into three groups.

The three group classifications for zone applications are based on the maximum experimental safe gap (MESG) or the minimum igniting current (MIC) ratio or both.

2.2.5.1.1 *MIC Ratio*

The MIC ratio is based on the use of methane gas as a base for comparison to the other gases in question.

The minimum ignition current is mostly used as a ratio relative to that of methane. Ignition on a hot surface occurs in a relatively large macroscopic part of the mixture. In contrast, the ignition from a spark spreads in a relatively small microscopic part of the volume. The discharge from a capacitor or the interruption of a predefined resistive/inductive electric circuit can be used for classifying gases and vapors or dusts according to their ease of ignition in the microscopic part of the mixture volume.

For the assessment of the ignition of gases and vapors in a circuit using a device defined in International Electrotechnical Commission (IEC) 60079-11, a comparative value with methane as reference in a standardized circuit is used. This comparative value is the MIC ratio. It is the means used for classifying gases and vapors within explosion group II in the subgroups IIA, IIB, and IIC.

2.2.5.1.2 MESG

The behavior of the explosive atmosphere after ignition is relevant for the design of equipment that will limit the damage caused by an explosion. To introduce the explosion groups, the relevant parameter is the MESG. Gas inside a test chamber is ignited. A gap exists between the cover and the chamber with a gap length of 25 mm. The hot burning gas is now forced through the narrow gap. If the escaping gas ignites the surrounding gas, the test will have to be repeated with a smaller gap. The gap, which prevents the ignition of the surrounding gas, is the MESG.

These group classifications are subdivided into Groups IIC, IIB, and IIA according to the nature of the gas or vapor and the type of protection technique.

Group IIC is an atmosphere containing acetylene, hydrogen, or flammable gas, flammable liquid-produced vapor, or combustible liquid-produced vapor mixed with air that may burn or explode when ignited, having either an MESG less than or equal to 0.50 mm or an MIC ratio less than or equal to 0.45.

Group IIB is an atmosphere containing acetaldehyde, ethylene, or flammable gas, flammable liquid-produced vapor, or combustible liquid-produced vapor mixed with air that may burn or explode when ignited, having either an MESG greater than 0.50 mm and less than or equal to 0.90 mm or an MIC ratio greater than 0.45 and less than or equal to 0.80.

Group IIA is an atmosphere containing acetone, ammonia, ethyl alcohol, gasoline, methane, propane, or flammable gas, flammable liquid-produced vapor, or combustible liquid-produced vapor mixed with air that may burn or explode when ignited, having either an MESG greater than 0.90 mm or an MIC ratio greater than 0.80.

Equipment can be listed for the specific gas or vapor, specific mixtures of gases or vapors, or any specific combination of gases or vapors. This permits electrical equipment to be designed, tested, and listed based on a specific anticipated usage. For example, electrical equipment that has been tested as a Group IIB would normally be required for ethylene-based chemical atmospheres but may also have been tested for hydrogen-based chemicals. This permits the equipment to be marked with a IIB marking instead of having to be subjected to the more rigid tests for acetylene, as would normally be required if a IIC equipment listing is used.

2.2.6 Explosion Group

The explosive characteristics of the air mixtures of gases, vapors, or dusts vary with the specific material involved. Materials have been placed in groups based on their ignition temperatures and explosion pressures. Electrical apparatus for potentially explosive atmospheres is divided into two groups per EN 50014:

- Group I: Electrical apparatus for mines susceptible to firedamp (for use underground).
- Group II: Electrical apparatus for places with a potentially explosive atmosphere other than mines susceptible to firedamp.

Electrical apparatus of Group II may be subdivided according to the nature of the potentially explosive atmosphere for which it is intended. The subdivision is based on the MESG for flameproof enclosures or the MIC for intrinsically safe electrical apparatus as required in the specific European Standards, as shown in Table 2.3.

A piece of apparatus classified as IIC may also be used in a IIA and IIB application, a piece of apparatus designated as IIB can also be used in a IIA but not in a IIC application, and apparatus designated as IIA can only be used in a IIA application.

2.2.7 Temperature Classification

Ignition temperature is influenced by various factors such as the size, shape, type, and composition of a surface. This is the minimum temperature

TABLE 2.3

Explosion Group

Explosion Group	Flammable Material (Gases and Vapors)	Maximum Experimental Safe Gap[a]	Minimum Ignition Current[a]
IIA	Propane	>0.9 mm	>0.8
	Acetone		
	Benzene		
	Butane		
	Methane		
	Gas		
	Hexane		
	Paint solvents		
IIB	Ethylene	0.5–0.9 mm	0.45–0.8
	Propylene oxide		
	Ethylene oxide		
	Butadiene		
	Cyclopropane		
	Ethyl ether		
IIC	Hydrogen	<0.5 mm	<0.45
	Acetylene		
	Carbon disulfide		

[a] IEC 60079-12 provides an overview of classification by MESG and MIC procedures.

required, at normal atmospheric pressures in the absence of spark or flame, to set afire; that is, to cause self-sustained combustion independently of the heating or heated element.

The maximum surface temperature of electrical or mechanical apparatus must always be lower than the ignition temperature of the surrounding explosive atmosphere. The ignition temperature of different gases varies considerably.

A mixture of air with hydrogen will ignite at 560°C but a mixture of air with diethyl ether will ignite at 170°C. To help manufacturers design their equipment, apparatus are given a temperature classification consisting of six temperatures ranging from 85°C (T6) to 450°C (T1). The six T classes are given in Table 2.4.

Apparatus will be marked according to the maximum surface temperature of any relevant part that might be in contact with the flammable gas. For flameproof and pressurized equipment the maximum surface temperature is on the outside of the enclosure, whereas for increased safety the hottest point is inside. The temperature classification for Group II electrical apparatus will be either

- T class as given in Table 2.4
- Actual maximum surface temperature
- Specific gas for which it is designed

Apparatus suitable in for example, the T3 temperature class can also be used in T1 and T2. Electrical apparatus should normally be designed for use in an ambient temperature of –20°C and +40°C. When designed for use in a different range, the ambient temperature must be stated by the manufacturer and specified in the certificate. The marking must include either the special

TABLE 2.4

Maximum Surface Temperature for Group II Electrical Apparatus

Temperature Class	Maximum Surface Temperature of Apparatus (°C)	Ignition Temperature of Flammable Substance (°C)
T1	450	>450
T2	300	>300 ≤ 450
T3	200	>200 ≤ 300
T4	135	>135 ≤ 200
T5	100	>100 ≤ 135
T6	85	>85 ≤ 100

temperature range; for example, $35°C \leq Ta \leq +55°C$ or the letter X after the certificate number.

Table 2.5 gives the classification of some gases in explosion groups and temperature classes.

If a hazard is present the equipment used within the installation must be given an appropriate T classification in order to maintain the integrity. If that hazard is, say, hydrogen, then all equipment used must meet the T6 rating. This means that all equipment used must not have a surface temperature greater than 85°C. Any equipment used that can generate a hotter surface temperature of greater than 85°C must not be used as this will then increase the likelihood of an explosion by igniting the hydrogen in the atmosphere.

Table 2.6 provides safety ratings such as ignition temperature, temperature class, and explosion group.

TABLE 2.5

Classification of Some Gases in Explosion Groups and Temperature Classes

	T1	T2	T3	T4	T5	T6
I	Methane					
IIA	Acetic acid	Butane	Aircraft fuel	Acetaldehyde		
	Acetone	Cyclohexane				
	Ammonia	Ethanol	Cyclohexane			
	Benzene	n-Butane	Diesel fuel			
	Carbon monoxide	n-Butyl alcohol	Fuel oil			
	Ethane Methanol	Propanol 2	Heptane			
	Propane Toluene		Kerosene			
			n-hexane			
			Petroleum			
			Turpentine			
			Pentane			
IIB	Coal gas	Ethylene	Ethylene glycol	Ethyl methyl ether		
		Ethylene oxide	Hydrogen sulfide			
		Methyl ethyl Ketone	Tetrahydrofuran			
		Propanol 1				
IIC	Hydrogen	Acetylene				Carbon disulfide

TABLE 2.6

Safety Ratings: Ignition Temperature, Temperature Class, and Explosion Group

Material	Ignition Temperature (°C)	Temperature Class	Explosion Group
1,2-Dichloroethane	440	T2	IIA
Acetaldehyde	155	T4	IIA
Acetic acid	485	T1	IIA
Acetic anhydride	330	T2	IIA
Acetone	535	T1	IIA
Acetylene	305	T2	IIC (also gas groups IIB + C2H2)
Ammonium	630	T1	IIA
Benzene	555	T1	IIA
Carbon disulfide	95	T6	IIC (also gas groups IIB + CS2)
Carbon monoxide	605	T1	IIA
Cyclohexanone	430	T2	IIA
Diethyl ether	175	T4	IIB
Diesel fuels	220	T3	IIA
Ethane	515	T1	IIA
Ethanol	400	T2	IIB
Ethene	440	T2	IIB
Ethyl chloride	510	T1	IIA
Ethyl ethanoate	470	T1	IIA
Ethyl glycol	235	T3	IIB
Ethylene oxide	435 (self-decomposing)	T2	IIB
Fuel oil EL, L, M, S	220 to 300	T3	IIA
Hydrogen	560	T1	IIC (also gas groups IIB + H2)
Hydrogen sulfide	270	T3	IIB
i-amyl acetate	380	T2	IIA
Methane	595	T1	IIA
Methanol	440	T2	IIA
Methyl chloride	625	T1	IIA
Naphthalene	540	T1	IIA
n-butane	365	T2	IIA
n-butanol	325	T2	IIB
n-hexane	230	T3	IIA
n-propyl alcohol	385	T2	IIB (the gas group for this substance has not yet been determined)
Petrol fuels	220 to 300	T3	IIA
Phenol	595	T1	IIA
Propane	470	T1	IIA
Toluene	535	T1	IIA

2.3 Area Classification

Area classification is a method of analyzing and classifying the environment where explosive gas atmosphere may occur to allow the proper selection of electrical apparatus to be installed in that environment. The object of the classification procedure is to enable electrical apparatus to be operated safely in that environment.

Installations in which flammable materials are manufactured, handled, or stored should be designed, operated, and maintained so that any releases of flammable material and the extent of hazardous areas are kept to a minimum.

In situations where there may be an explosive gas atmosphere, the following steps should be taken:

- Eliminate the likelihood of an explosive gas atmosphere occurring around the source of ignition or eliminate the source of ignition, or
- Limit the range of explosion flames and explosion pressures to a sufficient level of safety.

    Where this is not possible, protective measures, process equipment, systems, and procedures should be selected so the likelihood of both being present at the same time is acceptably small. In the first instance it is preferable to eliminate the presence of a flammable atmosphere. This is possible by

    - Substituting with a nonflammable substance or raising the flash point above the process temperature (e.g., adding water)
    - Lowering the process temperature (e.g., cooling)
    - Limiting the concentration below the LEL (e.g., dilution/ventilation or inerting)
    - Explosion-proof design (containment)

In practice however, it is very difficult to ensure that an explosive gas atmosphere will never occur. In this case, apparatus with special protective measures should be used.

Once an area has been classified, there is little difficulty in understanding the equipment requirements and associated wiring method because those can be determined from applicable publications.

Therefore, in situations where an explosive gas atmosphere has a high likelihood of occurrence, reliance is placed on using electrical apparatus that has an extremely low likelihood of creating a source of ignition. Conversely, where the likelihood of an explosive gas atmosphere occurring is reduced, electrical apparatus that has an increased likelihood of becoming a source of ignition may be used.

2.3.1 Class Definition

A hazardous area is defined by three main criteria:

1. The type of hazard (groups)
2. The AIT of the hazardous material (temperature or T rating)
3. The likelihood of the hazard being present in flammable concentrations (zones)

2.3.1.1 Type of Hazard

The type of hazard will be in the form of either a gas or vapor or a dust or fiber. The classification of these hazards is primarily divided into two groups depending on whether the hazard is in a mining or above-surface industry. These are defined below and in Table 2.7:

- *Group I:* electrical equipment for use in mines and underground installations but Group II and Group III comprise electrical equipment for use in surface installations.

- *Groups II and III:* are further subdivided depending on the hazard. Group II gases are grouped together based upon the amount of energy required to ignite the most explosive mixture of the gas with air. Group III dusts are subdivided according to the nature of the explosive atmosphere for which it is intended.

NFPA Publication 70, National Electrical Code (NEC) defines the type of hazardous substance that is or may be present in the air in quantities sufficient to produce explosive or ignitable mixtures.

TABLE 2.7

Classification of Hazards

Mining	Surface Industry			
Group I	Group II		Group III	
Electrical equipment for mines susceptible to firedamp	Electrical equipment for places with an explosive gas atmosphere		Electrical equipment for places with an explosive dust atmosphere	
	Subdivision	Ignition Energy	Subdivision	Explosive Atmosphere
	IIA	260 µJ	IIIA	Combustible flyings
	IIB	95 µJ	IIIB	Nonconductive dust
	IIC	18 µJ	IIIC	Conductive dust

- Class I locations are those in which flammable vapors and gases may be present
- Class II locations are those in which combustible dust may be found

The zone standard classifies the types of flammable substances present as such:

- Zone 0, zone 1, and zone 2 are zones where hazardous vapors and gases are present
- Zone 20, zone 21, and zone 22 are zones where hazardous dusts or fibers are present

To apply this approach, the first step is to assess the likelihood of an explosive gas atmosphere occurring in accordance with the definitions of zone 0, zone 1, and zone 2. The following sections give guidance on the classification of area in which there may be an explosive gas atmosphere into zones 0, 1, and 2.

2.3.2 Definitions of Zones

The hazardous location areas are defined by taking into account the different dangers presented by potentially explosive atmospheres. This enables protective measures to be taken that account for both cost and safety factors.

A hazardous place is a place in which an explosive atmosphere may occur in such quantities as to require special precautions to protect workers against explosion hazards. Such a quantity is termed a hazardous explosive atmosphere. As a basis for determining the extent of protective measures, any remaining hazardous places must be classified in terms of zones according to the likelihood of occurrence of such atmospheres. Tables 2.8 through 2.10 provide the details of zone classification. No exact correlation can be made between the Zone and Division designations.

- *Zone 0:* A place in which an explosive atmosphere consisting of a mixture with air of flammable substances in the form of gas, vapor, or mist is present continuously or for long periods or frequently. In this zone, flammable atmosphere is continuously present or present for long periods. Equipment suitable for zone 0 can be used in zones 0, 1, or 2.

TABLE 2.8

Summary of Flammable Mixture Present for Different Zones

Zone	Flammable Mixture Present
0	1000 or more hours/year (10%)
1	10 < hours/year <1000 (0.1%–10%)
2	1 < hour/year <10 (0.01%–0.1%)
Unclassified	Less than 1 hour per year (0.01%)

TABLE 2.9

Definition of Zones

Zone	Definitions (Per NEC Article 505-9, CEC Section 18, EN60079-10, IEC 60079-10)
0	In which ignitable concentrations of flammable gases or vapors are • Present continuously • Present for long periods of time
1	In which ignitable concentrations of flammable gases or vapors are • Likely to exist under normal operating conditions • May exist frequently because of repair, maintenance operations, or leakage
2	In which ignitable concentrations of flammable gases or vapors are • Not likely to occur in normal operation • Occur for only a short period of time • Become hazardous only in case of an accident or some unusual operating condition

TABLE 2.10

Zone Classification for Both Gas and Dust

Gas	Dust	Hazardous Area Characteristics
Zone 0	Zone 20	A hazardous atmosphere is highly likely to be present and may be present for long periods of time (>1000 hours per year) or even continuously
Zone 1	Zone 21	A hazardous atmosphere is possible but unlikely to be present for long periods of time (>10, <1000 hours per year)
Zone 2	Zone 22	A hazardous atmosphere is possible but unlikely to be present for long periods of time (>10, <1000 hours per year)

- *Zone 1:* A place in which an explosive atmosphere consisting of a mixture with air of flammable substances in the form of gas, vapor, or mist is likely to occur in normal operation occasionally. In this zone, flammable atmosphere is likely to occur in normal operation. Equipment suitable for zone 1 can be used in zones 1 or 2.

- *Zone 2:* A place in which an explosive atmosphere consisting of a mixture with air of flammable substances in the form of gas, vapor, or mist is not likely to occur in normal operation, but if it does occur, will persist for a short period only. In this zone, flammable atmosphere is not likely to occur in normal operation, and if it occurs, will exist only for short period. Equipment suitable for zone 2 can be used only in zone 2.

Table 2.9 shows the three zones for gases and vapors and dusts and fibers. It is recommended that plant and installation in which flammable materials are handled, stored, or processed be so designed that hazardous areas are kept to a minimum, in particular that zone 0 and zone 1 areas should be a kept to a minimum in both number and extent. In other words, the hazardous areas should be mainly zone 2.

Where release of flammable material is unavoidable, plant equipment items should be mainly limited to secondary-grade sources of release, or failing this (i.e., where primary or continuous sources of release are unavoidable), the source of release should be such as to have a very limited quantity or rate of release to atmosphere. In carrying out area classification, these principles should receive prime consideration and where necessary, the design, operation, or location of process equipment should be modified to meet these requirements.

Similarly, consideration should be given to the design and operation of process equipment to ensure that even when it is operating abnormally the amount of flammable material released to atmosphere is minimized in order to reduce the extent of the area (zone 2) made hazardous.

Once a plant has been classified and any necessary records made, it is important that no modification to equipment or operating procedures is made without discussion with those responsible for the area classification.

Determination may require inputs from various disciplines before the location classification drawing can be developed.

Note that the opening of parts of closed process systems (e.g., filter changing, batch filling) should also be considered as source of release in developing area classification.

Four principles ensure that electrical equipment does not become a source of ignition. The basic point is to ensure that parts to which a potentially explosive atmosphere has free access do not become hot enough to ignite an explosive mixture (Table 2.11).

TABLE 2.11

Principles and Protection Methods to Avoid Electrical Equipment as a Source of Ignition

Item Number	Principles	Protection Method
1	Explosive mixtures can penetrate the electrical equipment and be ignited. Measures are taken to ensure that the explosion cannot spread to the surrounding atmosphere.	Confine the explosion • Flameproof enclosure • Powder-filled enclosure
2	The equipment is provided with an enclosure that prevents the ingress of a potentially explosive mixture and/or contact with sources of ignition arising from the functioning of the equipment.	Isolate the hazard • Pressurized enclosure • Oil-filled enclosure • Potted enclosure
3	Potentially explosive mixtures can penetrate the enclosure but must not be ignited. Sparks and temperatures capable of causing ignition must be prevented.	Increased safety
4	Potentially explosive mixtures can penetrate the enclosure but must not be ignited. Sparks and raised temperatures must only occur within certain limits.	Limit the energy • Intrinsically safe

It is important that operators of hazardous location plants ensure that their personnel know when explosions are likely to happen and how to prevent them. A joint effort by the manufacturers of explosion-proof electrical equipment and the constructors and operators of industrial plants can help ensure the safe operation of electrical equipment in hazardous locations.

2.3.3 Examples of Zone Identification

The likelihood of the presence of an explosive gas atmosphere and hence type of zone depends mainly on the grade of source of release. In some cases the ventilation and other factors may also affect the type of zone.

Figure 2.6 shows a flammable liquid storage tank, situated outdoors, with fixed roof and no internal floating roof. Taking into account relevant parameters, the following are typical values that will be obtained for this example: $a = 3$ m from vent openings, $b = 3$ m above the roof, and $c = 3$ m horizontally from the tank.

Figure 2.7 shows a cone roof storage tank, Figure 2.8 shows a typical hazardous area classification for a hydrogen laboratory, and Figure 2.9 illustrates different zones for a floating roof storage tank.

2.3.4 Sources of Release

The basic elements for establishing the hazardous zone types are the identification of the source of release and the determination of the grade of release. Each item of process equipment (e.g., tank, pump, pipeline, vessel) should

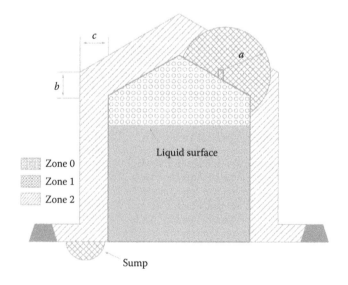

FIGURE 2.6
Flammable liquid storage tank.

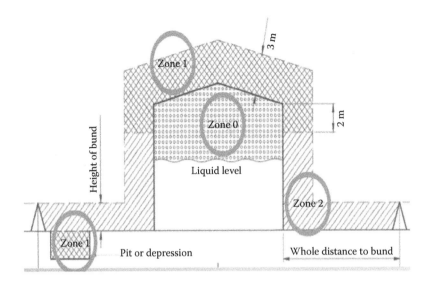

FIGURE 2.7
Cone roof storage tank.

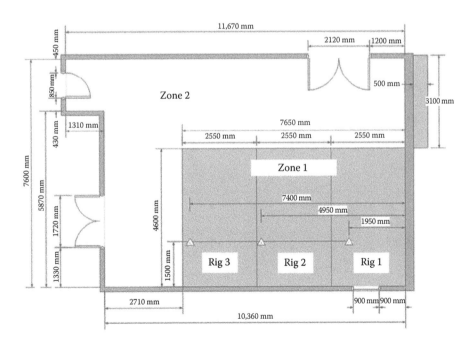

FIGURE 2.8
A typical hazardous area classification for a hydrogen laboratory. (Reprinted from Karri, V. et al., *International Journal of Hydrogen Energy* 33, pp. 2857–2867, 2008. With permission.)

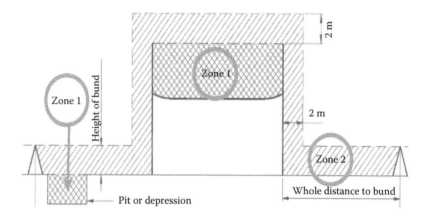

FIGURE 2.9
Floating roof storage tank.

be considered as a potential source of release. Items that contain flammable material but cannot release it to the atmosphere (e.g., all-welded pipeline) are not considered to be sources of release. Figure 2.10 shows a typical source of release. Figure 2.11 shows a typical source of ignition.

2.3.4.1 Grading of Source of Release

Each item of process equipment (e.g., tank, pump, pipeline, vessel) should be considered as a potential source of release of flammable material. If the item cannot contain flammable material it will clearly not give rise to a hazardous area around it. The same will apply if the item contains a flammable material but cannot release it to atmosphere (e.g., an all-welded pipeline is not a source of release).

FIGURE 2.10
Typical sources of release.

Electrical apparatus

FIGURE 2.11
Typical ignition source.

If it is established that the item may release flammable material to the atmosphere (as will be the case with most items) it is necessary, first of all, to determine the grade of the source of release in accordance with the definitions by establishing the likely frequency and duration of release.

By means of this procedure each item will be graded as *continuous, primary, secondary,* or *multigrade*.

The source grade should be estimated by considering operational openings to atmosphere and the possibility of releases under all (normal and abnormal) operating conditions of the plant, installation, or process.

Releases are categorized as follows:

Continuous grade of release: A release that is continuous or is expected to occur for long periods. Examples of continuous grade of release include

- Surface of a flammable liquid in a fixed roof tank
- Surface of an open reservoir (e.g., oil/water separator)

Primary grade of release: A release that can be expected to occur periodically or occasionally during normal operation. Examples of primary grade of release include

- Seals of pumps, compressors, or valves that are expected to release flammable material, particularly during start-up
- Water drainage points on vessels that contain flammable liquids
- Sample points from which analytical samples are drawn
- Relief valves, vents, and other openings, which that are expected to release during normal operation

Secondary grade of release: A release that is not expected to occur in normal operation, and if it does occur, is likely to do so only infrequently for short periods. Examples of secondary grade of release include

- Seals of pumps, compressors, or valves that are not expected to release flammable material during normal operation
- Flanges, connections, and pipe fittings where release of flammable materials is not expected during normal operation
- Relief valves, vents, and other openings that are not expected to release during normal operation

A continuous grade of release normally leads to a zone 0, a primary grade to zone 1, and a secondary grade to zone 2.

2.3.5 Extent of Zone

Quite a number of factors can influence the extent of the zone. If the gas is lighter than air, it rises on release and can become trapped in the roof space, or if the gas is heavier than air, it will fall and spread at ground level. This has an impact on the location of the site; is it on a hill or in a hollow?

The extent of zones depends mainly on the following parameters:

1. *Release rate of flammable material.* The extent may increase with increasing release rate (see release velocity).
2. *Release velocity.* Due to an improved dilution for release of flammable gases, vapors, and/or mists in air, the extent of a hazardous area may decrease if, with constant release rate, the release velocity increases that which causes turbulent flow.
3. *Concentration of flammable gases vapors and/or mists in the released mixture.* The extent may increase with increasing concentration at the release source.
4. *Ventilation.* With an increased rate of ventilation, the extent of a hazardous area may be reduced. The extent may also be reduced by an improved arrangement of the ventilation system.
5. *Obstacles.* Obstacles may impede the ventilation and thus may enlarge the extent of the zone. On the other hand some obstacles, for example dikes and walls, may limit the movement of a cloud of an explosive gas atmosphere and thus may reduce the extent.
6. *Boiling point of flammable liquids (initial boiling point for liquid mixtures).* For flammable liquids, the concentration of the released vapor is related to the vapor pressure at the relevant maximum liquid temperature. The lower the initial boiling point, the greater the vapor

pressure for a given liquid temperature and hence the greater the extent of the hazardous area.

7. *LEL.* The lower the LEL, the larger the extent of the hazardous area may be.

8. *Flash point.* An explosive gas atmosphere cannot exist if the flash point is significantly above the relevant maximum temperature of the flammable liquid. The lower the flash point, the larger the extent of the hazardous area may be. Some liquids (such as certain halogenated hydrocarbons) do not possess a flash point although they are capable of producing an explosive gas atmosphere; in these cases, the equilibrium liquid temperature corresponding to saturated concentration at LEL should be compared with the relevant maximum liquid temperature. (Note: In special conditions the mist of a flammable liquid may be released at a temperature below the flash point and may therefore produce an explosive gas atmosphere.)

9. *Relative density.* The horizontal extent of a hazardous area may increase with increasing relative density. Where the relative density is less than unity, the vertical extent may increase with a decrease in relative density.

10. *Liquid temperature.* The extent of a hazardous area may increase with increasing temperature of process liquid provided the temperature is above the flash point. Note that the liquid or vapor temperature after the release may be increased or decreased by the ambient temperature or other factors (e.g., a hot surface).

When sources of release are in an adjacent area, the migration can be prevented by

- Physical barriers
- Static overpressure in the area adjacent to the hazardous area
- Purging the area with a significant airflow

2.3.5.1 Procedure for Determining the Type and Extent of Zones

After establishing the grade of the release source, it is necessary to determine the release rate from the item, and furthermore, the other necessary factors that may influence the type and extent of the hazardous area, as mentioned in previous sections.

It is rarely possible by cursory examination of a plant or plant design to decide which parts of the plant can be matched to the three zonal definitions (zones 0, 1, and 2). A more objective approach is therefore necessary and involves the analysis of the basic possibility of an explosive gas atmosphere occurring. Since an explosive gas atmosphere can exist only if a flammable gas, vapor, or mist is present with air, it is necessary to decide if any of these flammable materials can exist in the area concerned. Generally speaking such materials (and also flammable liquids and solids that may give rise to them) are contained within process equipment that may or may not provide a totally enclosed containment. To discover the cause of an explosive gas atmosphere existing in an area, it is necessary therefore to determine how the items of process equipment containing flammable materials can release them to atmosphere.

Once the likely frequency and duration of release (and hence the grade of source of release), the release rate, concentration, velocity, ventilation, and other factors that affect the type of zone and/or extent of the hazardous area have been determined, there is then a firm basis on which to determine the likely presence of an explosive gas atmosphere in the surrounding areas. This approach therefore requires detailed consideration to be given to each item of process equipment that contains a flammable material and could therefore be a source of release. In some cases there may be other considerations (e.g., consequential damage) that may call for a different classification but these considerations are outside the scope of this book.

2.4 Ventilation

Gas or vapor released to the atmosphere can be diluted by dispersion or diffusion in the air until its concentration is below the lower explosive limit. Ventilation (i.e., air movement) will promote dispersion. The degree of ventilation, for example number of air changes per hour, may affect the type and/ or extent of zones.

The most important factor is that the design of ventilation is related to the source of release in order to achieve optimal ventilation conditions in the hazardous area. In considering the effect of ventilation, the relative density of the gases or vapors is important and should receive special consideration in determining the ventilation arrangements.

The following main types of ventilation are recognized:

- Natural ventilation
- General artificial ventilation
- Local artificial ventilation

Note that the above terms are related to the type and not to the degree of ventilation. It is also necessary to recognize another type:

- No ventilation

2.4.1 Natural Ventilation

Natural ventilation is created by the movement of air caused by the wind and/or by temperature gradients. Examples include

- Open-air situations typical of those in the chemical and petroleum industries that comprise open structures, pipe racks, pump bays, and the like
- An open building which, having regard to the relative density of the gases and/or vapors involved, has an opening in the wall and/or roof so dimensioned and located that the ventilation inside the building for the purpose of area classification can be regarded as equivalent to that in an open-air situation
- A building that is not an open building but has natural ventilation (generally less than that of an open building) provided by permanent openings made for ventilation purposes

2.4.2 General Artificial Ventilation

Artificial ventilation is provided by artificial means (e.g., fans or extractors). With the use of artificial ventilation it is possible to achieve

- Reduction in the extent of the zone
- Shortening of the time of persistence of an explosive atmosphere
- Prevention of the generation of an explosive atmosphere

Examples of artificial ventilation include

- A building that is equipped with fans in the walls and/or the roof to improve the general ventilation in the building
- An open-air situation provided with suitably located fans to improve the general ventilation of the area

2.4.3 Local Artificial Ventilation

Examples of local artificial ventilation include

- An air/vapor extraction system applied to a process machine or vessel that continuously or periodically releases flammable vapor
- A forced or extract ventilation system applied to a small inadequately ventilated local area where it is expected that an explosive atmosphere may otherwise occur

2.4.4 No Ventilation

An example of an area with no ventilation is an enclosure or room having no permanent opening. (Note that in a large building, the volume of which is large compared with the hazardous area around the source of release, that area need not necessarily be considered to have no ventilation.)

2.4.5 Degrees of Ventilation

There are three degrees of ventilation:

1. *High ventilation* (VH) can reduce the concentration at the source of release virtually instantaneously, resulting in a concentration below the LEL. A zone of negligible extent results.
2. *Medium ventilation* (VM) can control the concentration, resulting in a stable zone boundary while the release is in progress, and where the explosive gas atmosphere does not persist unduly after the release has stopped. The extent and type of the zone depend on the design parameters.
3. *Low ventilation* (VL) cannot control the concentration while release is in progress and/or cannot prevent undue persistence of a flammable atmosphere after release has stopped.

2.4.6 Availability of Ventilation

Three levels of availability of ventilation should be considered:

1. *Good:* Ventilation is present virtually continuously. Natural ventilation, which is generally obtained outdoors, is considered to be good when the wind speed is greater than 0.5 m/s (approximately 1.1 miles per hour).
2. *Fair:* Ventilation is expected to be present during normal operation. Interruptions are permissible provided that they occur infrequently and for short periods.
3. *Poor:* Ventilation, that does not meet the standard of fair or poor but interruptions are not expected to occur for long periods. Ventilation that is less than poor is ignored.

Table 2.12 shows influences of ventilation on type of zone.

TABLE 2.12

Influence of Ventilation on Type of Zone

Ventilation — Degree — Availability

Grade of Release	High			Medium			Low
	Good	Fair	Poor	Good	Fair	Poor	Good, Fair, or Poor
Continuous	(Zone 0 NE) Nonhaz (zone 0 NE, indicates a theoretical zone that would be of negligible extent)	(Zone 0 NE) Zone 2 (zone 0 NE, indicates a theoretical zone that would be of negligible extent)	(Zone 0 NE) Zone 1 (zone 0 NE, indicates a theoretical zone that would be of negligible extent)	Zone 0	Zone 0 (surrounded by zone 2)	Zone 0 (surrounded by zone 1)	Zone 0
Primary	(Zone 1 NE) Nonhaz (zone 1 NE, indicates a theoretical zone that would be of negligible extent)	(Zone 1 NE) Zone 2 (zone 1 NE, indicates a theoretical zone that would be of negligible extent)	(Zone 1 NE) Zone 2 (zone 1 NE, indicates a theoretical zone that would be of negligible extent)	Zone 1	Zone 1 (surrounded by zone 2)	Zone 1 (surrounded by zone 2)	Zone 1 or zone 0 (it will be zone 0 if the ventilation is so weak and the release is such that in practice an explosive atmosphere exists virtually continuously)
Secondary (the zone 2 area created by a secondary grade release may exceed that attributable to a primary or continuous grade release; in which case, the greater distance should be taken)	(Zone 2 NE) Nonhaz (zone 2 NE, indicates a theoretical zone that would be of negligible extent)	(Zone 2 NE) Nonhaz (zone 2 NE, indicates a theoretical zone that would be of negligible extent)	Zone 2	Zone 2	Zone 2	Zone 2	Zone 1 and even zone 0 (it will be zone 0 if the ventilation is so weak and the release is such that in practice an explosive atmosphere exists virtually continuously)

2.4.7 Relationships between Grades of Source of Release, Ventilation, and Type and Extent of Zone

2.4.7.1 Natural and General Artificial Ventilation

A continuous grade source of release may lead to a zone 0, a primary grade to a zone 1, and a secondary grade to a zone 2. In some cases, however, the degree of ventilation may be so good that the extent of the zone may be so small as to be negligible or the zone may have a higher number or in the limit may become nonhazardous. On the other hand, the degree of ventilation may be so poor that the zonal classification will have a larger extent and in some cases a lower zone number. In any case wind direction should not be overlooked.

2.4.7.2 Local Artificial Ventilation

Local artificial ventilation will usually be more effective than natural and general artificial ventilation in diluting explosive gas air mixtures. As a consequence, the extent of the zones will be reduced and in some cases will be so small as to be negligible or the zone will have a higher number or may become nonhazardous.

2.4.7.3 No Ventilation

Where a source of release is within an area with no ventilation, a continuous grade will, and a primary grade may, lead to a zone 0 and a secondary grade to a zone 1.

Where there are special factors, for example very infrequent release and/or monitoring of the release, higher zone number may apply.

2.4.7.4 Ventilation Restricted by Obstacles

Some obstacles that exist in the area may impede the ventilation local to the obstacles and may thus require a larger extent and/or a lesser zone number in the local area. In considering the effect of obstacles, especially in the case of pits and pockets, some of which may be inverted, particular attention should be paid to the relative density of the gases and vapors involved.

2.4.7.5 Consequences of Artificial Ventilation Failure

The area classification should be established assuming that the ventilation is in operation since this will be the normal condition. The risk of ventilation failure should then be established. If the risk of failure is negligible (e.g., because an automatic independent standby system is provided) the area classification determined with the ventilation operating will not need modification. However, should there be a risk of failure of the ventilation, the likely frequency and duration should be established together with the extent of spread of explosive gas atmosphere in the absence of artificial ventilation. This extent of spread may be greater than that of the area classification extent already determined with the ventilation operating. The zone number(s) of the whole area under consideration with ventilation off will depend on the likely frequency and duration of ventilation failure and on the classification determined with the ventilation on.

Where the ventilation is likely to fail infrequently and for short periods, the additional area due to ventilation failure needs to have a zone number no less than 2.

If provision is made to prevent release of flammable material when the ventilation has failed (e.g., by automatically closing down the process) the classification determined with the ventilation operating need not be modified.

The following items must be noted:

- The electrical installation in the additional area need not be suitable for the zone of this area if provision is made to switch off such electrical installation on ventilation failure.

- For detailed information about variation in ventilation conditions covering ventilation check for the classification of open sheltered or enclosed areas with released of heavier or lighter-than-air gases or vapors, see Chapter 6 of *IP Model Code of Safe Practice Part 15* (1990).

- While assessing ventilation condition, consideration should be given to abnormal topographical and or meteorological limitations (e.g., hollows, major unevenness, and sloping ground down where heavier-than-air gas/vapors or condensated vapor might accumulate).

- Natural ventilation is characterized by wind speed not less than 0.5 meters/second and frequently over 2 meters/second.

- Steady-state 12 fresh air changes per hour is referred to as adequate ventilation.

2.5 Equipment Protection Level

A risk assessment approach for the acceptance of Ex equipment has been introduced as an alternative method to the current prescriptive and relatively inflexible approach linking equipment to zones. To facilitate this, a system of equipment protection levels (EPLs) has been introduced to clearly indicate the inherent ignition risk of equipment no matter what type of protection is used.

The EPL is a level of protection assigned to equipment based on its likelihood of becoming a source of ignition and distinguishing the differences between explosive gas atmospheres, explosive dust atmospheres, and the explosive atmospheres in mines susceptible to firedamp.

- *EPL Ga:* Equipment for explosive gas atmospheres having a very high level of protection that is not a source of ignition in normal operation, during expected malfunctions, or during rare malfunctions
- *EPL Gb:* Equipment for explosive gas atmospheres having a high level of protection that is not a source of ignition in normal operation or during expected malfunctions
- *EPL Gc:* Equipment for explosive gas atmospheres having an enhanced level of protection that is not a source of ignition in normal operation and that may have some additional protection to ensure it remains inactive as an ignition source in the case of regular expected occurrences (e.g., failure of a lamp)

2.5.1 Zones

Hazardous areas are classified into zones. Zoning does not take account of the potential consequences of an explosion. In a relationship between EPLs and zones where only the zones are identified in the area classification documentation, then the relationship between EPLs and zones from Table 2.13 should be followed.

2.6 Gas Explosion Protection Concepts for Electrical Equipment

2.6.1 Zones of Use

To protect workers against explosion hazards, equipment and protective systems for all places in which explosive atmospheres may occur must be selected on the basis of the categories set based on local regulations if the explosion protection document on a risk assessment does not state otherwise.

TABLE 2.13

EPLs Where Only Zones Are Assigned

Zone	EPL
0	Ga
1	Ga or Gb
2	Ga, Gb, or Gc

In particular, the following categories of equipment must be used in the zones indicated, provided they are suitable for gases, vapors or mists, and/or dusts as appropriate:

- In zone 0, category 1 equipment
- In zone 1, category 1 or 2 equipment
- In zone 2, category 1, 2 or 3 equipment

Several methods may be used to make equipment safe for use in an explosive atmosphere. Tables 2.14 and 2.15 give an overview of the available concepts and their principles.

Essentially, protection methods fall under four main methods. These methods are detailed in Table 2.15 along with a brief description below of some of the concepts.

2.6.2 Protection Concepts

There are varying types of equipment that can be used within these zones to ensure that the potential for an explosion is removed or greatly reduced.

TABLE 2.14

Electrical Equipment for Gases, Vapors, and Mists (G)

Type of Protection	Symbol	Category	Suitable for Zones	Basic Concept of Protection
Increased safety	e	2	1, 2	No arcs, sparks, or hot surfaces
Nonsparking	nA	3	2	
Flameproof	d	2	1, 2	Contain the source of ignition, prevent flame propagation
Enclosed break	nC	3	2	
Quartz/sand-filled	q	2	1, 2	
Intrinsic safety	ia	1	0, 1, 2	Limit the energy of the spark and the surface temperature
Intrinsic safety	ib	2	1, 2	
Intrinsic safety	ic	3	2	
Energy limitation	nL	3	2	
Pressurized	p	2	1, 2	Keep the flammable gas out
Restricted breathing	nR	3	2	
Simple pressurization	nP	3	1, 2	
Encapsulation	ma	1	0, 1, 2	
Encapsulation	mb	2	1, 2	
Encapsulation	mc	3	2	
Oil immersion	o	2	1, 2	
Category 1G	–	1	0, 1, 2	Two independent methods of protection

TABLE 2.15

Protection Concepts and Brief Descriptions

Type of Protection Method	Equipment Code	Description	Suitable for Zones
Intended to prevent a potential ignition arising	Ex e	Increased safety: Precautions are applied to the installation to ensure increased security against the possibility of excessive temperatures and sparks from electrical equipment. Equipment that normally causes sparks is excluded from use within this method of protection.	1, 2
	Ex nA	Type-n protection: A type of protection where precautions are taken so that electrical equipment with the potential to arc is not capable of igniting a surrounding explosive atmosphere. This can be further categorized as follows: • Ex nA, where components used in construction are nonsparking • Ex nC, where components used in construction are nonincendive • Ex nR, where components used are tightly enclosed to restrict the breathing and prevent ignition • Ex nL, where components used in construction do not contain enough energy to cause an ignition	2
Intended to limit the ignition energy of the equipment	Ex ia	Intrinsic safety ia: A protection concept in which the electrical energy within the equipment is restricted to a level below that what may cause an ignition or to limit the heating of the surface of the equipment. There are two main subtypes to Ex i protection—ia and ib. Type ia protection allows for the occurrence of two faults during operation. Type ib protection allows for the occurrence of one fault during operation.	0, 1, 2
	Ex ib	Intrinsic safety ib is a protection concept in which the electrical energy within the equipment is restricted to a level below that what may cause an ignition or to limit the heating of the surface of the equipment. There are two main subtypes to Ex i protection—ia and ib. Type ia protection allows for the occurrence of two faults during operation. Type ib protection allows for the occurrence of one fault during operation.	1, 2

(continued)

TABLE 2.15 (Continued)

Protection Concepts and Brief Descriptions

Type of Protection Method	Equipment Code	Description	Suitable for Zones
Intended to limit the ignition energy of the equipment	Ex ic	Intrinsic safety ic is a protection concept in which the electrical energy within the equipment is restricted to a level below that what may cause an ignition or to limit the heating of the surface of the equipment. There are two main subtypes to Ex i protection—ia and ib. Type ia protection allows for the occurrence of two faults during operation. Type ib protection allows for the occurrence of one fault during operation.	2
	Ex nL	Type-n protection is a type of protection where precautions are taken so that electrical equipment that has the potential to arc is not capable of igniting a surrounding explosive atmosphere. This can be further categorized as follows: • Ex nA, where components used in construction are nonsparking • Ex nC, where components used in construction are nonincendive • Ex nR, where components used are tightly enclosed to restrict the breathing and prevent ignition • Ex nL, where components used in construction do not contain enough energy to cause an ignition	2
Intended to prevent the explosive atmosphere from contacting the ignition source	Ex p	Purge/pressurized protection: One process ensures that the pressure inside an enclosure is sufficient to prevent the entrance of a flammable gas, vapor, dust, or fiber and prevent a possible ignition. Another process maintains a constant flow of air (or an inert gas) to dilute to take away any potentially explosive atmosphere.	1, 2
	Ex px	Purge/pressurized protection px: One process ensures that the pressure inside an enclosure is sufficient to prevent the entrance of a flammable gas, vapor, dust, or fiber and prevent a possible ignition. Another process maintains a constant flow of air (or an inert gas) to dilute to take away any potentially explosive atmosphere.	1, 2

(continued)

TABLE 2.15 (Continued)

Protection Concepts and Brief Descriptions

Type of Protection Method	Equipment Code	Description	Suitable for Zones
Intended to prevent the explosive atmosphere from contacting the ignition source	Ex py	Purge/pressurized protection py: One process ensures that the pressure inside an enclosure is sufficient to prevent the entrance of a flammable gas, vapor, dust, or fiber and prevent a possible ignition. Another process maintains a constant flow of air (or an inert gas) to dilute to take away any potentially explosive atmosphere.	1, 2
	Ex pz	Purge/pressurized protection pz: One process ensures that the pressure inside an enclosure is sufficient to prevent the entrance of a flammable gas, vapor, dust, or fiber and prevent a possible ignition. Another process maintains a constant flow of air (or an inert gas) to dilute to take away any potentially explosive atmosphere.	2
	Ex m	Encapsulation: A protection concept whereby equipment that could potentially cause an ignition is encapsulated within a compound or resin so as to prevent contact with the explosive atmosphere. The concept also limits the surface temperature of the equipment under normal operating conditions.	1, 2
	Ex ma	Encapsulation: A protection concept whereby equipment that could potentially cause an ignition is encapsulated within a compound or resin so as to prevent contact with the explosive atmosphere. The concept also limits the surface temperature of the equipment under normal operating conditions.	0, 1, 2
	Ex mb	Encapsulation: A protection concept whereby equipment that could potentially cause an ignition is encapsulated within a compound or resin so as to prevent contact with the explosive atmosphere. The concept also limits the surface temperature of the equipment under normal operating conditions.	1, 2

(continued)

TABLE 2.15 (Continued)

Protection Concepts and Brief Descriptions

Type of Protection Method	Equipment Code	Description	Suitable for Zones
Intended to prevent the explosive atmosphere from contacting the ignition source	Ex o	Oil immersion: All equipment that has the potential to arc and potentially cause an ignition is immersed in a protective liquid or oil. The oil provides an insulating method to prevent ignition.	1, 2
	Ex nR	Type-n protection: A type of protection where precautions are taken so that electrical equipment with the potential to arc is not capable of igniting a surrounding explosive atmosphere. This can be further categorized as follows: • Ex nA, where components used in construction are nonsparking • Ex nC, where components used in construction are nonincendive • Ex nR, where components used are tightly enclosed to restrict the breathing and prevent ignition • Ex nL, where components used in construction do not contain enough energy to cause an ignition	2
Intended to prevent an ignition from escaping outside the equipment	Ex d	Flameproof protection: The equipment that may cause an explosion is contained within an enclosure that can withstand the force of an explosion and prevent transmission to the outside hazardous atmosphere. This method of protection also prevents the hazardous atmosphere from entering the enclosure and coming into contact with equipment.	1, 2
	Ex q	Sand/powder (quartz) filling: All equipment that has the potential to arc is contained within an enclosure filled with quartz or glass powder particles. The powder filling prevents the possibility of an ignition.	1, 2

(continued)

TABLE 2.15 (Continued)

Protection Concepts and Brief Descriptions

Type of Protection Method	Equipment Code	Description	Suitable for Zones
Intended to prevent an ignition from escaping outside the equipment	Ex nC	Type-n protection: A type of protection where precautions are taken so that electrical equipment that has the potential to arc is not capable of igniting a surrounding explosive atmosphere. This can be further categorized as follows: • Ex nA, where components used in construction are nonsparking • Ex nC, where components used in construction are nonincendive • Ex nR, where components used are tightly enclosed to restrict the breathing and prevent ignition • Ex nL, where components used in construction do not contain enough energy to cause an ignition	2
Special	Ex s	Special protection (Ex s special: This method of protection, as its name indicates, has no specific parameters or construction rules. In essence it is any method of protection that can provide a predetermined level of safety to ensure that there is no potential for an ignition. As such it does not fall under any specific protection method and may in fact be a combination of more than one).	0, 1, 2

This equipment must be designed and manufactured in accordance with particular construction parameters known as protection concepts.

2.6.2.1 Increased Safety e

Increased safety (IEC 60079-7—Increased Safety Marking EEx e in Accordance with EN 50 014 and EN 50 019) is intended for products in which arcs and sparks do not occur in normal or under fault conditions. The surface temperatures of the relevant parts are controlled below incentive values. Increased safety is achieved by reducing current ratings and enhancing insulation values and creepage and clearance distances above those required for normal service. Maximum voltage for the protection concept is 11 kV.

Additional measures are taken to achieve a higher degree of safety. This ensures reliable prevention of unacceptably high temperatures, sparks, or arcing, both on the inside and outside parts of electrical equipment whose normal operation does not involve unacceptably high temperatures, sparks, or arcing.

The protection concept provides a high level of safety, making it suitable for category 2 and gas group II.

Typical products are junction boxes, luminaries, induction motors, transformers, and heating devices.

The key design features for increased safety are

- Enclosures must be constructed such that they can withstand the mechanical impact test and provide a specified degree of ingress protection. Nonmetallic materials must comply with the following requirements:
  - Thermal endurance to heat.
  - Thermal endurance to cold.
  - Resistance to light.
  - Insulation resistance.
  - Thermal index (TI).
- Terminals must be generously dimensioned for the intended connections and ensure that the conductors are securely fastened without the possibility of self-loosening.
- Clearance between bare conductive parts must not be less than the values specified for the rated voltage.
- Creepage distances must not be less than the values specified for the rated voltage and the comparative tracking index (CTI) of the insulating material.
- Electrical insulating materials must have mechanical stability up to at least 20 K above the maximum service temperature.
- Temperatures of parts of equipment must be limited so as not to exceed values that could affect the thermal stability of the material and the temperature classification of the equipment
- Uninsulated, live components are subject to special protective requirements. Air and creepage gaps are wider than is generally the case in industry. Special conditions apply to the required IP enclosure protection.
- More stringent requirements apply to windings; mechanical strength, insulation, and resistance to high temperatures. Minimum cross sections are stipulated for winding wire, the impregnation and reinforcement of coils, and thermal monitoring devices.
- Applications: installation material such as marshalling and junction boxes, terminal compartments for heating systems, batteries, transformers, ballast, and squirrel cage motors.

2.6.2.2 Flameproof d

Flameproof enclosures (IEC 60079-1—Flame-Proof Enclosure Marking EEx d in Accordance with EN 50 014 and 50 018) are intended for equipment that produces arcs, sparks, or hot surfaces that may be incendive in normal operation or industrial components that cannot otherwise be made suitable for use in a hazardous area. The surrounding explosive atmosphere can enter the enclosure and internal explosions are expected during the life of the equipment.

The enclosure therefore has to be strong enough not to fracture or distort under the pressures generated. Any constructional joints in the enclosure are dimensioned such that they do not transmit the explosion from the inside to the surrounding atmosphere. These are called flame paths.

Typical products are electric motors and actuators, luminaries, loudspeakers, and switchgear. The key design features are

- Enclosures must be sufficiently strong to withstand the internal explosion.
- Joints and gaps have critical dimensions.
- Covers have warning labels if the enclosure contains parts that store energy or achieve temperatures in excess of the temperature classification.
- Fasteners must conform to dimensional and strength requirements.
- Enclosure materials must be fully specified and nonmetallic materials must be fully defined and have a suitable TI.
- Cable and conduit entries must meet constructional requirements so that the flameproof properties are maintained.
- Flameproof joints must be protected against corrosion. Flameproof joints must not be painted. The use of gaskets is only permitted when specified in the documentation. Nonsetting grease or anticorrosive agents can be applied. Nonhardening grease-bearing textile tape may also be employed outside of the joint with the following conditions:

    Where the enclosure is used in conjunction with gases allocated to group IIA, the tape must be restricted to one layer surrounding all parts of the flange joint with a short overlap, new tape must be applied whenever existing tape is disturbed.

    Where the enclosure is used in conjunction with gases allocated to group IIB, the gap between the joint surfaces must not exceed 0.1 mm, irrespective of the flange width. The tape must be restricted to one layer surrounding all parts of the flange joint with a short overlap. New tape must be applied whenever existing tape is disturbed.

Where the enclosure is used in conjunction with gases allocated to group IIC, the tape must not be applied.

This a type of protection in which components that could ignite a potentially explosive atmosphere are fitted in an enclosure that will contain the pressure of an explosion, preventing ignition of flammable gas outside the enclosure.

Technically, unavoidable gaps in the enclosure are so small and their lengths are restricted so that any hot gas released through them will have lost its power to cause ignition. If such gaps are only required by the production process they may be sealed with adhesive or gasket.

2.6.2.2.1 Important Design Parameters

1. Mechanical strength to withstand the pressure of explosion in accordance with a stipulated safety factor. The following guideline may be used:

   a. A pressure of approximately 0.8 MPa (8 bar) is produced within a sphere; for this sphere to be classified as an EEx d enclosure it would have to withstand a pressure of 1.2 MPa (12 bar).

   b. Gaps between two parts of an enclosure must be so small and their lengths restricted so that any hot gas released is unable to ignite a potentially explosive atmosphere that may be present in the hazardous area.

   c. The parameters for the spark ignition gap with regard to width and length are different in the explosion hazard subgroups IIA, IIB, and IIC. The most stringent requirements apply to enclosures in subgroup IIC.

2.6.2.2.2 Applications

Equipment whose operation normally involves sparks or arcing and/or hot surfaces such as switchgear, slip rings, collectors, rheostats, fuses, lamps, or heating cartridges.

2.6.2.2.3 Flame Paths, Gaps, Flanges, and Threaded Joints

A flame path is any small joint or gap in a flameproof enclosure through which the hot gases of an internal explosion might pass. When escaping through the gaps the hot gases are sufficiently cooled down that they do not ignite the surrounding atmosphere. The standard specifies the maximum permissible gaps for flanges, spigots, and other types of joints based experimental testing.

TABLE 2.16

Permissible Flange Gaps and Joints

Type of Joint	Maximum Length of Joint, L (mm)	For a Volume Less Than 100 cm³			For a Volume between 100 and 500 cm³			For a Volume between 500 and 2000 cm³			For a Volume Greater Than 2000 cm³		
		I	IIA	IIB	I	IIA	IIB	I	IIA	IIB	I	IIA	IIB
Flanged, cylindrical or spigot joints	6	0.3	0.3	0.2	–	–	–	–	–	–	–	–	–
	9.5	0.35	0.3	0.2	0.35	0.3	0.2	–	–	–	–	–	–
	12.5	0.4	0.3	0.2	0.4	0.3	0.2	0.4	0.3	0.2	0.4	0.2	0.15
	25	0.5	0.4	0.2	0.5	0.4	0.2	0.5	0.4	0.2	0.5	0.4	0.2
Cylindrical joints for shaft glands of rotating electrical machines (with sleeve bearings)	6	0.3	0.3	0.2	–	–	–	–	–	–	–	–	–
	9.5	0.35	0.3	0.2	0.35	0.3	0.2	–	–	–	–	–	–
	12.5	0.4	0.35	0.2	0.4	0.3	0.2	0.4	0.3	0.2	0.4	0.2	–
	25	0.5	0.4	0.25	0.5	0.4	0.25	0.5	0.4	0.25	0.5	0.4	0.2
	40	0.6	0.5	0.4	0.6	0.5	0.3	0.6	0.5	0.3	0.6	0.5	0.25
Cylindrical joints for shaft glands of rotating electrical machine (with Rolling element bearings)	6	0.45	0.45	0.3	–	–	–	–	–	–	–	–	–
	9.5	0.5	0.45	0.35	0.5	0.4	0.25	–	–	–	–	–	–
	12.5	0.6	0.5	0.4	0.6	0.45	0.3	0.6	0.45	0.3	0.6	0.3	0.2
	25	0.75	0.6	0.45	0.75	0.6	0.4	0.75	0.6	0.4	0.75	0.6	0.3
	40	0.8	0.75	0.6	0.8	0.75	0.45	0.8	0.75	0.45	0.8	0.75	0.4

Maximum Gap (mm)

Table 2.16 shows the values based on volume, gas subdivision, and type of joint. Cylindrical threads must have at least five full threads of engagement. In practice six threads are usually provided. If the thread has an undercut, then a nondetachable and noncompressible washer should be fitted to ensure the right thread engagement.

2.6.2.2.4 Cable Entry Devices

The design of the cable entry should be such that hot gases are not able to ignite the surrounding atmosphere following an internal explosion either through the gland or through the cable. Cable glands also have to conform to the requirements of threaded joints. Five fully engaged threads are required but six are usually provided.

Cables may be brought into the flameproof enclosure directly via a cable gland. This is called *direct entry*. All cable entry holes must be threaded. If the gas is IIC or the cable is not filled properly, a sealing compound must be used in the gland.

Alternatively the manufacturer might provide a terminating chamber and connect the components in the flameproof enclosure with the components in the terminating chamber through bushings. This is called *indirect entry*. The terminating chamber is usually an Ex e enclosure. For direct entry the cable entry system should comply with one of the following:

- Cable entry device in compliance with IEC 60079-1 and certified as part of the apparatus when tested with a sample of the particular type of cable.
- Thermoplastic, thermosetting, or elastomeric cable that is substantially compact and circular, has extruded bedding and the fillers, if any, are nonhygroscopic, may utilize flameproof cable entry devices incorporating a sealing ring selected in accordance with the selection chart for cable entry devices into flameproof enclosures. Compliance with the selection chart is not necessary if the cable gland complies with IEC 60079-1 and has been tested with a sample of specific cable to repeated ignitions of the flammable gas inside an enclosure and shows no ignition outside the enclosure (see Figure 2.12).
- Mineral-insulated metal-sheathed cable with or without plastic outer covering with appropriate flameproof cable gland complying with IEC 60079-1.
- Flameproof sealing device (for example a sealing chamber) specified in the equipment documentation or complying with IEC 60079-1 and employing a cable gland appropriate to the cables used. The sealing device should incorporate compound or other appropriate seals that permit stopping around individual cores. The sealing device should be fitted at the point of entry of cables to the equipment.

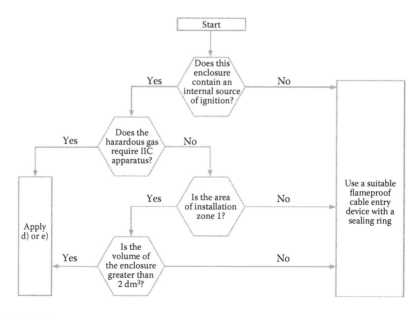

FIGURE 2.12
Selection chart for cable entry devices into flameproof enclosures for cables complying with item b. Internal sources of ignition include sparks or equipment temperatures occurring in normal operation that can cause ignition. An enclosure containing terminals only or an indirect entry enclosure is considered not to constitute an internal source of ignition. The term "volume" is defined in IEC 60079-1.

- Flameproof cable gland, specified in the equipment documentation or complying with IEC 60079-1, incorporating compound filled seals or elastomeric seals that seal around the individual cores or other equivalent sealing arrangements.

2.6.2.3 Intrinsic Safety i

Intrinsic safety (IEC 60079-11—Intrinsically Safe Marking EEx i in Accordance with EN 50 014 and EN 50 020) is intended for products in which the level of electrical energy circulating or stored in the product is insufficient to ignite a surrounding explosive atmosphere even under fault conditions.

Because of the method by which intrinsic safety is achieved it is necessary to ensure that not only the electrical apparatus exposed to the potentially explosive atmosphere, but also other electrical apparatus with which it is interconnected, is suitably constructed.

The equipment is suitable for gas group II, categories 1 (ia), 2 (ib), or 3 (ic).

Typical areas of use are control and instrumentation circuits with low voltage and current.

Depending on the design and purpose, apparatus are subdivided into two types:

1. Intrinsically safe electrical apparatus are apparatus in which all the circuits are intrinsically safe
2. Associated electrical apparatus are apparatus that contain both energy-limited and nonenergy-limited circuits and is constructed so that the nonenergy-limited circuits cannot adversely affect the energy-limited circuits. The associated electrical apparatus may be either
   a. Electrical apparatus that has an alternative standard type of protection suitable for its use in the appropriate potentially explosive atmosphere
   b. Electrical apparatus that is not protected and therefore cannot be used within a potentially explosive atmosphere

Intrinsically safe electrical equipment contains only circuits that meet the requirements of intrinsically safe circuits. Intrinsically safe circuits do not allow spark or thermal effect to occur under the test conditions of the potentially explosive atmosphere of subgroup IIA, IIB, or IIC.

2.6.2.3.1 *Important Design Parameters*
- Use of certain components for electrical and electronic circuits
- Lower permitted load on the components than in ordinary industrial applications with regard to
  - Voltage with regard to electric strength
  - Current with regard to heat

Voltage and current, including a safety margin, are kept so low that no impermissible temperatures can occur. Sparks and arcing (in the event of open-circuit or short-circuit) posses so little energy that they are unable to ignite a potentially explosive atmosphere.

2.6.2.3.1.1 *Applications*
- Instrumentation and control
- Sensors working on physical, chemical, or mechanical principles
- Actuators working on optical, acoustic, and, to a certain extent, mechanical principles

The limiting ignition curves for the different subdivisions are determined with the help of a spark test apparatus. Figure 2.13 shows the curves for a resistive circuit. Also the stored energy in a circuit has to be taken into consideration (e.g., capacitance or inductance). In the event of a short circuit, this energy could be released in addition to the energy from the associated apparatus.

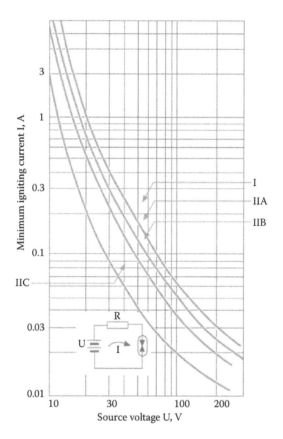

FIGURE 2.13
Curves for a resistive circuit.

2.6.2.3.2 Levels of Protection

Intrinsically safe apparatus and intrinsically safe parts of associated apparatus are placed in levels of protection ia, ib, or ic. In the determination of level of protection ia, ib, or ic, failure of components, connections, and separation should be considered.

2.6.2.3.3 Level of Protection ia

Intrinsically safe circuits in electrical apparatus of category ia must not be capable of causing an ignition during normal operation when two faults occur:

1. In normal operation and with the application of those noncountable faults that give the most onerous condition
2. In normal operation and with the application of one countable fault plus those noncountable faults that give the most onerous condition

3. In normal operation and with the application of two countable faults plus those noncountable faults that give the most onerous condition

In testing or assessing the circuits for spark ignition, the following safety factors should be applied:

- Safety factor 1.5 for both 1 and 2 above
- Safety factor 1.0 for 3 above

2.6.2.3.4 *Level of Protection ib*

Intrinsically safe circuits in electrical apparatus of level of protection ib should not be capable of causing ignition in each of the following circumstances:

- In normal operation and with the application of those noncountable faults that give the most onerous condition
- In normal operation and with the application of one countable fault plus the application of those noncountable faults that give the most onerous condition

In testing or assessing the circuits for spark ignition, a safety factor of 1.5 should be applied.

2.6.2.3.5 *Level of Protection ic*

Intrinsically safe circuits in electrical apparatus of level of protection ic should not be capable of causing ignition in normal operation. Where distances are critical for safety, they should meet the requirements of IEC 60079-11.

2.6.2.3.6 *Types of Interfaces*

There are two types of interfaces: the diode safety barrier and the galvanic isolator.

1. *Diode safety barrier.* This type of interface has been around for a long time. The assemblies incorporates shunt diodes or diode chains (including zener diodes) protected by fuses or resistors or a combination of these. The fuse restricts the fault power, the zeners restrict the voltage, and the resistor restricts the current.
2. *Galvanic isolation interface.* Figure 2.14 illustrates how this interface is constructed. The actual power limitation part of the isolator contains all the elements of the zener barrier. The power supply is via a transformer and the return signal can be via an optocoupler, transformer, or relay. The hazardous area circuit has effectively been isolated from the safe area circuit.

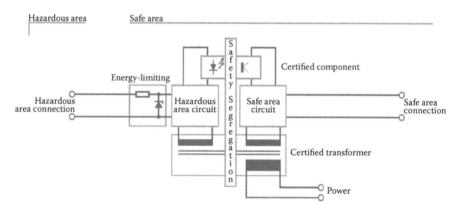

FIGURE 2.14
Galvanic isolation interface.

Discussions on the pros and cons of both interfaces are still ongoing. Earthing for zener barriers is essential for it to remain safe, unlike the isolator where earthing is not a requirement. Associated apparatus with galvanic isolation between the intrinsically safe and nonintrinsically safe circuits is preferred. Table 2.17 lists the relative merits of barriers and isolators. The significance depends on the installation.

2.6.2.3.7 Simple Electrical Apparatus and Components

Simple electrical apparatus and components (e.g., thermocouples, photocells, junction boxes, switches, plugs and sockets, resistors, light-emitting

TABLE 2.17

Relative Merits of Barriers and Isolators

Barriers	Isolators
Simple	Complex
Versatile	Application-specific
Low power dissipation	High power dissipation
Tightly controlled power supply	Wide range of power supply
High packing density	Lower packing density
Safety earth fundamental	Safety earth not essential
Reference 0 V imposed on system	Isolation between signals
Isolated from earth in hazardous area (500 V)	May be earth in hazardous areas
Accuracy and linearity (0.1%)	Lower accuracy and linearity (0.25%)
Lower cost	Increased cost
Good frequency response	Limited frequency response
Vulnerable to lightning and other surges	Less vulnerable to lightning and other surges
Cannot be repaired	Can be repaired

diodes [LEDs]) may be used in intrinsically safe systems without certification provided that they do not generate more than 1.2 V, 0.1 A, and 25 mW in the intrinsically safe system in the normal or fault conditions of the system.

Simple electrical apparatus and components should conform to all relevant requirements of the EN 60079-11. Simple electrical apparatus and components intended for use in areas where an explosive atmosphere is present continuously or for long periods (zone 0) require specific protective measures. For such apparatus, EN 60079-26 and EN 1127-1 apply in addition to EN 60079-0 and 60079-11. Special requirements (e.g., for electrostatic charging of plastic materials) as well as partition walls and the mounting of the apparatus are specified in this book. According to EN 1127-1, temperatures should not exceed 80% of the limit temperature of the temperature class.

Junction boxes and switches, however, may be awarded T6 (85°C) at a maximum ambient temperature of 40°C because, by their nature, they do not contain heat-dissipating components.

A wide variety of feed-through and disconnect terminals can be fitted in simple apparatus enclosures. Disconnect terminals that do not require the conductors to be removed from the terminals for test and calibration purposes are particularly useful during operational conditions.

It is important that the external terminal connections maintain 3-mm clearance between bare metal parts of the same intrinsically safe (IS) circuit and 6-mm clearance between bare metal parts of different IS circuits. Some users prefer Ex ia certified enclosures for IS circuits.

2.6.2.3.8 IS Electrical Systems

An IS system consists of one or more interfaces (zener barriers or isolators), one or more items of field equipment and interconnecting wiring in which any circuits intended for use in a potentially explosive atmosphere are IS circuits.

The requirements for IS systems are provided in EN 60079-25. The requirements for IS concepts for fieldbus are provided in EN 60079-27.

2.6.2.3.9 Individual Systems

Where the user or installer buys the components separately and builds his or her own system, he will be responsible to ensure that the combination of barrier and hazardous area equipment is safe. The system design engineer will be required to document the circuit with its interfaces, field equipment, and cable parameters. According to ATEX 95, this type of assembly is defined as an installation and as such does not require being CE marked.

If the field device(s) only include simple apparatus, the information needed to construct a safe system is included in within the certification of the barrier.

Where the field device is a certified item (e.g., a temperature transmitter or a solenoid valve), then extra checks are necessary.

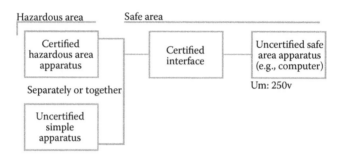

FIGURE 2.15
Individual systems.

The certificate of the field device will include its maximum input parameters that will specify one or more of the values Ui, Li, and Pi. Compatibility must be checked by ensuring that the maximum input figures of the field device are not exceeded by the maximum output values of the chosen barrier. If the system includes more than one item of certified apparatus, then compatibility with the barrier must be checked separately. The addition of simple apparatus will not affect the compatibility, except that the system temperature might be derated.

The system will be categorized according to the least favorable components of the barrier category and the apparatus category. For example, a barrier of [Ex ia] IIC with a field device of Ex ia IIC T6 will categorize the system as Ex ia IIC T6. The addition of afield device of Ex ia IIC T4 will change the system category to Ex ia IIC T4. Figure 2.15 provides a schematic of individual systems.

2.6.2.3.10 Cable Entry Devices

Cable entry devices in junction boxes of type of protection e or n, which contain only IS circuits, do not have to be certified and they do not have to maintain the e or n protection requirements of the enclosure.

Where e and i circuits are combined in one enclosure, the cable entry devices must comply with the requirements for increased safety. IS and non-IS circuits have to be separated by at least 50 mm.

2.6.2.4 Pressurization p

Pressurized apparatus (Pressurized Enclosure Marking EEx p in Accordance with EN 50 014 and EN 50 016) achieves separation of ignition sources from explosive atmospheres by purging the atmosphere inside the apparatus with air or an inert gas and then maintaining a positive pressure inside the equipment to prevent the ingress of the explosive atmosphere during operation. Failure of the pressurization leads to an alarm operating or the disconnection of the components having ignition capability.

The enclosure is filled with a pressurized gas (air, inert gas, or other suitable gas) in order to prevent the ingress of a surrounding atmosphere. The pressure within the enclosure is maintained with or without constant flushing of protective gas.

Typical products are electric motors, control cabinets, and gas analyzers. There are three types of pressurization:

1. *Static pressurization* involves charging the equipment with protective gas in a nonhazardous area and maintained only by the sealing of the enclosure. There is no protective gas supply in the hazardous area. When the overpressure drops below a set value, an alarm is raised or the equipment is switched off. The apparatus can only be recharged in the nonhazardous area.

2. *Pressurization with continuous flow of protective gas* involves an initial purge cycle followed by a continuous flow of protective gas through the enclosure while maintaining a positive pressure. The system can be used where cooling is required or dilution of an internal gas release.

3. *Pressurization with leakage compensation* involves an initial high purge with protective gas through the enclosure after which the outlet aperture is sealed and the protective gas supply is maintained to compensate for leakage from the enclosure. The minimum number of air changes before energization is usually 5. Pressurizing air should be drawn from a gas-free area. If it is not feasible to duct air from purged equipment into a safe area, a spark and flame arrestor may be required in the outlet air duct.

2.6.2.5 Equipment Protection Level Ga/Zone 0 Equipment

EN 60079-26 specifies the particular requirements for construction, test, and marking for electrical equipment that provides equipment protection level (EPL) Ga. This electrical equipment, within the operational parameters specified by the manufacturer, ensures a very high level of protection that includes rare faults related to the equipment or two faults occurring independently of each other.

This standard also applies to equipment mounted across a boundary where different protection levels may be required. It also applies to equipment installed in an area requiring a lower protection level but electrically connected to equipment with EPL Ga (associated apparatus).

To prevent ignition hazards by the electrical circuits of the apparatus, the very high level of safety required can be obtained by

- *Single apparatus*, which remains safe with two faults occurring independently from each other.
- *Two independent means of protection*. In the event that one protection fails, an independent second one is still available.

Individual concepts suitable for equipment protection level Ga or category 1 are

- Apparatus complying with the requirements of EN 60079-11, intrinsic safety ia
- Apparatus protected by encapsulation in accordance with EN 60079-18, encapsulation ma

2.6.2.5.1 *Two Independent Types of Protection Providing EPL Gb*

Electrical equipment shall comply with the requirements of two independent types of protection that provide EPL Gb. If one type of protection fails, the other type of protection shall continue to function. The independent types of protection shall not have a common mode of failure, except as specified in this section. Examples include

- Torchlight with Ex d and Ex e housing and Ex ib circuit
- Electric motor complying both with EN 60079-1, flameproof Ex d, EN 60079-7, and increased safety Ex e
- Measuring transducer with IS circuit type ib and flameproof enclosure Ex d
- Pressurized apparatus Ex px with increased safety enclosure Ex e

2.6.2.5.2 *Apparatus Mounted across the Boundary Wall*

Where apparatus are mounted across the boundary but are not IS type ia, they must contain a mechanical separation element inside the apparatus and comply with a means of protection.

Table 2.18 illustrates the possible combinations of separation elements and types of protection.

2.6.2.6 EPL Gc/Zone 2 Equipment

Equipment with EPL Gc is designed to comply with category 3 G equipment according to ATEX 95. They have a normal level of safety and are suitable for normal operation. Apparatus with an EPL of Ga or Gb may be used in an area requiring apparatus with an EPL Gc (zone 2).

The requirements for EPL Gc can be met by using the harmonized standard EN 60079-15 type of protection n. EPL Gc equipment does not have to be tested by a Notified Body (Atmosphères Explosibles [ATEX] only) such as Baseefa, PTB (German Notified Body), or KEMA (European Notified Body), but the manufacturer must be able to provide evidence that the product is safe.

Control boxes may house components with individual EC type examination certificates but no overall certification.

TABLE 2.18

Separation Elements and Types of Protection

	Requirements Depending on the Thickness, *t*, of Partition Wall		
	I: *T* > 3 mm: No Additional Requirements		
Type of Construction	**II: 3 mm *t* > *t* > 1**	**III: 1 mm > *t* > 0.2 ("X" marking required)**	**IV: *T* < 0.2 mm ("X" marking required)**
a. Partition wall 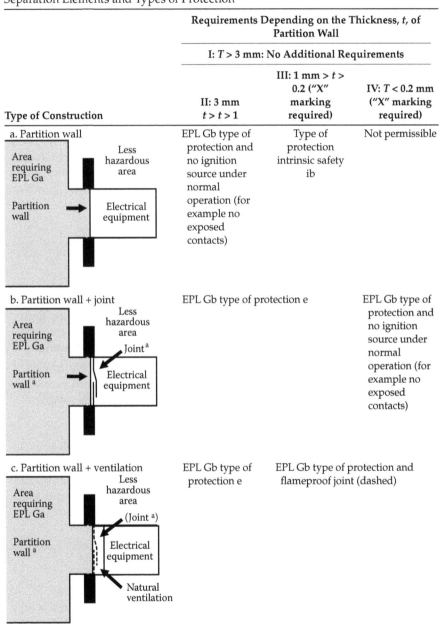	EPL Gb type of protection and no ignition source under normal operation (for example no exposed contacts)	Type of protection intrinsic safety ib	Not permissible
b. Partition wall + joint	EPL Gb type of protection e		EPL Gb type of protection and no ignition source under normal operation (for example no exposed contacts)
c. Partition wall + ventilation	EPL Gb type of protection e	EPL Gb type of protection and flameproof joint (dashed)	

Note: a. Flameproof joint and partition wall are exchangeable in sequence of order.

The manufacturer must establish the maximum surface temperature and provide the necessary documentation. The Ex n standard makes a distinction between apparatus that does not produce arcs, sparks, or hot surface nonsparking apparatus, an apparatus that produce arcs, sparks, or hot surface in normal operation sparking apparatus.

2.6.2.6.1 Nonsparking Apparatus

The risk of the occurrence of arcs, sparks, or hot surface during normal operation has been minimized by constructional means. The equipment is marked with "nA." Examples of apparatus are motors, luminaries, junction boxes, and control boxes. Fuse terminals are considered to be nonsparking provided they are not opened under load. Fuses must be nonrewireable. When mounted in an enclosure and built to protection type Ex nA the manufacturer must ensure the internal or external surface temperature is within the T classification.

2.6.2.6.2 Sparking Apparatus

In this case, arcs, sparks, or hot surfaces do occur during normal operation. The following protection concepts are allowed:

- *Equipment with protected contacts nC.* These include enclosed-break devices, nonincendive components, hermetically sealed devices, and sealed devices. Note that the requirements for encapsulated devices are now part of the EN 60079-18 as mc with a transition period running.
- *Restricted-breathing enclosures nR.* Restricted-breathing equipment shall be limited in dissipated power such that the temperature measured on the outside does not exceed the maximum surface temperature requirements.

These types of protection may be applied to enclosures containing sparking contacts but with a limitation in dissipated power such that the temperature measured on the outside of the enclosure does not exceed the external ambient temperature by more than 20 K. When applied to enclosures without sparking contacts, the only limitation is the outside temperature. This temperature should not exceed the marked temperature class.

2.6.2.7 Powder Filling q

In powder filling q (IEC 60079-5—Powder-Filled Enclosure Marking EEx q in Accordance with EN 50 014 and EN 50 017), the enclosure is filled with a fine powder to prevent arcing inside the enclosure from igniting a potentially explosive atmosphere outside the enclosure. There must be no risk of ignition by flame or the surface heat of the enclosure. The filling, which may be sand, glass beads, or similar, is subject to special requirements, as is the design of the enclosure. The filling must not leak out of the enclosure, either during normal operation or as a result of arcing or other events inside the enclosure.

Protection is provided by immersing the ignition-capable parts in a fine powder, usually quartz. The arc is quenched before it can ignite the surrounding gas. Current is limited to a safe level. The equipment is suitable for areas requiring equipment protection level Gb and equipment category 2G.

2.6.2.7.1 Applications

Applications are capacitors, electronic subassemblies, or transformers used in hazardous areas, as well as a wide variety of components whose operation involve sparks or hot surfaces but is not impeded by the powder filling.

2.6.2.8 Oil Immersion o

In oil immersion *(IEC 60079-6—Oil-Filled Enclosure Marking EEx o in Accordance with EN 50 014 and EN 50 015)*, protection is provided by immersing the apparatus in oil so that an explosive atmosphere cannot be ignited by the arcs and sparks generated under the oil.

Parts that could ignite a potentially explosive atmosphere are immersed in oil or another noncombustible, insulating liquid to an extent that gas or vapor above the liquid or outside the enclosure cannot be ignited by arcing, sparks, hot components (such as resistors), or hot residual gases from switching operations beneath the surface of the liquid.

The equipment is designed for areas requiring equipment protection level Gb, equipment category 2G.

2.6.2.8.1 Important Design Parameters
- Stipulated insulating liquids (e.g., oil)
- Assurance that the liquid remains in good condition with regard to dirt and viscosity
- Assurance and possibility of checking on safe oil level

    On heating and cooling

    Leak detection
- Restriction to nonportable equipment

2.6.2.8.2 Applications

Applications include large transformers, switchgear, starting resistors, and complete starting controllers.

2.6.2.9 Encapsulation m

In encapsulation *(IEC 60079—Encapsulation Marking EEx m in Accordance with EN 50 014 and EN 50 028)*, protection is provided by encapsulating any hot or sparking components with a material that prevents the ingress of explosive gas and cools any heat produced by the components. The equipment is suitable for areas requiring equipment protection levels Ga (ma), Gb (mb), and Gc (mc), equipment categories 1G, 2G, and 3G.

Parts that could ignite a potentially explosive atmosphere by means of sparks or heat are potted so as to prevent ignition. This is done by encapsulating the components in a compound, which is proof against physical—especially electrical, thermal, and mechanical—and chemical influences.

2.6.2.9.1 *Important Design Parameters*

- Encapsulation:

      Breakdown strength

      Low water absorption

      Resistance to various influences

      Potting must be of stipulated thickness all around

      Cavities are only permitted to a limited extent

      Potting is only penetrated by cable entries

- Load on components is limited or reduced

- Increased clearance between live parts

2.6.2.9.2 *Applications*

Applications include static coils in ballast, solenoid valves or motors, relays and other control gear of limited power, and complete printed circuit boards (PCBs) with electronic circuits.

2.6.3 Summary of Types of Protection for Electrical Apparatus in Explosive Gas Atmosphere

Only explosion-protected equipment may be used in areas in which an explosive atmosphere may still be expected despite the implementation of prevention measures. Electrical explosion-protected equipment can have various types of protection according to the construction regulations of the standards.

The type of protection employed by the manufacturer depends mainly on the kind and function of the apparatus. Various safety levels exist for some types of protection. These correspond to the equipment categories.

Tables 2.19 and 2.20 give an overview of the standardized types of protection and describe the basic principles as well as the usual applications.

TABLE 2.19

Types of Protection for Electrical Apparatus in Explosive Gas Atmosphere

Type of Protection in Accordance with IEC, EN, UL, FM, and NFPA	Representation (Diagram)	Basic Principle	Main Application
Increased safety e EN 60079-7 UL 60079-7 IEC 60079-7 FM 3600		Additional measures are applied to increase the level of safety, thus preventing the possibility of excessive temperatures and the occurrence of sparks or electric arcs within the enclosure or on exposed parts of electrical apparatus, where such ignition sources would not occur in normal service.	Terminal and connection boxes, control boxes for installing Ex components (which have a different type of protection), squirrel cage motors, light fittings
Flameproof enclosure d EN 60079-1 UL 60079-1 IEC 60079-1 FM 3600		Parts that can ignite a potentially explosive atmosphere are surrounded by an enclosure that withstands the pressure of an explosive mixture exploding inside the enclosure, and prevents the transmission of the explosion to the atmosphere surrounding the enclosure.	Switchgear and control gear and display units, control systems, motors, transformers, heating equipment, light fittings
Pressurized enclosure p EN 60079-2 NFPA 496 IEC 60079-2 FM 3620		The formation of a potentially explosive atmosphere inside a casing is prevented by maintaining a positive internal pressure of protective gas in relation to the surrounding atmosphere and, where necessary, by supplying the inside of the casing with a constant flow of protective gas acting to dilute any combustible mixtures.	Switchgear and control cabinets, analyzers, large motors px = use in zone 1, 2 py = use in zone 1, 2 pz = use in zone 2

(continued)

TABLE 2.19 (Continued)

Types of Protection for Electrical Apparatus in Explosive Gas Atmosphere

Type of Protection in Accordance with IEC, EN, UL, FM, and NFPA	Representation (Diagram)	Basic Principle	Main Application
Intrinsic safety i EN 60079-11 UL 60079-11 IEC 60079-11 FM 3610		Apparatus used in a potentially explosive area contain intrinsically safe electric circuits only. An electric circuit is intrinsically safe if no sparks or thermal effects produced under specified test conditions (which include normal operation and specific fault conditions) is not capable of causing ignition of a given explosive atmosphere.	Measurement and control technology, communication technology, sensors, actuators ia = use in zone 0, 1, 2 ib = use in zone 1, 2 [Ex ib] = associated apparatus—installation in safe area
Oil immersion o EN 60079-6 UL 60079-6 IEC 60079-6 FM 3600		Electrical apparatus or parts of electrical apparatus are immersed in a protective fluid (such as oil), such that a potentially explosive atmosphere existing over the surface or outside of the apparatus cannot be ignited.	Transformers, starting resistors
Powder filling q EN 60079-5 UL 60079-5 IEC 60079-5 FM 3600		Filling the casing of an electrical apparatus with a fine granular packing material has the effect of making it impossible for an electric arc created in the casing under certain operating conditions to ignite a potentially explosive atmosphere surrounding the casing. Ignition must not result either from flames or from raised temperature on the surface of the casing.	Sensors, display units, electronic ballast, transmitter

Encapsulation m EN 60079-18 UL 60079-18 IEC 60079-18 FM 3600		Parts that are capable of igniting an explosive atmosphere by either sparking or heating are enclosed in a compound in such a way as to avoid ignition of an explosive atmosphere.	Switchgear with small breaking capacity, control and signaling units, display units, sensors ma = use in zone 0, 1, 2 mb = use in zone 1, 2
Type of protection n_ EN 60079-15 UL 60079-15 IEC 60079-15 FM 3600		Electrical apparatus cannot ignite a explosive atmosphere surrounding them (in normal operation and under defined abnormal operating conditions).	All electrical equipment for zone 2 nA = nonsparking apparatus nC = sparking apparatus in which contacts are protected conveniently nL = energy-limited apparatus nR = purged/pressurized apparatus nZ = purged pressurized apparatus, n
Optical radiation "op_" EN 60079-28 IEC 60079-28		Appropriate measures prevent ignition of an explosive atmosphere by optical radiation.	Optical fiber There are three different methods: 1. Ex op is = intrinsically safe optical radiation 2. Ex op pr = protected optical radiation 3. Ex op sh = blocking optical radiation

TABLE 2.20

Types of Protection for Nonelectrical Apparatus in Explosive Gas Atmosphere and Used in the Presence of Combustible Dust

Type of Protection in Accordance with IEC or EN	Representation (Diagram)	Basic Principle	Main Application
Constructional safety "c" EN 13463-5		Proven technical principles are applied to equipment types that do not have any ignition source under normal operating conditions, so that the risk of mechanical failure that cause ignitable temperatures and sparks is reduced to a minimum degree.	Couplings, pumps, gearing, chain drives, belt conveyors
Flameproof enclosure d EN 13463-3		Parts that can ignite a potentially explosive atmosphere are surrounded by an enclosure that withstands the pressure of an explosive mixture exploding inside the enclosure, and prevents the transmission of the explosion to the atmosphere surrounding the enclosure.	Brakes, couplings
Pressurized enclosure p EN 13463-7		The formation of a potentially explosive atmosphere inside a casing is prevented by maintaining a positive internal pressure of protective gas in relation to the surrounding atmosphere and, where necessary, by supplying the inside of the casing with a constant flow of protective gas acting to dilute any combustible mixtures.	Pumps

Ignition source monitoring b EN 13463-6	 Sensors are integrated in the equipment to detect hazardous conditions to come, and to take steps against them before potential ignition sources become effective. The measures can be initiated automatically by means of a direct connection between the sensors and the ignition protection system or manually by issuing a warning message intended for the operator of the equipment.	Pumps, belt conveyors
Liquid immersion k EN 13463-8	 Ignition sources are rendered inactive by immersion in a protective liquid or by constant moistening using a liquid film.	Submerged pumps, gears, liquid immersion
Restricted breathing fr EN 13463-2	The effective sealing of the enclosure can reduce penetration of explosive atmosphere to an extent that no potentially explosive atmosphere can form in it. Pressure differences between the interior and the exterior atmosphere have to be taken into account. Application is limited to equipment category 3.	Equipment exclusively for zone 2 or zone 22

2.6.4 Selection of Equipment

In order to select the appropriate electrical equipment for hazardous areas, the following information is required:

- Classification of the hazardous area including the EPL requirements where applicable
- Where applicable, gas, vapor, or dust classification in relation to the group or subgroup of the electrical equipment
- Temperature class or ignition temperature of the gas or vapor involved
- Minimum ignition temperature of the combustible dust cloud, minimum ignition temperature of the combustible dust layer, and minimum ignition energy of the combustible dust cloud
- External influences and ambient temperature

2.6.5 Inspection and Maintenance

Electrical installations in hazardous areas have features specifically designed to make them suitable for use in such atmospheres. ATEX 137 stipulates that it is the operator's responsibility to maintain the integrity of those special features.

Electrical installations in hazardous areas have features specifically designed to make them suitable for use in such atmospheres. The operator must ensure that electrical equipment is

- Installed and operated correctly
- Monitored on a regular basis
- Maintained with due regard to safety

Electrical apparatus for explosive gas atmospheres, "Inspection and maintenance of electrical installations in hazardous areas (other than mines)." Depending on the country and industry (e.g., offshore or gas stations) additional national standards might have to be complied with.

2.6.6 General Technical Requirements

The general temperature range for the use of explosion-protected electrical equipment is given as from –20°C to +40°C. Permissible extensions or restrictions of the temperature range are to be indicated.

The rules for marking the electrical equipment are uniformly laid down in the standards relating to general technical requirements. The marking must indicate the following:

- The manufacturer who has put the item of electrical equipment on the market and who must be able to identify it

- The type or types of protection the item of electrical equipment conforms to
- The temperature class for which it is suitable
- The explosion hazard group or subgroup applicable to the item of electrical equipment
- The test center issuing the test certificate
- Any special conditions that have to be observed
- The standard or revision of the standard applicable to the item of electrical equipment

Depending on the country certifications, slight variations to the marking is shown in Table 2.21.

The zone standard discussed in this book is not the only standard that exists for hazardous location applications. There is another standard known as the Class/Division standard that is predominantly used in North America.

The comparisons between these two systems are not easily accomplished. Both systems are good and are developed independently from each other.

Each has its own approach to area classification and advocates and approval organizations. No one system is better than the other as neither has been proven to be safer than the other.

Each has its own merits. Which system is preferred depends on the user preference, how the areas are classified, and the wiring system used in the facility. Currently the zone system has wider use throughout the world in the chemical and petrochemical industries.

The class/division method is the dominant method used in the North America with requirements set by NEC/CEC. This method is very straightforward, with little interpretation as to the classification and what electrical materials can or cannot be used.

On the other hand, the zone method offers more choices on how to handle a particular application, which may make it seem more complicated.

TABLE 2.21

Variations of the Markings in Some Countries

	United States	Canada	IEC	Europe
Class	Class I	Class I (optional)	–	–
Zone	Zone 0, 1, or 2	Zone 0, 1, or 2 (optional)	Zone 0, 1, or 2	Zone 0, 1, or 2
Explosion Protection Symbol	AEx	Ex	Ex	EEx
Protection Method Symbol(s)	i.e., d = flameproof	i.e., d = flameproof	i.e., d = flameproof	i.e., d = flameproof
Gas Group	IIA, IIB, or IIC	IIA, IIB, or IIC	IIA, IIB, or IIC	IIA, IIB, or IIC
Temperature Class	T1...T6	T1...T6	T1...T6	T1...T6

This is because the division equipment for hazardous location is marked in accordance to the area that it is classified to use, whereas equipment within the zone method is marked in accordance with the type of protection used by the equipment.

It is then the responsibility of the user to apply the proper method of protection in each zone. However, under the new approach, Directive 94/9/EC requires additional markings to specify exactly which categories and zones the product may be used in. Both methods are meant to serve all hazardous areas from oil to sewage treatment to paint spray to everyday gas stations, as deemed appropriate by the user.

Standards for electrical installations have been established and are governed by a variety of organizations throughout the world to ensure safe electrical systems in hazardous locations. In Europe, the European Committee for Electrotechnical Standardization (CENELEC) has developed standards called EuroNorm (EN) Standards to which many European countries work. Other countries either work to their standards based on the international standards governed by the IEC or accept products and systems certified to European and/or North American standards.

For a simplified side-by-side comparison between the NEC (class/division) standard and IEC (zone), NEC (class/zone), Canadian Electrical Code (CEC) Section 18 Zone Standards, please refer to Appendix A.

2.6.7 Documentation

Up-to-date information of the following items must be made available:

- Site drawings outlining the hazardous areas with the required EPL
- For gas: the required equipment group (IIA, IIB, or IIC) and the temperature class
- For dust: the required equipment group (IIIA, IIIB, or IIIC) and the maximum surface temperature
- Characteristics of the apparatus: rated temperatures, type of protection, IP rating, corrosion resistance
- Records sufficient to enable the explosion-protected equipment to be maintained in accordance with its type of protection
- Copies of previous inspection records

2.6.8 Qualification of Personnel

The inspection and maintenance of installations should only be carried out by experienced personnel whose training has included instruction on the various types of protection and installation practices, the relevant rules and regulations, and the general principles of area classification. Appropriate continuing training should be undertaken on a regular basis. Evidence of the relevant experience and training claimed must be available.

2.6.9 Permit-to-Work

A permit-to-work system is a formal written system used to control certain types of work that are potentially hazardous. A permit-to-work is a document that specifies the work to be done and the precautions to be taken. Permits-to-work form an essential part of safe systems of work for many inspection and maintenance activities. They allow work to start only after safe procedures have been defined and they provide a clear record that all foreseeable hazards have been considered.

A permit is needed when inspection or maintenance work can only be carried out if normal safeguards are dropped or when new hazards are introduced by the work. Examples are entry into vessels, hot work, and pipeline breaking. The precise format of a work permit will vary from site to site.

2.6.10 Inspections

Before a plant is brought into service for the first time, it must be given an initial inspection. This work can be done by the operator or an outside company (third party). To ensure that the installation is maintained in a satisfactory condition it is necessary to carry out either regular periodic inspections or continuous supervision by skilled personnel, and where necessary, maintenance must be carried out.

If, at any time, there is a change in the area classification or if any apparatus is moved from one location to another, a check must be made to ensure that the type of protection, apparatus group, and temperature class, where appropriate, are suitable for the revised conditions.

2.6.10.1 Types of Inspection

1. Initial inspections are used to check that the selected type of protection and its installation are appropriate. Example checklists are shown in Appendix B.

2. Periodic inspections are carried out on a routine basis. They may be visual or close but could lead to a further detailed inspection.

   The type of equipment, manufacturer's guidance, deterioration of the apparatus, zone of use and/or the EPL requirements of the installation area, and the result of previous inspections determine the grade and the interval between periodic inspections. The interval between periodic inspections should not exceed three years without seeking expert advice or the use of extensive inspection data. Movable electrical apparatus are particularly prone to damage or misuse and therefore the interval between inspections should be set accordingly.

3. Sample inspections can be visual, close, or detailed. The size and composition of all samples depends on the purpose of the inspection.

4. Continuous supervision is based on the frequent attendance, inspection, service, care, and maintenance of the electrical installation by skilled personnel who have experience in the specific installation and its environment in order to maintain the explosion-protection features of the installation in satisfactory condition.

Where the installation falls outside the capability of continuous supervision it will be subject to periodic inspection.

2.6.10.2 Grades of Inspection

- Visual inspections identify, without the use of access equipment or tools, those defects, such as missing bolts, which will be apparent to the eye.

- Close inspections include those aspects covered by a visual inspection, and in addition, identifies those defects, such as loose bolts, which will be apparent only by the use of access equipment, for example steps (where necessary) and tools. Close inspections do not normally require the enclosure to be opened or the equipment to be de-energized.

- Detailed inspections include those aspects covered by a close inspection, and in addition, identifies those defects, such as loose terminations, which will only be apparent by opening the enclosure, and/or using, where necessary, tools and test equipment.

2.6.11 Regular Periodic Inspections

To set accurately an appropriate inspection interval is not easy, but it should be fixed taking into account the expected deterioration of the equipment. Major factors affecting the deterioration of apparatus include susceptibility to corrosion, exposure to chemicals or solvents, likelihood of accumulation of dust or dirt, likelihood of water ingress, exposure to excessive ambient temperatures, risk of mechanical damage, exposure to undue vibration, training and experience of personnel, likelihood of unauthorized modifications or adjustments, and likelihood of inappropriate maintenance (e.g., not in accordance with manufacturer's recommendation).

Once intervals have been set, the installation can be subjected to interim sample inspections to support or modify the proposed intervals or inspection grades. Where inspection grades and intervals have been established for similar apparatus, plants, and environments, this experience can be used to determine the inspection strategy.

2.6.12 Continuous Supervision by Skilled Personnel

The objective of continuous supervision is to enable the early detection of arising faults and their subsequent repair. It makes use of existing personnel who are in attendance at the installation in the course of their normal work (e.g., erection work, modifications, inspections, maintenance work, checking for faults, cleaning, control operations, functional tests, and measurements).

Therefore it may be possible to dispense with the regular periodic inspection and utilize the more frequent presence of the skilled personnel to ensure the ongoing integrity of the apparatus.

A technical person with executive function will be responsible for each installation and its skilled personnel. He will assess the viability of the concept and define the scope of equipment to be considered under continuous supervision.

He will also determine the frequency and grade of inspection as well as the content of reporting to enable meaningful analysis of apparatus performance.

2.6.13 Maintenance

Appropriate remedial measures might have to be taken following an inspection report. Care must be taken to maintain the integrity of the type of protection provided for the apparatus; this may require consultation with the manufacturer.

When necessary, the area of work should be confirmed gas-free prior to commencement of work.

Maintenance requires more detailed knowledge than when the equipment is first installed. Defect parts should only be replaced by manufacturers' authorized replacement parts and modifications that might invalidate the certificate or other documents should not be made.

For equipment that is manufactured and certified according to ATEX 95, the maintenance requirements, including the need for special tools, can be found in the operating instructions supplied with each piece of equipment.

Some maintenance tasks are listed below.

Flameproof flanges should not be broken without justification. When reassembling flameproof enclosures, all joints should be thoroughly cleaned and lightly smeared with a nonsetting grease to prevent corrosion and to assist weatherproofing. Only nonmetallic scrapers and noncorrosive cleaning fluids should be used to clean flanges.

The gasket on increased safety enclosures should be checked for damage and replaced if necessary. Terminals might have to be tightened.

Any discoloration could indicate a rise in temperature and the development of a potential hazard. Cable glands and stopping plugs should be checked for tightness. When replacing lamps in luminaries the correct rating and type should be used or excessive temperatures may result. If it is necessary for maintenance purposes to withdraw the equipment, the exposed conductors must be correctly terminated in an appropriate enclosure (e.g., Ex e) or isolated from all sources of supply and either insulated or earthed.

2.6.14 Repair

Ideally, repair work on explosion-proof electrical equipment should only be carried out by the manufacturer. In this case the manufacturer will test the equipment.

Repaired products are products whose functionality has been restored following a defect without adding new features or any other modification. As this occurs after the product has been placed on the market, the product is not to be sold as a new product. This does not preclude that national regulations of the member states on the working environment may require some kind of assessment of the repaired product as well.

A typical repair operation would be replacement by a spare part. The manufacturer of the spare part is normally not required to comply with Directive 94/9/EC unless the spare part represents an equipment or component as defined by the Directive. If so, all obligations laid down in the Directive have to be fulfilled.

If the manufacturer of the original spare part offers a new, different one in its place (due to technical progress, discontinued production of the old part, etc.), and it is used for the repair, the repaired product (as long as no substantial modification of the repaired product takes place) does not need to be brought into conformity at this time with Directive 94/9/EC as the repaired product is not then placed on the market and put into service. The repairer should be aware of certain specific requirements in the relevant national legislation that may govern the repair and overhaul operation.

2.7 Combustible Dusts

Installation in which combustible dust is handled, produced, or stored should be designed, operated, and maintained so that any releases of combustible dust, and consequently the extent of classified areas, are kept to a minimum.

In situations where explosive dust/air mixtures are possible, the following steps should be taken: eliminate the likelihood of an explosive dust/air mixture and combustible dust layers or eliminate the likelihood of any ignition source.

If this cannot be done, measures should be taken to avoid that either or both exist at the same time. If it not possible to eliminate the probability of an explosive dust/air mixture and a source of ignition at the same time, then explosion-protective systems should be considered to halt an incipient explosion immediately or to mitigate the effects (e.g., dust explosion venting systems).

However, in order to avoid unnecessary and costly plant downtime, measures would still be put in place to minimize the possibility of an ignition occurring.

The concept for area classification is similar to that used for flammable gases and vapors. However, combustible dusts, unlike flammable gases and vapors, will not necessarily be removed by ventilation or dilution after release has stopped. Very diluted and therefore nonexplosive dust clouds could form, and in time, thick dust layers.

Dust layers present three risks:

1. A primary explosion within a building may raise dust layers into clouds and cause secondary explosions more damaging than the primary event. Dust layers should always be controlled to reduce this risk.
2. Dust layers may be ignited by the heat flux from equipment on which the layer rests. The risk is of fire, rather than explosion, and this may be a slow process.
3. A dust layer may be raised into a cloud, ignite on a hot surface, and cause an explosion. In practice, dust cloud ignition temperatures are often much higher than layer ignition temperatures.

The likelihood of a layer causing a fire can be controlled by the correct selection of equipment and effective housekeeping.

2.7.1 Dust Characteristics

As part of the explosion risk assessment when dealing with dust, three basic questions should be answered:

1. Is it flammable?
2. How easily can it be ignited?
3. How violent will the explosion be?

Some dusts will glow when in contact with a heat source but extinguish immediately when removed, while others will burn fiercely and sustain a fire, which could ignite a dust cloud. If the combustibility of a product is required at high ambient temperature, the sample should be tested at the anticipated high temperature (e.g., drying temperature). Sometimes there can be a big difference in the combustion behavior.

The ease of ignition is addressed by the measurement of the minimum ignition temperature of a dust layer and dust cloud and the minimum ignition energy.

The ignition temperature of a dust layer is the lowest temperature of a hot surface at which ignition occurs in a dust layer of specified thickness on a hot surface.

Combustible dusts, when deposited in heaps or layers, may under certain circumstances develop internal combustion and high temperatures. Mostly this occurs when the dust deposit or layer rests on a heated surface that supplies the heat needed to trigger self-ignition in the dust. Such surfaces include overheated bearings, heaters in workrooms, lightbulbs, and walls in dryers. If disturbed and dispersed by an air blast or a mechanical action, the burning dust can easily initiate a dust explosion if brought in contact with a combustible dust cloud. Sometimes the dust in the deposit that has not yet burnt forms the dust cloud.

The ignition temperature of a dust cloud is the lowest temperature of the hot inner wall of a furnace at which ignition occurs in a dust cloud in air contained therein.

Hot surfaces capable of igniting dust clouds exist in a number of situations in industry, such as in furnaces, burners, and dryers or by overheated bearings. The minimum ignition temperature is not a true constant for a given dust cloud, but depends on the geometry of the hot surface and the dynamics of the cloud.

If the dust cloud is kept at a high temperature for a long period of time (e.g., in a fluidized bed), ignition can occur at temperatures below the experimentally determined minimum ignition temperature.

2.7.2 Apparatus Groups and Temperature Classes for Common Flammable Dusts and Fibers: Group III

When considering installations that are at risk of a potential explosion due to dust ignition, the equipment used is classified in much the same way as with gases. No equipment should be installed where the surface temperature of the equipment is greater than the ignition temperature of the given hazard. Table 2.22 shows some common dust hazardous and their minimum ignition temperature.

The minimum ignition energy of a dust cloud is the lowest energy value of a high-voltage capacitor discharge required to ignite the most ignitable dust/air mixture at atmospheric pressure and room temperature. Powders show quite a broad spectrum of ignition sensitivity and the vast majority need a very energetic ignition source. On plants where powders and solvents

TABLE 2.22

Ignition Temperatures for Common Flammable Dusts and Fibers

	Ignition Temperature (°C)	
Material	Cloud	Layer
Coal dust	380	225
Polythene	420	(melts)
Methyl cellulose	420	320
Starch	460	435
Flour	490	340
Sugar	490	460
Grain dust	510	300
Phenolic resin	530	>450
Aluminum	590	>450
PVC	700	>450
Soot	810	570

are handled, the risk assessment will normally be centered on the solvent characteristics.

The explosion violence is determined from the explosion pressure characteristics. The maximum explosion pressure, the maximum rate of pressure rise and the lower explosion limit are determined in a standard test apparatus with a content of 20 liters.

The maximum rate of pressure rise $(dp/dt)_{max}$ measured in the 20 liter-sphere is used to obtain the Kst value.

The maximum explosion pressure and the Kst value describe the explosion behavior of a combustible dust in a closed system. Table 2.23 gives the characteristics of dust explosions.

Most process equipment is normally far too weak to withstand the pressures exerted even by only partly developed, confined dust explosions. Consequently a primary objective of fighting an explosion after it has been initiated is to prevent the buildup of destructive overpressures. Explosion protective systems such as venting, suppression, and isolation can be used.

TABLE 2.23

Dust Characteristics

Class	Ks Value (bar m/s)	Characteristics	Example
St0	0	No explosion	–
St1	0–200	Weak/moderate explosion	Coal dust, flour
St2	200–300	Strong explosion	Epoxy resin
St3	>300	Very strong explosion	Aluminum

The explosion limits describe the range of dust concentrations in air within which an explosion is possible. Generally only the lower explosion limit is determined.

Other factors affecting dust flammability are particle size, moisture content, solvent content and temperature.

Having obtained the relevant information regarding the process, plant, and material characteristics, the next step is to locate the flammable atmospheres and identify any potential sources of ignition.

2.7.3 Area Classification

2.7.3.1 Definitions of Zones

The concept of zones for dusts is based on the classification of areas where combustible dust may be present, either as a layer or a cloud of combustible dust, mixed with air. The area where there is a possibility of combustible dust being present is divided into three zones dependent on the probability of a release and the presence of the dust.

2.7.3.2 Zone 20

2.7.3.2.1 Definition of Zone 20

Zone 20 is a place in which an explosive atmosphere in the form of a cloud of combustible dust in air is present continuously or for long periods or frequently.

2.7.3.2.2 Examples of Typical Zone 20 Locations

In general, these conditions arise only inside containers, pipes, vessels, and so forth (i.e., usually only inside the plant).

- Inside hoppers, silos, etc.
- Inside cyclones and filters
- Inside dust transport systems, except some parts of belt and chain conveyors
- Inside blenders, mills, dryers, bagging equipment, etc.
- Outside the containment, where bad housekeeping allows layers of dust of uncontrollable thickness to be formed

2.7.3.3 Zone 21

2.7.3.3.1 Definition of Zone 21

Zone 21 is a place in which an explosive atmosphere in the form of a cloud of combustible dust in air is likely to occur in normal operation occasionally.

FIGURE 2.16
Example of dust Ex zone 21 for explosion-protected plug and socket and terminal box in the field.

Figure 2.16 shows an example of dust Ex zone 21 for an explosion-protected plug and socket and terminal box in the field.

2.7.3.3.2 Examples of Typical Zone 21 Locations

Zone 21 can, for example, include places in the immediate vicinity of, for example, powder filling and emptying points and places.

- Areas outside dust containment and in the immediate vicinity of access doors subject to frequent removal or opening for operation purposes when internal explosive dust/air mixtures are present.
- Areas outside dust containment in the proximity of filling and emptying points, feed belts, sampling points, truck dump stations, belt dump-over points, etc., where no measures are employed to prevent the formation of explosive dust/air mixtures.

- Areas outside dust containment where dust accumulates and where due to process operations the dust layer is likely to be disturbed and form explosive dust/air mixtures.

- Areas inside dust containment where explosive dust clouds are likely to occur (but not continuously, not for long periods, and not frequently), such as silos (if filled and/or emptied only occasionally) and the dirty side of filters if large self-cleaning intervals are occurring.

2.7.3.4 Zone 22

2.7.3.4.1 Definition of Zone 22

Zone 22 is a place in which an explosive atmosphere in the form of a cloud of combustible dust in air is not likely to occur in normal operation, but if it does occur, will persist for a short period only.

2.7.3.4.2 Examples of Typical Zone 22 Locations

Example locations are places in the vicinity of the plant containing dust, if dust can escape at leaks and form deposits in hazardous quantities.

- Outlets from bag filter vents, because in the event of a malfunction there can be emission of explosive dust/air mixtures.

- Locations near equipment that has to be opened at infrequent intervals or equipment that from experience can easily form leaks, where due to pressure above atmospheric, dust will be blow out (pneumatic equipment, flexible connections that can become damaged, etc.).

- Storage of bags containing dusty product. Failure of bags can occur during handling, causing dust leakage.

- Areas that normally are classified as zone 21 can fall into zone 22 when measures are employed to prevent the formation of explosive dust/air mixtures. Such measures include exhaust ventilation. The measures should be used in the vicinity of (bag) filling and emptying points, feed belts, sampling points, truck dump stations, belt dump-over points, etc.

- Areas where controllable dust layers are formed that are likely to be raised into explosive dust/air mixtures. The area is designated as nonclassified only if the layer is removed by cleaning before hazardous dust/air mixtures can be formed.

Figure 2.17 is a sample zone identification.

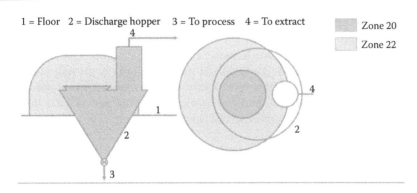

1 = Floor   2 = Discharge hopper   3 = To process   4 = To extract

Zone 20

Zone 22

FIGURE 2.17
Zone identification.

2.7.4 Grades of Release, Extent of Zones, and Housekeeping

2.7.4.1 Grades of Release

The conditions need to be identified in which process equipment, process steps, or other actions that can be expected in plants can form explosive dust/air mixtures or create combustible dust layers. It is necessary to consider separately the inside and outside of a dust containment.

Inside a dust containment area, dust is not released to the atmosphere but as part of the process continuous dust clouds may be formed. These may exist continuously or may be expected to continue for long periods or for short periods that occur frequently depending on the process cycle.

Outside the dust containment many factors can influence the area classification. Where higher than atmospheric pressures are used within the dust containment, dust can easily be blown out of leaking equipment. In the case of negative pressure within the dust containment, the likelihood of formation of dusty areas outside the equipment is very low. Dust particle size, moisture content, and where applicable, transport velocity, dust extraction rate, and fall height can influence release rate potential.

There are three grades of release:

1. *Continuous presence of dust cloud:* If a dust cloud is constantly present or is expected to persist for long periods or for short periods that occur frequently. Examples are the insides of process equipment such as silos, blenders, and mills in which dust is introduced or formed.

2. *Primary grade of release:* Release, by which it can be expected that it arises regularly or occasionally during the intended enterprise.

Examples are the close vicinity around an open bag filling or emptying point.

3. *Secondary grade of release:* Release, by which it is not expected that
   it arises in the intended enterprise, and if it does occur, will persist
   rarely and for a short time only. Examples are manholes that need to
   be opened occasionally and only for a very short period, or a dust-
   handling plant where deposits of dust are present.

Based on the likelihood of the formation of potentially explosive dust/air
mixtures, the areas can be designated according to Table 2.24.

2.7.4.2 Extent of Zones

The extent of a zone for explosive dust atmospheres is defined as the distance in any direction from the edge of a source of dust release to the
point where the hazard associated with that zone is considered to exist
no longer.

Consideration should be given to the fact that fine dust can be carried
upwards from a source of release by air movement within a building. The
extent of an area formed by a source of release also depends on several dust
parameters such as dust amounts, flow rate, particle size, product moisture
content, and so forth. In case of areas outside buildings (open air), the boundary of the zone can be reduced because of weather effects such as wind, rain,
and so forth.

2.7.4.3 Housekeeping

Inside a dust containment where powders are handled or processed, layers
of dust of uncontrolled thickness often cannot be prevented because they are
an integral part of the process.

In principle, the thickness of dust layers outside equipment can be limited. The limitation is by housekeeping and during the consideration of
sources of release it is essential to agree on the nature of the housekeeping
arrangements for the plant. The frequency of cleaning alone is not enough
to determine whether a layer contains sufficient dust to control these risks.

TABLE 2.24

Designation of Zones Depending on Presence of Combustible Dust

Presence of Combustible Dust	Resulting Zone Classification
Continuous presence of dust cloud	20
Primary grade of release	21
Secondary grade of release	22

The rate of deposition of the dust has an effect; for example, a secondary grade of release with a high deposition rate may create a dangerous layer much more quickly than a primary grade with a lower deposition rate. The effect of cleaning is therefore more important than frequency. Three levels of housekeeping can be described:

1. *Good.* Dust layers are kept to negligible thickness, or are nonexistent, irrespective of the grade of release. In this case the risk of the occurrence of explosive dust clouds from layers and the risk of fire due to layers has been removed.
2. *Fair.* Dust layers are not negligible but are short-lived (e.g., less than 1 shift). Depending on the thermal stability of the dust and the surface temperature of the equipment, the dust may be removed before any fire can start.
3. *Poor.* Dust layers are not negligible and persist for more than, for example, 1 shift. The fire risk may be significant. When a planned level of housekeeping is not maintained, additional fire and explosion risks are created. Some equipment may no longer be suitable.

2.7.4.4 Removal of Dust Deposits

Hazardous dust deposits can be avoided by regular cleaning of work and technical rooms. A proven approach is the use of cleaning schedules prescribing the nature, extent, and frequency of cleaning and the responsibilities of those concerned. These instructions can be tailored to the specific case. Particular attention should be paid to (e.g., elevated) surfaces that are difficult to inspect or reach, where considerable amounts of dust may be deposited over time.

Where appreciable quantities of dust are released as a result of operational malfunctions (e.g., damage to or bursting of containers, leakage) additional steps should be taken to remove the dust deposits with as little delay as possible. Wet cleaning and exhausting of dust deposits (using central extraction systems or mobile industrial vacuum cleaners containing no ignition sources) has proved to have safety advantages. Cleaning processes in which dust is raised into suspension should be avoided.

Bear in mind that wet cleaning can create extra problems of disposal. Where light-metal dusts are collected in wet scrubbers, hydrogen may be formed. The practice of blowing away deposited dust should be avoided.

The cleaning arrangements can be laid down as part of operational instructions for working with flammable substances. Note that only vacuum cleaners that do not contain ignition sources may be used for flammable dusts.

2.7.5 Dust Explosion Protection Measures

Combustible dusts and fibers can be ignited by several types of ignition sources, such as

- Hot surfaces
- Flames and hot gases
- Mechanically generated sparks
- Electrical apparatus
- Stray electrical currents, cathodic corrosion protection
- Static electricity
- Lightning
- Electromagnetic fields in the frequency range from 9 kHz to 300 GHz
- Electromagnetic radiation in the frequency range from 300 GHz to 3×10^6 GHz or wavelength range from 1000 to 0.1 μm (optical spectrum)
- Ionizing radiation
- Ultrasonics
- Adiabatic compression, shock waves, gas flows
- Chemical reactions

In order to avoid effective ignition sources or mitigate their effect, a number of explosion protection measures can be applied.

2.7.5.1 Protective Systems

Explosion protection measures means all measures that

- Prevent the formation of hazardous explosive atmospheres
- Avoid the ignition of hazardous explosive atmospheres
- Mitigate the effects of explosions so as to ensure the health and safety of workers

2.7.5.1.1 Explosion-Resistant (Containment)

An explosion-resistant design ensures that the explosion is contained inside the vessel. This also means that connecting and isolating equipment have to meet the same requirements. Two types of design exist:

1. Explosion-pressure-resistant vessels or apparatus are able to withstand the expected pressure of the explosion without permanent deformation

2. Explosion-pressure shock-resistant vessels or apparatus are able to withstand the expected explosion pressure without destruction but may be permanently deformed

2.7.5.1.2 Venting System

An explosion vent is a relief device that ruptures at a predetermined pressure to allow the fireball and explosive pressure to vent into a safe area. The vents fit into the walls of a process volume and are available in a variety of sizes, configurations, and materials to ensure fast, reliable operation in an explosion situation. Typically, vents are installed in conjunction with an isolation system.

2.7.5.1.3 Suppression System

In a matter of milliseconds, an explosion suppression system detects the buildup of pressure in an explosion and discharges an explosion suppressant into the enclosed space before destructive pressures are created.

The suppressant works in two ways:

1. *Chemically,* by interfering with the explosion's reaction
2. *Thermally,* by removing heat from the deflagration's flame front and thereby lowering its temperature below that needed to support combustion

The explosion suppressant also creates a barrier between the combustible particles to prevent the further transfer of heat.

2.7.5.1.4 Isolation System

Isolation systems are designed to detect incipient explosions and react instantly to keep the deflagration from spreading to unprotected areas or interconnected equipment.

The chemical type isolation method discharges an explosion suppressant into the pipeline to suppress the fireball and prevent it from reaching other plant areas or equipment.

The mechanical type isolation method produces the same results by triggering the release of a high speed valve that forms a mechanical barrier in the pipeline.

2.7.5.2 Protection by Enclosures—t

The type of protection "protection by enclosures" is based on limiting the maximum surface temperature of the enclosure and restricting the ingress of dust by using dust-tight or dust-protected enclosures. The equipment inside the enclosure can be sparking or at a higher temperature than the surface temperature. Only when gas and dust are present at the same time will the type of enclosure and/or content be restricted. The term enclosure is used for boxes, motor housings, luminaries, and so forth.

The degrees of dust protection are defined as

1. *Dust-tight enclosure:* an enclosure that prevents the ingress of all observable dust particles (IP 6X). Usable for zone 20 (category 1D), zone 21 (category 2D), and even zone 22 (in the presence of conductive dust) (category 3D).

2. *Dust-protected enclosure:* an enclosure in which the ingress of dust is not totally prevented, but dust does not enter to interfere with the safe in sufficient quantities to operation of the equipment. Dust should not accumulate in a position within the enclosure where it is liable to cause an ignition hazard (IP 5X) for zone 22 (in the presence of nonconductive dust) (category 3D). The requirement for category 1D and 2D enclosure and gasket materials is basically the same as for increased safety enclosures. For electrical equipment of the category 3D is sufficient, if plastic materials have a TI-value, which is at least 10 K higher as the temperature in the hottest place of the enclosure. However the demands on nonmetallic materials with regard to static electricity are more onerous. Propagating brush discharges have to be avoided and this can be achieved by using plastic material with one or more of the following characteristics:

Insulation resistance ≤109 Ω

Breakdown voltage ≤4 kV

Thickness ≥8 mm of the external insulation on metal parts

Marking usually should include

> Manufacturer's name and address (logo)
> Type identification
> Serial number (if required)
> Year of manufacture
> Ex
> Equipment group II
> "D" for dust
> Category 1, 2, or 3
> Certificate number (if required)
> Maximum surface temperature
> IP rating
> Relevant electrical information
> CE mark

The label in Figure 2.18 is an example of the marking. The assembly has been certified for gas as well as for dust atmospheres. However, when used

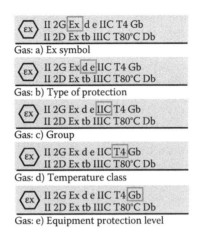

Gas: a) Ex symbol

Gas: b) Type of protection

Gas: c) Group

Gas: d) Temperature class

Gas: e) Equipment protection level

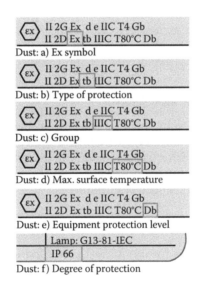

Dust: a) Ex symbol

Dust: b) Type of protection

Dust: c) Group

Dust: d) Max. surface temperature

Dust: e) Equipment protection level

Dust: f) Degree of protection

FIGURE 2.18
Example of the markings for both gas and dust atmospheres.

in an environment where gas and dust is present at the same time, additional precautions must be taken before use.

2.7.5.3 Protection by Pressurization pD

The concept of protection by pressurization pD is basically the same as for gas. Special attention is placed on the presence of dust when opening and closing the enclosure.

Before switching on the pressurization of the inside of the enclosure, it should be cleaned from dust that could have settled during switched-off foreign ventilation. This type of protection can be used for zone 21 (category 2D) and zone 22 (category 22) only.

The temperature classification of the unit is determined by the higher of the following temperatures:

- The maximum external surface temperature of the enclosure
- The maximum surface temperature of internal parts that are protected and remain energized when the supply of protective gas for pressurization is removed or fails

2.7.5.4 Protection by Encapsulation mD

The concept of protection by encapsulation mD is basically the same as for gas. With this type of protection, a piece of electrical equipment that generates sparks or heats up excessively can be encapsulated in a casting compound (thermosets or thermoplastics with or without fillers) so that it is shielded from an external explosive dust atmosphere.

2.7.5.5 Protection by Intrinsic Safety iD

The concept of protection by intrinsic safety iD is basically the same as for gas atmospheres. It specifies requirements for the construction and testing of IS apparatus intended for use in potentially explosive dust cloud or dust layer environments and for associated apparatus that is intended for connection to IS circuits that enter such environments.

2.7.6 Selection of Apparatus

When selecting apparatus for use in dust atmospheres, the following information should be available:

- Zones in which the equipment will be used
- Characteristics of the dust present, such as

    Ignition temperature of 5-mm dust layer

    Ignition temperature of the dust cloud

The equipment category suitable for the zones is selected in accordance with Table 2.25.

The maximum surface temperature for apparatus operating in any zone is calculated by deducting a safety margin from the minimum ignition temperatures of a dust cloud and a dust layer up to 5 mm thick.

TABLE 2.25

Types of Dust

Type of Dust	Zone 20	Zone 21	Zone 22
Conductive	Category 1D, excessive and uncontrollable dust layers. Test under simulated working conditions.	Category 1D or category 2D	Category 1D or category 2D
Nonconductive	Category 1D, excessive and uncontrollable dust layers. Test under simulated working conditions.	Category 1D or category 2D	Category 1D, category 2D, or category 3D

1. Maximum permissible surface temperature in case of dust clouds

$$T_{max} = 2/3 \ T_{cloud}$$

(T_{cloud} is the ignition temperature of a cloud of dust.)

2. Maximum permissible surface temperature in case of dust layer (max 5 mm)

$$T_{max} = T_5 \ mm - 75K$$

(T_5 mm is the ignition temperature of a 5-mm dust layer.)

Example: Milk powder, skimmed spray dried

$$T_5 \ mm = 340°C \ \text{and} \ T_{cloud} = 540°C$$

$$T_{max}(1) = 2/3 \times 540°C = 360°C$$

$$T_{max}(2) = 340 \times C - 75K = 265°C$$

According to this, the maximum surface temperature of the apparatus must not exceed 265°C. Where it is possible that dust layers in excess of 5 mm up to 50 mm are formed on top of the apparatus, the maximum permissible surface temperature must be reduced in accordance with the graph in Figure 2.19.

In our example, the ignition temperature of the 5-mm layer is between 320°C and 400°C, therefore the middle curve (320°C) should be used. For a layer, say 20 mm thick, the maximum surface temperature derived from Figure 2.19 is

$$T_{max} = 160°C$$

Where it cannot be avoided that a dust layer in excess of 50 mm is formed on top of the apparatus or around the sides, or where the apparatus is totally

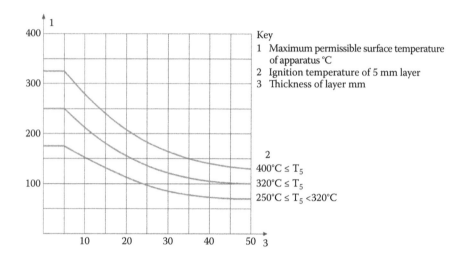

FIGURE 2.19
Reduction in maximum permissible surface temperature for increasing depth of dust layers.

submerged in the dust (typical zone 20 applications), a much lower surface temperature may be required. This should be investigated under simulated working conditions.

The special requirements for zone 20 can be met by a system of power limitation, with or without temperature control. Power engineering apparatus (e.g., motors, luminaries, plug and sockets) should wherever practicable be placed outside zone 20 areas, or if used at all, be submitted for special testing.

Table 2.26 gives an overview of the standardized types of protection in the presence of combustible dust, and describes the basic principle as well as the usual applications.

2.7.6.1 Installations

The installation requirements are similar to those in areas free of combustible dust. Installations in dust atmospheres should be designed and apparatus installed with a view to providing ease of access for cleaning.

2.7.6.2 Types of Cables

All common types of cable can be used if they are drawn into screwed, solid drawn or seamed welded conduit. It is also possible to use cables that are inherently protected against mechanical damage and are impervious to dust; for example,

- Thermoplastic or elastomer insulated, screened, or armored cable with a PVC, PCP, or similar sheath overall

TABLE 2.26

Electrical Apparatus for Use in the Presence of Combustible Dust

Type of Protection in Accordance with IEC or EN	Representation (Diagram)	Basic Principle	Main Application
Protected by enclosures tD EN 61241-1 IEC 61241-1		Thanks to the tightness, dust cannot ingress the apparatus at all or its quantity is limited to a safe degree. For this reason, ignitable apparatus can be mounted into the enclosure. The surface temperature of the enclosure must not ignite the surrounding atmosphere.	Switchgear and control gear, control, connection, and terminal boxes, motors, light fittings td A21 = according to method A for zone 21 td B21 = according to method B for zone 21
Pressurized enclosure pD EN 61241-4 IEC 61241-4		The formation of a potentially explosive atmosphere inside a casing is prevented by maintaining a positive internal pressure of protective gas in relation to the surrounding atmosphere and, where necessary, by supplying the inside of the casing with a constant flow of protective gas that acts to dilute any combustible mixtures.	Switchgear and control cabinets, motors

(continued)

TABLE 2.26 (Continued)

Electrical Apparatus for Use in the Presence of Combustible Dust

Type of Protection in Accordance with IEC or EN	Representation (Diagram)	Basic Principle	Main Application
Intrinsic safety iD EN 61241-11 IEC 61241-11		Apparatus used in a potentially explosive area contain intrinsically safe electric circuits only. An electric circuit is intrinsically safe if no sparks or thermal effects produced under specified test conditions (which include normal operation and specific fault conditions) is not capable of causing ignition of a given explosive atmosphere.	Measurement and control technology, communication technology, sensors, actuators iaD = use in zones 20, 21, 22 ibD = use in zones 21, 22 [Ex ibD] = associated electrical apparatus—installation in safe area
Encapsulation mD EN 61241-18 IEC 61241-18		Parts that are capable of igniting an explosive atmosphere by either sparking or heating are enclosed in a compound in such a way as to avoid ignition of a dust layer or cloud.	Switchgear with small capacity, control and signaling units, display units, sensors maD = use in zones 20, 21, 22 mbD = use in zone 21, 22

- Cables enclosed in a seamless aluminum sheath with or without armor
- Mineral insulated cables with metal sheath
- Cables externally provided with protection or where there is no danger of mechanical damage, thermoplastic or elastomer insulated with a PVC, PCP, or similar overall sheath are allowed

Note that the current inside cables may need to be derated to limit surface temperature.

2.7.6.3 Cable Installation

- Cable runs should be arranged so that they are not exposed to the friction effects and buildup of electrostatic charge due to the passage of dust.
- Cable runs should be arranged insofar possible that they collect the minimum amount of dust and are accessible for cleaning. Wherever possible, cables that are not associated with the hazardous areas should not pass through them.
- Where layers of dust are liable to form on cables and impair the free circulation of air, consideration should be given to reduce the current carrying capacity of the cables, especially if low ignition temperature dust is present.
- When cables pass through a floor, partition, or a ceiling that forms a dust barrier, the hole that is provided should be made good to prevent the passage or collection of combustible dust.
- When metal conduit is used, care should be taken to ensure that no damage might occur to the connecting points, that they are dust-proof, that the dustproofing of connected equipment is maintained, and that they are included in the potential equalization.
- Cable should not to be led through potentially explosive dust areas if they do not stand with these ranges in connection.

2.7.6.4 Cable Entry Devices

The requirements for the entries in category 1D and 2D dust-explosion protection equipment are basically the same as for increased safety. The only difference is the IP rating—IP 6X for zone 20 and zone 21—instead of IP54 for zone 1 and zone 2.

2.7.7 Inspection and Maintenance

2.7.7.1 Inspection

Inspection criteria are still undergoing extensive revision. The procedures are similar to those for gas atmospheres but special consideration should be given to

- Presence of accumulations of dust on the outside of apparatus is to be noted and recorded. Excessive dust layers will cause apparatus to overhead, which may lead to premature failure.
- Presence of any dust within the equipment and enclosures when carrying out detailed inspections. Note and record presence of dust.

2.7.7.2 Maintenance

Maintenance procedures follow very closely those for gas atmospheres. The principle requirement is to ensure that no excessive accumulations of dust remain on the electrical equipment or are able to cause friction in mechanical equipment.

Where significant dust layers are allowed to settle and remain for a long period of time, this could lead to serious deterioration of the equipment or could become a combustible atmosphere when disturbed.

2.7.7.3 Repair

Repair procedures follow those for gas atmospheres.

2.8 Class-Division Schemes

North American installation requirements for hazardous areas, the well-known class-division schemes, have historically differed from their IEC zone counterparts.

The growing trend in North America, however, toward the use of IEC equipment has led to the introduction of parallel hazardous area codes that allow for IEC-recognized protection techniques and wiring methods. The result is simpler and lower-cost electrical installations that are easier to maintain and modify.

The choice of classification scheme is not simply a matter of whether to name an area a zone rather than a division. Rather, each scheme carries with it a whole approach to design and execution—a philosophy incorporating a set of techniques, wiring methods, and even aesthetics.

In addition, because North American installations may now be executed according to one scheme or the other, particularly in the United States, and both may even appear in the same facility, the requirements must now provide for a hybrid approach.

For example, they must answer questions such as, May a division touch a zone? Is equipment marked for a division permitted in a zone, or vice versa? In the United States, electrical requirements are set forth in the NEC. The NEC itself is not a law but a standard that jurisdictions (i.e., counties and states) typically adopt by reference and modify according to local preferences.

The NEC is updated on a 3-year cycle, the most recent edition being 2008. The edition in force in any jurisdiction is left to the discretion of the jurisdiction and is not always the most recent. Thus, there is not strict uniformity among jurisdictions, because no single authority has national enforcement oversight.

In Canada, electrical requirements are set forth in the CEC. The Canadian code cycle is also 3 years; the most recent edition is 2009.

Both the NEC and CEC hazardous location requirements differ from the European in a number of ways, particularly in the classification schemes for hazardous areas and in the types of protection allowed.

In the traditional North American schemes, there are two classification levels: a class designation of either I, II, or III, identifying the type of hazardous substance encountered, and a division or zone designation, identifying the level of hazard. The IEC scheme provides only a single zone designation, 0, 1, or 2 for gases and 20, 21, or 22 for dusts, which indicates both the type of substance and the level of hazard.

There also exist differences between the NEC and the CEC. For example, the NEC recognizes both class-division and class-zone for class I. The CEC recognizes only class-zone for new class I installations, although existing class I installations using divisions are allowed to continue doing so. As another example, the zone designations 20, 21, and 22, recognized in the NEC as an alternative to classes II and III, have no provision in the CEC.

Since the North American zone schemes and their associated equipment requirements are identical to those in the IEC, this section will not recapitulate discussions found elsewhere in this guide. Rather, this section will address two new subjects:

1. The traditional class-division scheme used in North America (primary in the United States and secondary in Canada)
2. The hybrid requirements that recognize and reconcile two schemes in current use

2.8.1 United States: The NEC

The NEC provides for the classification of hazardous locations according to class-division or according to class-zone, and for the use of both schemes in

TABLE 2.27

NEC Articles Covering Hazardous Locations

Article 500	Hazardous (classified) locations, classes I, II, and III, divisions 1 and 2
Article 501	Class I locations
Article 502	Class II locations
Article 503	Class III locations
Article 504	Intrinsically safe systems
Article 505	Class I, zones 0, 1, and 2 locations
Article 506	Zones 20, 21, and 22 locations for combustible dusts or ignitible fibers/flyings

a single facility. In the United States, the NEC includes seven articles regarding hazardous locations in general (see Table 2.27).

In addition to the articles listed in Table 2.27, Articles 510 through 516 contain the requirements for specific types of facilities such as commercial garages, service stations, bulk storage plants, spray application, and aircraft hangars.

2.8.2 Canada: The CEC

The CEC provides for classifying class I locations according to class-zone and classes II and III locations according to class-division. The class-division scheme for class I locations is also preserved in an appendix for facilities historically employing that scheme. Interestingly, unlike the NEC and the IEC, the CEC contains no provision for the two-digit zone designations indicating dust-bearing atmospheres. In Canada, Section 18 of the CEC contains the numerous rules regarding hazardous locations in general (see Table 2.28). In addition, Section 20 contains the requirements for specific types of facilities such as commercial garages, service stations, bulk storage plants, spray application, and aircraft hangars.

2.8.3 Class I, II, and III Hazardous Locations

Locations are classified according to the properties of vapors, liquids, gases, dusts, or fibers/flyings that may be present, and the likelihood that a flammable or combustible concentration or quantity is present.

TABLE 2.28

CEC Rules Covering Hazardous Locations

Rule 18-000	Classification of hazardous locations
Rule 18-090	Class I, zone 0 locations
Rule 18-100	Class I, zones 1 and 2 locations
Rule 18-200	Class II, divisions 1 and 2 locations
Rule 18-300	Class III, divisions 1 and 2 locations

2.8.3.1 Class I Locations

Class I locations are those in which flammable gases, flammable liquid-produced vapors, or combustible liquid produced vapors are or may be present in the air in quantities sufficient to produce explosive or ignitable mixtures.

2.8.3.1.1 Class I, Division 1 Locations

Class I, division 1 locations are those where

1. Ignitable concentrations of flammable gases, flammable liquid-produced vapors, or combustible liquid-produced vapors can exist under normal operating conditions
2. Ignitable concentrations of such gases or vapors above their flash points may exist frequently because of repair or maintenance operations or because of leakage
3. Breakdown or faulty operation of equipment or processes might release ignitable concentrations of such gases or vapors and might also cause simultaneous failure of electrical equipment in such a way as to directly cause the electrical equipment to become a source of ignition

2.8.3.1.2 Class I, Division 2 Locations

Class I, division 2 locations are those

1. In which volatile flammable gases, flammable liquid-produced vapors, or combustible liquid-produced vapors are handled, processed, or used, but in which the liquids, vapors, or gases will normally be confined within closed containers or closed systems from which they can escape only in case of accidental rupture or breakdown of such containers or systems or in case of abnormal operation of equipment
2. In which ignitable concentrations of such gases or vapors are normally prevented by positive mechanical ventilation and might become hazardous through failure or abnormal operation of the ventilating equipment adjacent to a Class I, Division 1 location, and to which ignitable concentrations of such gases or vapors above their flash points might occasionally be communicated unless such communication is prevented by *adequate positive-pressure ventilation from a source of clean air and effective safeguards against ventilation failure are provided*

2.8.3.2 Class II Locations

Class II locations are those that are hazardous because of the presence of combustible dust.

2.8.3.2.1 Class II, Division 1 Locations

Class II, division 1 locations are those where

1. Combustible dust is in the air under normal operating conditions in quantities sufficient to produce explosive or ignitable mixtures
2. Mechanical failure or abnormal operation of machinery or equipment might cause such explosive or ignitable mixtures to be produced, and might also provide a source of ignition through simultaneous failure of electrical equipment through operation of protection devices or from other causes
3. Group E combustible dusts may be present in quantities sufficient to be hazardous

2.8.3.2.2 Class II, Division 2 Locations

Class II, division 2 locations are those where

1. Combustible dust due to abnormal operations may be present in the air in quantities sufficient to produce explosive or ignitable mixtures
2. Combustible dust accumulations are present but are normally insufficient to interfere with the normal operation of electrical equipment or other apparatus, but could as a result of infrequent malfunctioning of handling or processing equipment become suspended in the air
3. Combustible accumulations on, in, or in the vicinity of the electrical equipment could be sufficient to interfere with the safe dissipation of heat from electrical equipment, or could be ignitable by abnormal operation or failure of electrical equipment

2.8.3.3 Class III Locations

Class III locations are those that are hazardous because of the presence of easily ignitable fibers or materials producing combustible flyings are handled, manufactured, or used, but in which such fibers/flyings are not likely to be in suspension in the air in quantities sufficient to produce ignitable mixtures.

2.8.3.3.1 Class III, Division 1 Locations

Class III, division 1 locations are those where easily ignitable fibers/flyings are handled, manufactured, or used.

2.8.3.3.2 Class III, Division 2 Locations

Class III, division 2 locations are those where easily ignitable fibers are stored or handled other than in the process of manufacture.

2.8.3.3.3 Protection Techniques

Most of the protection techniques available in the zone scheme are also available in the division scheme, but the indication of a zone protection technique does not indicate suitability in a division.

Equipment in general-purpose enclosures may be installed in division 2 locations provided the equipment does not constitute a source of ignition under normal operating conditions. Typically, UL-listed rail-mounted terminals fitted in a type 4X enclosure may be installed in a division 2 location.

Table 2.29 lists the protection techniques under class-division.

2.8.3.4 Enclosure Types to IP Designations

Table 2.30 shows the conversion from common enclosure types (sometimes referred to as National Electrical Manufacturers Association [NEMA] types) to the corresponding IP designation.

2.8.4 Degrees of Protection According to NEMA Standards

Table 2.31 shows degree of protection provided by enclosures according to NEMA (Publication No. 250, Enclosures for Electrical Equipment 1000 Volts Maximum).

TABLE 2.29

Protection Techniques under Class-Division

	Class		
Division	I	II	III
1	• Explosion-proof • Intrinsic safety • Purged/pressurized (type X or Y)	• Dust-ignition proof • Intrinsic safety • Purged/pressurized	• Dust tight • Intrinsic safety • Hermetically sealed • Purged/pressurized
2	• Hermetically sealed • Nonincendive circuits • Nonincendive components • Nonincendive equipment • Nonsparking devices • Oil immersion • Purged/pressurized (type Z) • Any class I, division 1 method • Any class I, zones 0, 1, or 2 method	• Dust tight • Hermetically sealed • Nonincendive circuits • Nonincendive components • Nonincendive equipment • Nonsparking devices • Any class II, division 1 method	• Nonincendive circuits • Nonincendive components • Nonincendive equipment • Nonsparking devices • Any class III, division 1 method

TABLE 2.30

Enclosure Types to IP Designations

Type	Meaning	IP Designation	Remarks
1	General purpose	IP10	The enclosure type implies
3R	Rainproof	IP14	materials and methods of
4 and 4X	Watertight	IP56	construction as well as function.
6P	Submersible	IP67	The IP designation indicates a
7	Class I explosion-proof	–	degree of ingress protection only,
9	Class II explosion-proof	–	without reference to materials,
12	Dust tight	IP52	methods, or environmental
13	Oil tight	IP54	suitability. It is therefore not
			possible to convert directly from IP
			designation to enclosure type. Note
			that types 7 and 9 explosion-proof
			enclosures have no corresponding
			IP designation.

2.8.4.1 Equipment

Suitability of equipment is determined by any of the following (NEC 500.8):

1. Equipment listing or labeling
2. Evidence of equipment evaluation from a qualified testing laboratory or inspection agency concerned with product evaluation
3. Evidence acceptable to the authority having jurisdiction, such as a manufacturer's self-evaluation or the owner's engineering judgment

Evidence of suitability may include certificates demonstrating compliance with applicable equipment standards, indicating special conditions of use and other pertinent information.

2.8.4.2 Material Groups for Class-Division

Material groups in the NEC are identified by a letter code and are different from the groups in the European system (Table 2.32). NEC material groups are shown in Table 2.32 followed by a typical material from that group. Class III does not have material groups. In addition, the NEC allows equipment to be used only for those groups for which it is explicitly labeled. (The IEC material groups imply suitability for less stringent groups.)

2.8.4.3 Temperature Ratings for Class-Division

Equipment marked for classes I and II is not permitted to have any exposed surface that operates at a temperature above the ignition temperature of the hazardous substance. Equipment is marked with a T code, indicating its

TABLE 2.31

Degree of Protection Provided by Enclosures According to NEMA Publication No. 250, Enclosures for Electrical Equipment 1000 Volts Maximum

Digit	Degree of Protection	Use
Type 1	Protection against incidental contact with the enclosed equipment.	Indoor
Type 2	Protection against limited amounts of falling water and dirt.	Indoor
Type 3	Protection against rain, sleet, windblown dust, and damage from external ice formation.	Outdoor
Type 3R	Protection against rain, sleet, and damage from external ice formation.	Outdoor
Type 3S	Protection against rain, sleet, windblown dust, and for operation of external mechanisms when ice-laden.	Outdoor
Type 4	Protection against, rain, splashing water, hose-directed water, and damage from external ice formation.	Indoor or outdoor
Type 4X	Protection against, rain, splashing water, hose-directed water, and damage from external ice formation. Indoor or outdoor protection against corrosion.	Indoor or outdoor
Type 5	Protection against settling airborne dust, falling dirt, and dripping noncorrosive liquids.	Indoor
Type 6	Protection against hose-directed water, penetration of water during occasional temporary submersion at a limited depth, and damage from external ice formation.	Indoor or outdoor
Type 6P	Protection against hose-directed water, penetration of water during prolonged submersion at a limited depth, and damage from external ice formation.	Indoor or outdoor
Type 7	For use in locations classified as class I, groups A, B, C, or D as defined in the NEC.	Indoor
Type 8	For use in locations classified as class I, groups A, B, C, or D as defined in the NEC.	Indoor or outdoor
Type 9	For use in locations classified as class II, groups E, F, or G as defined in the NEC.	Indoor
Type 10	Constructed to meet the applicable requirements of the Mine Safety Health Administration.	Mining
Type 11	Protection against the corrosive effects of liquids and gases by oil immersion.	Indoor
Type 12, 12K	Protection against circulating dust, falling dirt, and dripping noncorrosive liquids.	Indoor
Type 13	Protection against dust, splashing water, oil, and noncorrosive liquids.	Indoor

maximum surface temperature (see Table 2.33). Lower maximum temperatures and higher T codes indicate more stringent requirements.

Class III does not have T codes. For class III, the maximum surface temperatures under operating conditions should not exceed 165°C (329°F) for equipment that is not subject to overloading, and 120°C (248°F) for equipment (such as motors or power transformers) that may be overloaded.

TABLE 2.32

Material Groups for Class-Division

Class I	Class II	Class III
D (propane)	E (metal dusts)	No material groups
C (ethylene)	F (coal dusts)	–
B (hydrogen)	G (grain dust)	–
A (acetylene)	–	–

TABLE 2.33

Temperature Ratings for Class-Division

Maximum Temperature	Temperature Class
450°C (842°F)	T1
300°C (572°F)	T2
280°C (536°F)	T2A
260°C (500°F)	T2B
230°C (446°F)	T2C
215°C (419°F)	T2D
200°C (392°F)	T3
180°C (356°F)	T3A
165°C (329°F)	T3B
160°C (320°F)	T3C
135°C (275°F)	T4
120°C (248°F)	T4A
100°C (212°F)	T5
85°C (185°F)	T6

2.8.4.4 Equipment Marking

Equipment is to be marked according to its suitability for use in the class and division or class and zone where it is installed. In addition, division equipment may also be marked for use in a zone. The marking requirements are similar to those existing prior to the ATEX directive.

2.8.4.5 Equipment Marking for Class-Division

Equipment must be marked to show the class, group, and operating temperature or temperature class referenced to a 40°C ambient temperature. Electrical equipment suitable for ambient temperatures exceeding 40°C must be marked with both the ambient temperature and the operating temperature or temperature class at that ambient temperature. Table 2.34 provides sample equipment marking.

Equipment of the nonheat-producing type, such as junction boxes, conduits, and fittings, is not required to have the operating temperature or temperature class marked. The coding is identical for the United States and Canada.

TABLE 2.34

Sample Equipment Marking: Class I, Division 1, Groups A and B, T4

Marking	Meaning
Class I	Flammable gas or vapor environments
Division 1	Explosive atmosphere can exist under normal operating conditions
Groups A & B	A = acetylene, B = hydrogen
T4	Surface temperature will not exceed 135°C (275°F) under specified conditions

2.8.4.6 Equipment Marking for Class-Zone

Because not all division 1 equipment is suitable for zone 0, it is necessary to pay special attention when marking division-based equipment for use in zone-classified areas. Markings are identical for the United States and Canada.

IS products suitable with certain class I, division 1 requirements may be marked as meeting zone 0 requirements, as shown in Table 2.35.

In general, products suitable for class I, division 1 locations may be marked for use in zone 1, as shown in Table 2.36, while products suitable for class I, division 2 locations may be marked for use in zone 2 as shown in Table 2.37.

Table 2.38 provides examples of markings for division-based zone certification, and Table 2.39 provides examples of markings for IEC-based zone certification.

The markings between the United States and Canada are different. The Canadian markings are identical to the IEC markings (see Tables 2.39 and 2.40).

A summary of the classification of hazardous locations for class-division schemes is presented in Table 2.41.

TABLE 2.35

Zone 0 Group Markings Permitted for Class I, Division 1

Intrinsically Safe Equipment Marked as Class I, Division 1...	May Also Be Marked as Class I, Zone 0...
Group A *and* group B	Group IIC
Group B	Group IIB + H2
Group C	Group IIB
Group D	Group IIA

TABLE 2.36

Zone 1 Group Markings Permitted for Class I, Division 1

Equipment Marked as Class I, Division 1...	May Also Be Marked as Class I, Zone 1...
Group A *and* group B	Group IIC
Group B	Group IIB + H2
Group C	Group IIB
Group D	Group IIA

TABLE 2.37

Zone 2 Group Markings Permitted for Class I, Division 2

Equipment Marked as Class I, Division 2...	May Also Be Marked as Class I, Zone 2...
Group A *and* group B	Group IIC
Group B	Group IIB + H2
Group C	Group IIB
Group D	Group IIA

TABLE 2.38

Example Markings: Class I, Division 1, Group IIB + H2, T6

Marking	Meaning
Class I	The equipment is suitable for use in environments containing flammable gas or vapor
Zone 1	The equipment is suitable for use in environments where an explosive atmosphere is likely to exist under normal operating conditions
Group IIB + H2	The equipment is suitable for use in environments consisting of ethylene and/or hydrogen
T6	The surface temperature of the equipment will not exceed 85°C (185°F)

TABLE 2.39

Example Marking (Canada): EX ia IIC T4

Marking	Meaning
EX	Approved to Canadian standards
ia	Intrinsic safety
IIC	Acetylene and hydrogen atmospheres
T4	135°C maximum surface temperature

TABLE 2.40

Example Markings (United States): Class I, Zone 0, AEX ia IIC T4

Marking	Meaning
Class I	Flammable gas or vapor
Zone 0	Explosive atmosphere always present
AEx	Approved to U.S. standards
ia	Intrinsic safety
IIC	Acetylene and hydrogen
T4	135°C maximum surface temperature

TABLE 2.41

Classification of Hazardous Locations for Class-Division Schemes

Gases, Vapors, or Mists Classification Class I	Dusts	Fibers and Flyings
NEC 500-5 CEC J18-004	NEC 500-7 CEC 18-006	Fibers and Flyings Classification Class III
Division 1: locations where ignitable concentrations of flammable gases or vapors can exist under normal operating conditions as well as frequently because of repair or maintenance operations or because of leakage.	Zone 0: locations where ignitable concentrations of flammable gases or vapors are present continuously or for long periods of time.	NEC 500-6 CEC 18-008
		Division 1: locations where easily ignitable fibers or materials producing combustible flyings are handled, manufactured, or used.
—	Zone 1: locations where ignitable concentrations of flammable gases or vapors are likely to exist under normal operating conditions or may exist frequently because of repair or maintenance or because of leakage.	Division 1: locations where ignitable concentrations of combustible dust is in the air under normal operating conditions.
		—
Division 2: locations where ignitable concentrations of flammable gases or vapors can exist under abnormal operating conditions.	Zone 2: locations where ignitable concentrations of flammable gases or vapors are not likely to occur in normal operation, and if they do, will exist only for a short period.	Division 2: locations where ignitable concentrations of combustible dust is in the air under abnormal operating conditions.
		Division 2: locations where easily ignitable fibers and materials producing combustible flyings are stored or handled other than in the process of manufacture.
Class I Groups NEC 500-3 CEC J18-050	Class II Groups NEC 505-7 CEC J18-050	NEC 500-3 CEC J18-050
		Class III
		—

(*continued*)

TABLE 2.41 (Continued)

Classification of Hazardous Locations for Class-Division Schemes

Gases, Vapors, or Mists Classification Class I		Dusts	Fibers and Flyings
Division 1 and 2:	Zones 0, 1, and 2:	Division 1 and 2:	Division 1 and 2:
• A (acetylene)	• IIC (acetylene + hydrogen)	• E (metal)	None
• B (hydrogen)	• IIB (ethene)	• F (coal)	
• C (ethene)	• IIA (propane)	• G (grain)	
• D (propane)			
Class I Temperature Classes, Division 1 and 2	Zones 0, 1, and 2	Class II Temperature Classes, Divisions 1 and 2	
T1 (≤450°C, 842°F)	T1 (≤450°C)	T1 (≤450°C, 842°F)	
T2 (≤300°C, 572°F)	T2 (≤300°C)	T2 (≤300°C, 572°F)	
T2A, T2B, T2C, T2D (≤280°C, ≤260°C, ≤230°C, ≤215°C) (536°F, 500°F, 446°F, 419°F)	—	T2A, T2B, T2C, T2D (≤280°C, ≤260°C, ≤230°C, ≤215°C) (536°F, 500°F, 446°F, 419°F)	
T3 (≤200°C, 392°F)	T3 (≤200°C)	T3 (≤200°C, 392°F)	
T3A, T3B, T3C (≤180°C, ≤165°C, ≤160°C) (356°F, 329°F, 320°F)	—	T3A, T3B, T3C (≤180°C, ≤165°C, ≤160°C) (356°F, 329°F, 320°F)	
T4 (≤135°C, 275°F)	T4 (≤135°C)	T4 (≤135°C, 275°F)	
T4A (≤120°C, 248°F)	—	T4A (≤120°C, 248°F)	
T5 (≤100°C, 212°F)	T5 (≤100°C)	T5 (≤100°C, 212°F)	
T6 (≤85°C, 185°F)	T6 (≤85°C)	T6 (≤85°C, 185°C)	

References

Directive 94/9/EU of the European Parliament and the Council of 23 March 1994 on the approximation of the laws of the member states concerning equipment and protective systems intended for use in potentially explosive atmospheres. Official Journal of the European Communities, No. L 100/1.

Directive 99/92/EC on the minimum requirements for improving the health and safety protection of the worker at risk from explosive atmospheres 16/12/1999. *Official Journal of the European Communities*, L23/57–64.

EN 50014:1999 Electrical apparatus for potentially explosive atmospheres - General requirements, Amendment A1, European Union.

IEC 60079 or EN 60079 series, *Electrical Apparatus for Explosive Gas Atmospheres*, 2004, VDE-Verlag GmbH, Berlin.

IEC 61241 and EN 61241 series, *Electrical Apparatus for Use in the Presence of Combustible Dust*, 2004, VDE-Verlag, GmbH, Berlin.

The Equipment and Protective Systems for Use in Potentially Explosive Atmospheres Regulations, 1996 (EPS), ATEX 95 (UK).

Further Readings

Babiarz, P. 1998. How will you classify your hazardous areas? *InTech* 45 (2), 34–36.

Babiarz, P.S., Liggett, D.P. and Wellman, C.M. 1998. Adapting to the dual hazardous area classification system. *IEEE Industry Applications Magazine* 4 (2), 16–24.

Danen, G.W.A. 1988. Electrical safety of process instrumentation in areas with potentially explosive gas atmospheres. *IEE Conference Publication* (296), 67–69.

Dust explosions ISSA Prevention Series No. 2044 (G) The basics of dust explosion protection R. Stahl Schaltgeräte GmbH.

EN 60529 Specification for degrees of protection provided by enclosures (IP code) EN 13463 Part 1–Part 8.

Ennis, T. 2006. Experiences and issues in ATEX and DSEAR compliance. *Institution of Chemical Engineers Symposium Series* (151), 309–321.

Garside, R.H. 1994. Area classification—An alternative approach. *IEE Conference Publication* (390), 16–21.

Glynn, K.J. 1999. Risk based approach to hazardous area classification. *IEE Conference Publication* (469), 17–23.

Hartwell, F.P. 1996. Be aware of zone system equipment markings. *EC and M: Electrical Construction and Maintenance* 95 (8), 3.

Hattwig, M. and Steen, H. 2004. *Handbook of Explosion Prevention and Protection*, Wiley VCH, Weinheim, Germany.

Herres, D. 2011. Wiring methods for hazardous locations. *EC and M: Electrical Construction and Maintenance* 110 (8).

LeBlanc, J.A. and Lawrence Jr., W.G. 2000. Benefit from the three-zone National Electrical Code. *Chemical Engineering Progress* 96 (12), 75–82.

Lionetto, P.F., Ostano, P. and Testa, A. 1988. Italian experience on hazardous area classification and selection of electrical installations and apparatus in industrial plants. *IEE Conference Publication* (296), 9–12.

Moodie, T.W. 1971. Measurement comes to hazardous dust area classification. *ISA Transactions* 10 (3), 224–230.

Nieuwmeyer, H. 1997. Hazardous area classification, whose job is it? Rules and regulations are nothing but the scar tissue of past mistakes. *Elektron* 14 (7), 87–89.

Niu, Y. 2002. Hazard analyses of fire and explosion with the electrical equipment in CNG motor-repair-shop. *Process in Safety Science and Technology Part B 3*, 933–936.

Patel, M. 2007. Hazardous area classification of flammable dust operations. *Chemical Engineering World* 42 (5), 63–65.

The Dangerous Substances and Explosive Atmospheres Regulations 2002, Statutory Instrument 2002 No. 2776 (UK).

Thurnherr, P. and Schwarz, G. 2009. Selection of electrical equipment for hazardous areas. *IEEE Industry Applications Magazine* 15 (1), 50–55.

Tommasini, R., Pons, E. and Toja, S. 2005. The Italian approach to the classification of areas where combustible dusts may be present, Series on Energy and Power Systems, No. 468-180, 512–517.

Toney, M.K., Griffith, T., Hyde, W., Schram, P.J. and Soffrin, D.E. 2000. History of electrical area classification in the United States, Record of Conference Papers—Annual Petroleum and Chemical Industry Conference, 273–280.

3

Layout and Spacing

3.1 Introduction

Equipment in a process unit can be arranged in many ways. Safety, economy, operability, and ease of maintenance should be considered in locating each item within the unit. Adequate spacing between equipment will help in minimizing the spread of fire. Consideration should be given to access for firefighting. This chapter specification covers the basic requirements of the plant layout and spacing of oil and gas refineries and petrochemical and similar chemical plants to ensure safety and fire prevention together with ease of operation and maintenance.

Hydrocarbon processing and handling plants are inherently hazardous. Today's trend of large and complex plants presents substantial risk potential. At times plants are modified to operate at higher capacities or efficiencies, necessitating larger storage requirements than originally contemplated. For these reasons, initial site analysis for a proposed new construction or addition should be done carefully while considering the space allocation to the various facilities.

3.2 Layout in Oil, Gas, and Chemical Facilities

To prepare a layout, information can be collected on the following aspects, as applicable:

- Process units, utility requirements, storage tanks, LPG storage vessels, and other pressurized storage vessels
- Product receipt/dispatch and mode of transport (rail, road, and pipeline)
- Warehouses, storage areas for solid products such as petroleum coke, petroleum wax, sulfur, bitumen/asphalt, etc., and other open storage areas like scrap yards and dumping grounds
- Chemical/toxic chemical storage, hazardous waste storage/disposal
- Flares

- Service buildings, fire station, and fire training ground
- Site topography including elevation, slope, and drainage
- Meteorological data
- Bathymetric data (high tide level, surge wave height, etc.) for installations in coastal areas
- Seismic data
- Highest flood level in the area, water table, natural streams/canals
- Approach roads to main plant areas
- Aviation considerations
- Risk to and from adjacent facilities
- Environmental considerations
- Statutory obligations

3.2.1 General Classification of Petroleum Products

Petroleum products are classified according to their closed cup flash points as given below:

- *Class-A petroleum:* Liquids with a flash point below 23°C
- *Class-B petroleum:* Liquids with a flash point of 23°C and above but below 65°C
- *Class-C petroleum:* Liquids with a flash point of 65°C and above but below 93°C
- *Excluded petroleum:* Liquids with a flash point of 93°C and above

Liquefied gases including LPG do not fall under this classification but form a separate category.

In the following cases, the above classification does not apply and special precautions should be taken as required:

i. Where ambient temperatures or the handling temperatures are higher than the flash point of the product
ii. Where product handled is artificially heated to a temperature above its flash point

HAZARDOUS AREA

An area will be deemed to be hazardous where

i. Petroleum with a flash point below 65°C or any flammable gas or vapor in a concentration capable of ignition is likely to be present

ii. Petroleum or any flammable liquid with a flash point above 65°C is likely to be refined, blended, or stored at above its flash point

For classification and extent of a hazardous area, refer to Chapter 2.

3.2.2 Terrain

First, considerations should be given to the physical setting. It should not automatically be assumed that it is necessary to level the site. There may instead be ways that the process can take advantage of whatever slopes are present.

With respect to terrain, an assessment should be made as to whether there is adequate space in general. If not, ingenuity will be required to meet requirements such as those for flares. Available space can help to govern whether the plant can be located on one floor or instead occupy several stories. The physical setting shall also be considered in light of the transportation requirements for raw materials, products, wastes, and supplies.

Block layout should be adopted as far as possible. Plant layout arrangement should follow the general route of raw material to process unit(s) with tankages interposed as required, followed by storage and dispatch facilities. The entire area should be subdivided into blocks.

3.2.3 Safety and Environment

The hydrocarbon industry over the years has learned lessons from fires and explosions throughout the world and has been updating plant safety norms including interdistances between facilities and their relative locations. The minimum distances recommended many years ago need review in the context of today's environment in the industry.

Familiarization with pertinent environmental regulations (local, national, and international) and how they might change is essential prior to the conclusion of preproject studies.

Attention should be given to the pertinent safety regulations, including health and welfare needs. Hazardous and flammable materials require special handling, which can take up layout space.

If the process fluids are especially toxic, layout is affected by the need for close chemical sewers and other protection measures. Security requirements may necessitate a special layout design when the plant produces a high-value product.

If a plant site is governed by particular building, piping, plumbing, electrical, and other codes, these can affect plant layout. Similar governing standards and regulations for plant sites will also affect the layout concept.

3.2.4 Throughput

It is important not only to know the initial capacity but also to have a good feel for how much the plant might be expanded in the future, as well as how likely the process technology will be modernized. These factors indicate how much space should be left for additional equipment.

Multiple processing lines (trains) are often required for the plant. Pairs of trains can either be identical or be mirror images. The former option is less expensive. But the mirror image approach is sometimes preferable for layout reasons. Two such reasons are

1. For operator access via a central aisle
2. The need for the outlet sides of two lines of equipment (pumps, for instance) to point toward each other so that they can be readily hooked to one common line

3.3 Basic Considerations

a. Process flow sequence and operating procedures should be thoroughly understood so that equipment arrangement in the plot plan is functional. Equipment should be arranged in a logistic process sequence for optimum piping runs and operational and maintenance ease. Spacing between equipment should be adequate for undertaking maintenance jobs.

b. The unit pipe rack should be kept in the center, thereby splitting the unit into two or more areas of equipment. Pumps may be arranged in two rows close to and on either side of the pipe rack. Heat exchangers and vessels should be grouped together forming outer rows on both sides of the rack.

c. Heat exchangers should be located perpendicular to the pipe rack on the outer row to facilitate pulling of tube bundles with mobile crane or by other means. Shell and tube heat exchangers should have a longitudinal clearance of at least 1 m plus the length of removable bundles.

d. Air fin coolers should be installed above the pipe rack/technological structures/independent structure.

e. Vessels having large liquid holdup should be installed at lower heights and preferably at grade. Adequate drainage should be provided around such vessels. Where process requirement dictates their installation above grade, these should be located in an open area.

f. Towers/columns should be located along the pipe rack toward open areas for unobstructed erection as well as maintenance of internals at grade. Tall towers requiring frequent operating attention at upper levels may be located at one place so that common connecting platforms can be provided.

g. Thermosiphon reboilers should preferably be placed close to their associated towers.

h. Vessels, columns, reactors with internals and/or containing catalysts, chemicals, and so forth should have a dropout area for removing/installing the internals and/or for loading/unloading of catalysts and chemicals.

i. Heaters should be located upwind at one corner of the unit. Space should be provided for removal and cleaning of heater tubes and for an approach for a crane. Areas around the heaters should be graded for guiding spills away from process equipment. Forced draft fans should be located away from process equipment where they are likely to suck hydrocarbon vapors.

j No trenches or pits that might hold flammables should extend under the furnace and connections with underground drain system should be sealed over an area 15 m from the furnace walls.

k. The local control panel for a soot blower control and flue gas analyzer should only be located on and near the process heater. The rest of controls should be taken to the control room.

l. Gas compressors should be located downwind from heaters so that leaked gases will not drift toward the heater. Gas compressors should have roofing and open from sides to avoid accumulation of heavier vapors/gases on the floor of the compressor house. The compressor house should be located near the battery limits to facilitate ease in maintenance and operation. A dropout area should be provided for maintenance.

m. No other tankage except day tanks/process chemicals should be provided within battery limits of any process unit.

n. Process chemical storage tanks should be provided with a curb wall that is a minimum of 300 mm in height. Hydrocarbon day tanks should be provided with a dike in line with Section 7.0 of this standard.

o. Cold boxes should be located on grade or on separate elevated structures. Adequate space should be provided around cold boxes for ease of operation and maintenance.

p. Flare knockout drums for the process units should be located at the battery limit of the unit.

q. Blowdown facilities/buried drums should be located at one corner of the plant farthest from a furnace or any fired equipment and on the leeward side of the unit. The vent from the blowdown facility should be a minimum of 6 m above the highest equipment falling within radius of 15 m from the vent stack.

r. An operator's cabin may be provided in the process unit. The cabin should be located on the upwind side of the unit in a nonhazardous area and away from draining/sampling facilities. The cabin should be for minimum occupancy of the shift operators of the respective facilities only.

s. Stairways should be provided for the main access.

t. Minimum headroom under vessels, pipes, cable racks, and so forth should be 2.1 m.

u. Equipment should be spaced to permit use of mobile equipment and power tools or servicing and maintaining equipment during turn-around periods.

v. The plant layout should be arranged for

1. Maximization of safety

2. Prevention of the spread of fire and also ease of operation

3. Maintenance consistent with economical design and future expansion

Figure 3.1 shows the layout of different equipment in a typical plant.

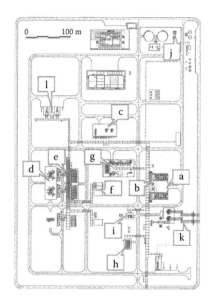

FIGURE 3.1
Layout of different systems: (a) three-phase separators, (b) booster pumps, (c) pumps, (d) compressors, (e) dehydration, (f) Triethylene Glycol (TEG) regenerator, (g) water separators, (h) vapor recovery, (i) drain treatment, (j) diesel tank, (k) inlet, and (l) outlet. (Reprinted from *Journal of Hazardous Materials*, 191, A. Di Padova et al., Identification of fireproofing zones in oil and gas facilities by a risk-based procedure, 83–93, Copyright 2011, with permission from Elsevier.)

3.3.1 Blocking

The plant site should be blocked in consideration of hazards attendant to plant operation in the area. All blocked areas should be formed as square as possible by divided access roads and/or boundary lines.

3.3.2 Location and Weather

The plant layout should be arranged in consideration of geographic location and weather in the region of the site.

Where the prevailing wind is defined, the administration and service facilities and directly fired equipment, and so forth, should be located windward of process units, storage tanks, and so forth.

3.3.3 Layout Indication

The basic requirements to be met in the appropriate diagram when making a piping and equipment layout are

- All equipment, ladders, structures, davits, and trolley beams, should be indicated
- All instruments should be located and indicated
- All valving and handwheel orientations should be indicated
- Drip funnel locations for underground drains should be indicated
- All electrical switch gears and lighting panels should be indicated
- All sample systems should be indicated

3.4 Plant Layout

While locating the various facilities/blocks, the following should be considered:

a. Layout of blocks/facilities should be in sequential order of process flow.
b. Process unit(s), tank farm, loading gantry, solid storage, utilities, effluent treatment plant (ETP), and approach roads should be located on high ground to avoid flooding.
c. In cases when process units are operated in an integrated way and shutdowns are taken simultaneously, then it may be considered as a single block.

d. The control room should be located in a nonhazardous area upwind of process plant/hydrocarbon storage and handling facilities. It should not be located on a lower level than surrounding plants and tank farms. There should be no structure that would fall on the control room in case of a blast.

e. Utility block(s) should preferably be located adjacent to unit blocks.

f. Power generation facilities that also supply steam for process requirements should be located near the process unit block. When external power grid is interconnected with plant power generation facilities, either the power plant should be located at the side of the boundary wall or the external power transmission lines should be taken underground up to the interconnection grid.

g. Overhead power transmission lines should not pass over the installation including the parking areas. High tension (HT) substation(s) should be located close to major load centers.

h. Low tension (LT) substation(s) should be located at load centers in such a way that the distance between the distribution transformer and farthest motor is minimum.

i. Cooling towers should be located downwind of process equipment and substation(s) so that if fog develops it will not cause corrosion or obstruct vision or short-circuiting.

j. Storage tanks should be grouped according to product classification. In undulating areas, storage tanks should be located at lower elevations.

k. Truck loading/unloading facilities should be located close to the product movement gate and should be oriented to provide a one-way traffic pattern for entrance and exit.

l. Rail loading facilities should be located along the periphery of the installation.

m. The sulfur recovery unit and sulfur loading area should be located close to the product movement gate and away from the process units and hazardous and populated areas. Equipment drawing air (e.g., air compressors or air blowers) should be located away from the sulfur recovery unit/sulfur handling facility. A minimum separation distance of 50 m is recommended between sulfur storage/handling and any facility or boundary wall.

n. Petroleum coke storage and handling facilities should be located as far away as possible from process units, air separation plants, and populated and hazardous areas.

o. A separate collection system should be provided for different types of waste generated in the process plant such as oily water, caustic, acid effluents, fecal, and so forth. ETPs should be located at minimum one block away from the process unit area, and downwind of process units and important areas with consideration of odors and emissions of volatile organic compound. This should be closer to the disposal point by the side of the boundary and at a lower grade to facilitate the gravity flow of effluent.

p. Flares should be located upwind of process units and the area around flares should be paved. Main pipe racks/pipe track should not be routed through process units. Overhead clearance for vehicles over roadways and railroads should be provided.

q. Roads should be provided in a symmetric manner to serve all process areas requiring access for the operation, maintenance, and firefighting. These roads should encircle the process blocks/process units.

r. Smoking booths should not be provided in the hydrocarbon industry. However, if required, these should be located at a minimum distance of 60 m from any hydrocarbon source.

s. A fire station, firewater storage, and firewater pump house should be located at a safe place away from hazardous areas. A fire station should be upwind of process units and hydrocarbon storage areas with a straight approach to process units/other critical areas.

3.4.1 Area Arrangement

Classified blocked areas, such as process areas, storage areas, utilities areas, administration and service areas, and other areas should be arranged as follows:

1. The process area should be located in the most convenient place for operating the process unit.

2. The storage area should be located as far away as possible from buildings occupied by personnel at the site, but should be located near the process area for ready operation of the feed stocks and product rundowns.

3. The utilities area should be located beside the process area for ready supply of utilities.

4. The loading and unloading area should be located on a corner of the site with a capable connection to public road directly, for inland traffic. For marine transportation, the area should be located on the seaside or riverside in the plant site.

5. The administration and service area should be located at a safe place on the site in order to protect personnel from hazards. It should preferably be located near the main gate alongside the main road of the plant.

6. Flare and burn pits should be located at the end of the site with sufficient distance to prevent personnel hazard.

7. The wastewater treating unit should be located near the lowest point of the site so as to collect all effluent streams from the processing unit.

8. The process unit to which the feed stock is charged first should be located on the side near the feed stock tanks to minimize the length of the feed line.

9. The process unit from which the final product(s) is (are) withdrawn, should be located on the side near the products tanks to minimize the length of the product rundown line.

10. Process units in which large quantities of utilities are consumed should be preferably located on the side near the utility center.

3.4.2 Roadways

1. Road and access ways should offer easy access for mobile equipment during construction and maintenance, firefighting, and emergency escape in a fire situation.

2. Access roads should be at least 3 m from processing equipment between road edges to prevent vehicle collisions.

3.4.3 Pipe Racks and Sleepers

In general, the pipe racks for process units and pipe sleeps for the off-site facilities should be considered as the principal support of the pipeway. Run pipe lines overhead should be grouped in pipe racks in a systematic manner.

Pipe rack runs oriented in the same direction should be at consistent elevations. Pipe rack runs oriented opposite to these runs should be at other elevations to accommodate crossing of lines at pipe rack junctions and to accommodate branch line intersections.

Single-level pipe racks are preferred. If more than one level is required, the distance between levels oriented in the same direction should be adequate for maintenance but not less than 1.25 m.

Maximum pipe rack widths should be 10 m. If widths larger than 10 m are required, the pipe rack should be designed to be two-stage. Actual widths should be 110% of the required widths or the required widths plus 1 m. In cases where air fin coolers are to be placed on the pipe racks, the pipe rack widths should be adjusted based on the length of the air coolers.

Avoid flat turns. When changing direction, change elevation. Allow ample space for routing instrument lines and electrical conduit. Provide 25% additional space for future instrument lines and electrical conduit adjacent to that required.

Provide 20% additional space on the pipe rack for future piping. This space should be continued and clear on each level for the full length of the rack. The width allocation may be split in no more than two sections.

Allow a continuous clear area of 4 m high by 4 m wide below main racks in process units for maintenance access ways.

Pipe racks outside process areas should have the following minimum overhead refinery/plant clearances: main roadway, 5 m; access roads, 4.5 m; and railroads, 6.7 m above the top of the rail.

A typical layout of a pipe rack for process plants, depending on the number of process units incorporated and the process complexities, are given in Figures 3.2 through 3.5 with reference descriptions as follows:

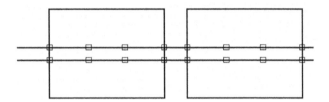

FIGURE 3.2
Single-rack type layout.

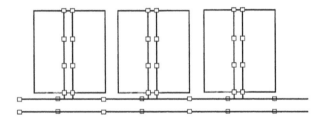

FIGURE 3.3
Comb type layout.

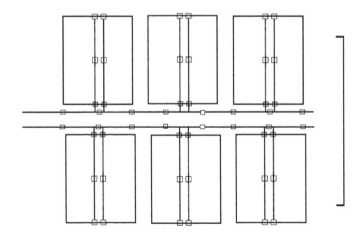

FIGURE 3.4
Double-comb type layout.

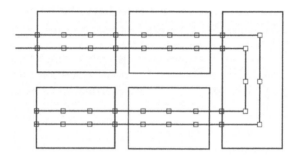

FIGURE 3.5
U type layout.

a. A single-rack type layout is suitable for a small-scale process complex consisting of two to three process units. It is economical without requiring any large area (Figure 3.2).

b. The comb type layout shown in Figure 3.3 is recommended for use in a process complex consisting of three or more process units. The single rack type in this case will not be suitable since separate maintenance and utility administration in normal operation would be difficult because of the utility and flare line that are placed on the common rack.

c. A double-comb type layout is an expansion of the comb type, which is recommended for the use in large-scale process complexes where five to 10 process units are to be arranged. This layout as shown in Figure 3.4, can be conveniently utilized.

d. The U type layout shown in Figure 3.5 is recommended to be used in the case of process units whose maintenance cannot be conducted separately within the complex. This type can be regarded as an expansion of the single rack type. Even process complexes of this nature can be regarded as one process unit in the planning of their layout.

Location of pipe racks should be in general agreement with the plot plan.

3.5 Layout of Control Room and Electrical Substation

General recommendations for spacing of a control room and substations in refineries, petrochemical plants, utilities, pump stations, and so forth should be in accordance with the Oil Insurance Association (OIA) recommendation as given in Tables 3.1 through 3.4.

TABLE 3.1

OIA Recommended Spacing at Refineries, Chemicals

Minimum Distance (m)	Petrochemicals and Gas Plants for Buildings					
	Fire-Resistive Construction with Noncombustible Contents	Noncombustible Construction with Noncombustible Contents	Ordinary Construction, IC, and SIC, with Noncombustible Contents	Fire-Resistive Construction with Combustible Contents	Noncombustible Construction with Combustible Contents	Ordinary Construction with Combustible Contents
Fire-resistive construction with noncombustible contents	None	9	9	9	9	12
Noncombustible construction with noncombustible contents	9	None	9	9	12	15
Ordinary construction, IC, and SIC, with noncombustible contents	9	9	None	12	15	18
Fire-resistive construction with combustible contents	9	9	12	None	12	15
Noncombustible construction with combustible contents	9	12	15	12	None	15
Ordinary construction with combustible contents	12	15	18	15	15	None

TABLE 3.2

OIA General Recommendations for Spacing in Refineries

Minimum Distance (m)	Service Building	Process Units	Boilers, Utilities, and Electrical Generating Equipment, etc.	Fired Process Heaters	Process Vessels, Fractionating Equipment, etc.	Gas Compressor Houses
Service buildings	10 (see building chart)	—	—	—	—	—
Process units	30	15–30	—	—	—	—
Boilers, utilities, and electrical generating equipment, etc.	30	30	—	—	—	—
Fired process heaters	30	15	30	8	—	—
Process vessels, fractionating equipment, etc.	30	—	30	15	—	—
Gas compressor houses	30	—	30	30	9	—
Large oil pump houses	30	—	30	30	6	9
Control rooms	—	—	30	15	15	15
Cooling towers	15–30	30	30	30	30	15–30
Dropout controls, steam snuffing, and water spray controls	—	—	—	15	15	15
Blowdown drums and flare stacks	60–90	60–90	60–90	60–90	60–90	60–90
Product storage tanks	60	75	75	75	75	75
Rundown tanks	30	60	60	60	60	60
Blending tanks	60	60	60	60	60	60
Hazardous loading and unloading facilities, including docks	60	60	60	60	60	60
Fire pumps	15–30	75	—	75	75	30

TABLE 3.2 (Continued)

OIA General Recommendations for Spacing in Refineries

Minimum Distance (m)	Large Oil Pump Houses	Control Rooms	Cooling Towers	Controls for Dropout Steam Snuffing and Spray Dropout	Blowdown Drums and Flare Stacks	Product Storage Tanks	Rundown Tanks
Service buildings	–	–	–	–	–	–	–
Process units	–	–	–	–	–	–	–
Boilers, utilities, and electrical generating equipment, etc.	–	–	–	–	–	–	–
Fired process heaters	–	–	–	–	–	–	–
Process vessels, fractionating equipment, etc.	–	–	–	–	–	–	–
Gas compressor houses	–	–	–	–	–	–	–
Large oil pump houses	–	–	–	–	–	–	–
Control rooms	9	–	–	–	–	–	–
Cooling towers	15–30	15–30	8–15	–	–	–	–
Dropout controls, steam snuffing, and water spray controls	6	–	15	–	–	–	–
Blowdown drums and flare stacks	60–90	60–90	60–90	60–90	–	–	–
Product storage tanks	75	75	75	75	60–90	–	–
Rundown tanks	60	60	60	60	60–90	–	–
Blending tanks	60	60	60	60	60–90	–	–
Hazardous loading and unloading facilities, including docks	60	60	60	60	60–90	75	75
Fire pumps	30	–	–	–	90	90	90

(continued)

TABLE 3.2 (Continued)

OIA General Recommendations for Spacing in Refineries

Minimum Distance (m)	Blending Tanks	Hazardous Loading and Unloading Facilities, Including Docks	Fire Pumps	Turret Nozzles	Fire Hydrants	Fire Equipment Houses
Service buildings	–	–	–	15	15–75	15–75
Process units	–	–	–	15–30	15–75	30
Boilers, utilities, and electrical generating equipment, etc.	–	–	–	15–30	15–75	30
Fired process heaters	–	–	–	15–30	15–75	30
Process vessels, fractionating equipment, etc.	–	–	–	15–30	15–75	30
Gas compressor houses	–	–	–	15	15–75	30
Large oil pump houses	–	–	–	15	15–75	30
Control rooms	–	–	–	15	15–75	30
Cooling towers	–	–	–	15–30	15–75	30–60
Dropout controls, steam snuffing, and water spray controls	–	–	–	–	–	–
Blowdown drums and flare stacks	–	–	–	30	30	75
Product storage tanks	–	–	–	15–30	15–75	90
Rundown tanks	–	–	–	15–30	15–75	90
Blending tanks	–	–	–	15–30	15–75	75
Hazardous loading and unloading facilities, including docks	75	15–75	–	15–30	15–75	75
Fire pumps	90	90	–	–	–	–

TABLE 3.3

OIA General Recommendations for Spacing in Petrochemical Plants

Minimum Distance (m)	Process Unit, High Hazard	Process Unit, Low Hazard	Tank Frames, High Hazard	Tank Frames, Low Hazard	Process Warehouse, Low Hazard	Shipping and Receiving, High Hazard	Shipping and Receiving, Low Hazard
Process unit, high hazard	60	–	–	–	–	–	–
Process unit, low hazard	30	15	–	–	–	–	–
Tank frames, high hazard	75	75	1.5 (larger diameter)	–	–	–	–
Tank frames, low hazard	60	30	Larger diameter	0.5 (larger diameter)	–	–	–
Process warehouse, low hazard	45	45	75	30	15	–	–
Shipping and receiving, high hazard	60	60	45	30	45	15	–
Shipping and receiving, low hazard	45	30	30	15	6	15	–
Service buildings	60	30	60	30	30	45	30
Boiler area	60	45	60	45	30	60	30

(continued)

TABLE 3.3 (Continued)

OIA General Recommendations for Spacing in Petrochemical Plants

Minimum Distance (m)	Service Buildings	Fire Pumps	Emergency Controls	Water Spray Controls	Future Nozzles	Pilot Plants	Large Cooling Towers	Fired Process Heaters
Process unit, high hazard	–	75	30	15	15–30 to center of target	60	45	15–30
Process unit, low hazard	–	45	15	–	15–30 to center of target	60	30	15
Tank frames, high hazard	–	75	–	30	15–30 to center of target	75	75	60
Tank frames, low hazard	–	60	–	–	15–30 to center of target	60	60	60
Process warehouse, low hazard	–	60	–	–	15–30 to center of target	60	45	30
Shipping and receiving, high hazard	–	45	30	15	15–30 to center of target	60	60	60
Shipping and receiving, low hazard	–	30	15	–	15–30 to center of target	45	45	30
Service buildings	–	30	–	–	15–30 to center of target	60	30	30
Boiler area	30	–	–	–	15–30 to center of target	60	30	30

TABLE 3.4

OIA General Recommendations for Spacing in Gas Plants

Minimum Distance (m)	Service Building	Gas Compressor House	Large Process Oil Pump House	Distillation and Fractionation	Utilities	Pressure Tanks	Atmospheric Tanks	Loading Racks
Service building	–	–	–	–	–	–	–	–
Gas compressor house	30	–	–	–	–	–	–	–
Large process oil pump house	30	15	–	–	–	–	–	–
Distillation and fractionation	30	15	9	–	–	–	–	–
Utilities	15	30	30	30	–	–	–	–
Pressure tanks	45	60	60	60	45	–	–	–
Atmospheric tanks	30	60	60	60	30	15	–	–
Loading racks	30	60	60	60	30	30	30	15–30
Fired heaters	30	30	30	30	15	45–60	30	30
Cooling towers	15–30	15–30	15–30	30	30	75	60	60
Skid units for package plant	30	15	15	12	30	30	30	60

(continued)

TABLE 3.4 (Continued)

OIA General Recommendations for Spacing in Gas Plants

Minimum Distance (m)	Main Gas Control Valve	Fire Pumps	Open Flames	Ordinary Electrical	Turret Nozzles	Fire Equipment House	Steam Snuff and/or Blowdown Control	Hydrants	Lean Oil Pumps
Service building	15	30	–	–	–	15	–	15–30	–
Gas compressor house	75–150	60	30	15	–	15	15	15–30	15
Large process oil pump house	75–150	60	30	–	15	15	15	15–30	–
Distillation and fractionation	75–150	60	30	–	15	15	15	15–30	15
Utilities	75–150	–	–	–	–	15	–	15–30	30
Pressure tanks	30	75	30	–	15	30	–	15–30	60
Atmospheric tanks	30	75	30	–	15	30	–	15–30	60
Loading racks	30	45	30	–	15	30	–	15–30	60
Fired heaters	30	45	–	–	–	15	15	15–30	30
Cooling towers	30	–	30	30	15–30	15–30	–	15–30	15–30
Skid units for package plant	75–150	45	30	–	15	30	–	–	15

The following basic requirements should also be met when designing the layout of a control room:

1. The control room and substation should be located as close as possible to the plant equipment, maintaining a minimum distance from viewpoint of noise and safety requirements.

2. The control room and substation should be spaced at least 15 m, from the nearest process equipment surface.

3. The control room and substation should be located with consideration to convenience in daily operation.

4. The control room and substation should be located from an economical standpoint so as to minimize the length of electrical and instrument cables entering and leaving from them.

5. The control room should be positioned so that the operator can command a view of the whole system that is under his or her control. Large buildings or equipment should not be placed in front of the control room.

3.6 Firefighting Requirements

Each individual process unit should be provided with sufficient open spaces around so that fire trucks can be run and operated thereat. The width of an access way thereto should be 6 m minimum.

Process units consisting of large hazardous material storage tanks should be located ideally in an outer area in the complex site.

Figure 3.6 shows a sample layout and footprint of the fireproofing zones, the internal area shows the impingement zone, and the outer area is the radiative heat zone.

3.7 Building Requirements

Service buildings include offices, control rooms, laboratories, houses, shops, warehouses, garages, cafeterias, and hospitals. These structures and areas require protection of personnel from possible fires and explosions of major plant equipment and may require additional spacing from high-risk facilities.

The service buildings should be located near the entrance to the plant and be readily accessible to a public road or highway.

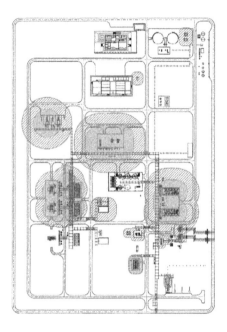

FIGURE 3.6
A sample layout and footprint of the fireproofing zones. Internal area: impingement zone, outer area: radiative heat zone. (Reprinted from *Journal of Hazardous Materials*, 191, A. Di Padova et al., Identification of fireproofing zones in oil and gas facilities by a risk-based procedure, 83–93, Copyright 2011, with permission from Elsevier.)

3.8 Layout in Process Units

In cases where process fluid is run by a gravity head, elevated layouts must be considered. Unless there are any such limitations as indoor arrangement and confined locations, equipment should be placed at grade as a rule except in cases where the gravity flow is specifically required for any reason.

Since the directions of the incoming feed stocks, outgoing products, and utility supply are determined on the overall plot plan, the direction of the pipe rack in the unit should be decided first.

Equipment should be arranged to minimize piping runs (particularly for large-sized piping and alloy piping) as far away as possible. Clear access ways with a minimum width of 600 mm should be provided for the operator's access around equipment.

Large-capacity storage tanks containing flammable and explosive fluids should be located in outer areas as far away as possible.

Space should be allowed for the provision of future spare equipment. Consideration should also be given to future plant expansion.

3.8.1 Process Requirements

Equipment should be laid out along the flows on the process flow diagram. In particular, the fractionator and its reboilers, condensers, and overhead receivers should be collectively located.

Gravity flow lines should be laid out with consideration given to related elevations so that their lengths will become minimum. In particular, in the case of lines in which liquids will flow near their boiling points, related equipment should be located close to each other so that the lines need not be elevated.

Pieces of equipment that are to be connected by large-size piping or alloy piping should be located close to each other.

Air coolers should be laid out so that no heated air may be recirculated.

3.8.2 Safety Requirements

All process equipment should be kept at least 15 m from fired heaters. Exception are permitted for certain pieces of equipment where the heater in question is being used to heat the process flow of the equipment and where any leakage from the equipment would probably ignite instantly, thus creating no additional fire hazard. Such an exception is permitted when locating reactors on a platforming unit, for example. Each exception must be individually investigated as to its potential hazards and must not violate any process practices in oil and gas producer (OGP) plants and governing codes.

Locate fire heaters on the side of the process unit from which the prevailing wind blows. This is done to blow gas away from the heaters instead of toward them. Fired equipment should not be located in a hazardous area classification.

Generally, fired equipment should be located at a distance of more than 15 m from any sources of hazards (hot oil pumps, light end pumps, compressors, etc.).

Emergency showers, if required in a certain process, should be located as near the hazard as possible and indicated in the plot plan.

Adequate and easy access and egress must be considered not only for safety, but also operation and maintenance requirements.

Control rooms and their roadways should not be located in a hazardous area classified in the project specification; generally, they should be located at a distance of more than 15 m from the nearest equipment.

High-pressure gas compressors should be located at leeward locations.

Large-capacity hazardous material storage tanks should be located in outer areas as far away as possible.

Pumps intended to handle flammable materials (that fall under the control of the hazardous area classification), should be located as outlined in Table 3.5.

Fired heater stacks should be located at such points that the performance of the air coolers and operators on tower tops may not be adversely affected

TABLE 3.5

Basis for Pump Location to Handle Flammable Materials

Use of Pump	Under Pipe Rack	Under Air-Cooled Exchangers
Cold oil pumps	Acceptable	Acceptable
Hot oil pumps (hot oil means the oil whose operating temperature is higher than autoignition point)	Unacceptable (a space of minimum 3 m should be provided from pipe rack, hot oil pumps can be located under pipe rack when the proper devices for fire protection such as fire detector, water spray, etc., are provided)	Unacceptable
Light end pumps (light ends means the fractional distillates with a BP range of 110°C–120°C, consisting of mixtures of benzenes, toluenes, xylenes, pyridine, phenole, cresol, etc.)	Acceptable	Unacceptable (light end pumps should not be located under air-cooled exchangers except where concrete decks are provided under the air-cooled exchangers as shown in Figure 3.5. Light end pumps can be located under air coolers without concrete decks, when the proper devices for fire protection are provided)

by hot flue gas emitted through them; the direction of the prevailing wind should be taken into consideration in the determination of the stack locations. If this problem cannot be settled in the equipment layout, the heights of the heater stacks should be increased or a common stack should be provided.

In layout design for safety and fire protection requirements, consideration of NFPA codes, standards, and recommendations is strongly recommended along with the following requirements:

- All units in which heaters exist should be collectively located on the windward side of their process units. If this is impracticable the leeward heater should be located at a distance of more than 15 m from the windward equipment.

- Process units in which poisonous gas is to be handled should be located leeward.

- Ideally, process units in which high-pressure gas is to be handled should be located leeward. Figure 3.7 shows a typical layout plan for a plant.

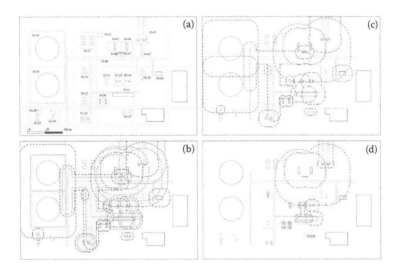

FIGURE 3.7
A typical layout considered in a case study and footprint of the fireproofing zones. (a) Layout
and location of isolable sections, (b) fireproofing zones for structural elements: Internal area,
outer area, (c) fireproofing zone for atmospheric vessel targets (AV), (d) fireproofing zone for
pressurized vessel targets (PV). (Reproduced from A. Tugnoli et al., *Reliability Engineering and
System Safety* 105, 25–35. 2012. With permission.)

3.8.3 Basic Design Considerations

The equipment spacing distances put forth here are general recommendation for minimum requirements and are based on the following:

a. To permit access for firefighting

b. To permit access for normal operation and maintenance

c. To permit access for operation to perform emergency shutdown
action in a fire situation

d. To ensure that critical emergency facilities are not subject to fire
damage

e. To separate continuous ignition sources from possible sources of
release of flammable materials

f. To allow for likely additional equipment that may be added at some
time during the life of the plant

g. To consider recommendations of NFPA codes, standards, and regulations for safety and fire protection

Where spacing is reduced significantly below the recommended distances,
it is usually necessary to compensate for the increased degree of risk by providing additional safety features such as emergency shutdown facilities,

fireproofing, water sprays, and so forth. Reduction below the recommended spacing should require the company's approval.

Location of equipment such as fired heaters, heat exchangers, towers, vessels, drums, pumps, and compressors should be in general agreement with the plot plan.

Space around equipment must be left for items like pipe supports, control valve manifolds, and hose stations where required.

Consideration of the following factors that affect spacing and layout may justify deviations from the basic requirement:

- Physical limitations of the available site
- Special hazards
- Flexibility space requirements for future expansion
- Topography and prevailing winds
- Environmental considerations
- Location conditions
- Association with the type of adjacent property

3.8.4 Erection and Maintenance Requirements

Special maintenance requirements for each item of equipment in a given process should be considered. For example, most machines that consist principally of a stirring mechanism require space for removal of the impeller shaft, and large compressors and crashers need floor or ground area for laying down components during maintenance.

Process units consisting of large equipment should be located in outer areas in the complex site so that sufficient spaces will be available for their haulage and erection.

Sufficient access areas should be provided around each individual unit for use in the operation of mobile cranes.

Process units that will require maintenance separately from other units should be provided with a minimum spacing of 6 m between them and the nearest unit.

Space requirements for maintenance access should be in conformity with the codes, standards, and regulation set forth in Table 3.6.

Table 3.7 shows the proximity of refrigerated storage vessels to boundaries and other facilities.

Sufficient access spaces should be provided around large-sized equipment (e.g., main columns, reactors, coke chambers) for use in haulage and erection. In cases where gin poles are to be used, spaces will be required to haul the equipment and to assemble and disassemble the gin poles and to provide guy wire anchors. A construction engineer should be consulted regarding these access spaces.

TABLE 3.6

Codes and Standards Affecting Maintenance Requirements

For Vessels	For Mechanical Equipment
• American Society of Mechanical Engineers (ASME): Boiler and Pressure Vessel Code: Section 1, Power Boilers. Section VIII, Pressure Vessels	• API, Occupational Safety and Health Administration (OSHA), Environmental Protection Agency (EPA)
• American Petroleum Institute (API): Standard 620, On Large, Welded, Low-Pressure Storage Tanks. Standard 650, On Welded Steel Tanks for Oil Storage.	• National Fire Code
	• NFPA
	• Crane Manufacturers Association of America
• Tubular Exchanger Manufacturers Association Standards	• Monorail Manufacturer's Association
• American Society of Civil Engineers: Minimum Design Loads for Structures	• Conveyer Equipment Manufacturers Association
• Uniform Building Code, from International Conference of Building Officials	• Hydraulic Institute
	• NEC

An open space should be provided on one side of the equipment for crane access. In the case of air coolers placed on pipe racks, if equipment is installed on both sides of the air coolers, access ways (whose minimum width should be 5 m) should be provided at a proper interval on one side of the air coolers. If no such access can be provided, trolley beams should be installed over the air coolers for use in bundle maintenance.

An open space should be provided under the tower top davit as a tower internal maintenance area.

A working area required for catalyst loading and unloading should be provided.

Shell-and-tube heat exchangers should be located collectively at one point as far away as possible and their tube bundle pulling area (tube bundle length + min 2 m) should be provided thereat.

Horizontal coil arrangement (fired) heaters require a mechanical cleaning and tube pulling area.

Generally, pumps should be located collectively under pipe racks. However, this requirement need not be applied to pumps whose suction line lengths must be minimized in connection with the process performance.

Compressors of large capacity should be sheltered unless otherwise requested by the company. Permanent gantry cranes may be provided for the compressors if necessary and an open space should be provided on one side of the shelter. Figure 3.8 shows the layout considered in the case study and footprint of the fireproofing zones: (a) according to the pool-fire scenario and (b) according to API 2218.

TABLE 3.7

Proximity of Refrigerated Storage Vessels to Boundaries and Other Facilities

Boundary Lines or Other Facilities	Minimum Spacing of Dome Roof Tanks	Minimum Spacing of Spheres or Spheroids
Property lines adjacent to land that is developed or could be built upon public highways, and mainline railroads	60 m (distance from boundary line or facility to centerline of peripheral dike wall surrounding the storage vessel shall not be less than 30 m at any point).	60 m (distance from boundary line or facility to centerline of peripheral dike wall surrounding the storage vessel shall not be less than 30 m at any point)
Utility plants, buildings of high occupancy (offices, shops, labs, warehouses, etc.)	One and a half vessel diameter but not less than 45 m, not to exceed 60 m (distance from boundary line or facility to centerline of peripheral dike wall surrounding the storage vessel shall not be less than 30 m at any point)	60 m (distance from boundary line or facility to centerline of peripheral dike wall surrounding the storage vessel shall not be less than 30 m at any point)
Process equipment (or nearest process unit limits if firm layout not available)	One vessel diameter, but not less than 45 m need not exceed 60 m (distance from boundary line or facility to centerline of peripheral dike wall surrounding the storage vessel shall not be less than 30 m at any point)	60 m (distance from boundary line or facility to centreline of peripheral dike wall surrounding the storage vessel shall not be less than 30 m at any point)
Nonrefrigerated pressure storage facilities	One vessel diameter, but not less than 30 m need not exceed 60 m	Three-quarter vessel diameter but not less than 30 m need not exceed 60 m
Atmospheric storage tanks (stock closed cup flash point under 55°C)	One vessel diameter, but not less than 30 m need not exceed 60 m	One vessel diameter, but not less than 30 m need not exceed 60 m
Atmospheric storage tanks (stock closed cup flash point 55°C or higher)	One-half vessel diameter, but not less than 30 m need not exceed 45 m	One-half vessel diameter, but not less than 30 m need not exceed 45 m

3.8.5 Operational Requirements

Clear access ways (whose minimum widths should be 600 mm) should be provided as the operator's access around each individual item of equipment. No auxiliaries and accessories such as piping and instruments should be present in the access ways.

Entrances of structure ladders should be located on the side near the control room.

Stages and platforms through which patrol personnel will frequently pass should be provided with stairways.

Patrol routes should be considered in the preparation of the plot plan.

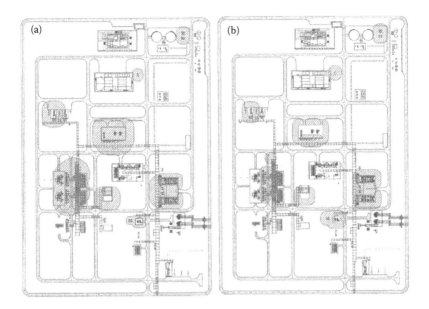

FIGURE 3.8
Layout considered in the case study and footprint of the fireproofing zones: (a) according to the pool–fire scenario, (b) according to API 2218. (Reprinted from *Journal of Hazardous Materials*, 191, A. Di Padova et al., Identification of fireproofing zones in oil and gas facilities by a risk-based procedure, 83–93, Copyright 2011, with permission from Elsevier.)

3.8.6 Economic Requirements

The following should be considered for cost reduction purposes:

- Minimize piping lengths
- Minimize pipe rack lengths
- Minimize common duct lengths
- Minimize cable lengths

3.9 Distances/Clearances Requirements for Storage Tanks

3.9.1 Diked Enclosures

a. Petroleum storage tanks should be located in diked enclosures with roads all around the enclosure. Aggregate capacity of tanks located in one diked enclosure should not exceed following values:

- 60,000 m³ for a group of fixed roof tanks
- 120,000 m³ for a group of floating roof tanks

Fixed m³ floating roof tanks should be treated as fixed roof tanks. However, if these tanks are provided with windows opening on the shell and these windows will not get blocked, then these may be considered as floating roof tanks.

If a group of tanks contains both fixed and floating roof tanks, then it should be treated as a group of fixed roof tanks for the purpose of the above limits.

b. A diked enclosure should be able to contain the complete contents of the largest tank in the dike in case of any emergency. Enclosure capacity should be calculated after deducting the volume of tanks (other than the largest tank) and the tank pads within the dike up to the height of the enclosure. A free board of 200 mm above the calculated liquid level should be considered for fixing the height of the dike.

c. The height of a tank enclosure dike (including free board) should be at least 1.0 m and should not be more than 2.0 m above the average inside grade level. A dike wall made up of earth, concrete, or solid masonry should be designed to withstand the hydrostatic load. An earthen dike wall should have no less than a 0.6 m wide flat section on top for stability of the dike wall.

d. For excluded petroleum, the capacity of the diked enclosure should be based on spill containment and not for containment on tank rupture. The minimum height of a dike wall in the case of excluded petroleum should be 600 mm.

e. Separation distances between the nearest tanks located in separate dikes should not be less than the diameter of the larger of the two tanks or 30 m, whichever is more.

f. Process equipment should not be located inside the dike. Pump stations and piping manifold should be located outside dike areas by the sides of roads.

g. Tanks located overhead should meet safety distances and should also have a diked enclosure of roller-compacted concrete (RCC) construction and be provided with an efficient drainage system for the dike enclosure.

3.9.2 Grouping

a. Grouping of petroleum products for storage should be based on the product classification. Class A and/or class B petroleum may be stored in the same diked enclosure. Class C petroleum should preferably be stored in separate enclosure. However, where class C petroleum is stored in a common dike along with class A and/or class B petroleum, all safety stipulations applicable for class A and/or class B, respectively, will apply.

b. Excluded petroleum should be stored in a separate diked enclosure and should not be stored along with class A, class B, or class C petroleum.

c. Tanks should be arranged in a maximum of two rows so that each tank is approachable from the road surrounding the enclosure. This stipulation need not be applied to tanks storing excluded petroleum class.

Tanks having a 50,000 m³ capacity and above should be laid in a single row.

3.9.3 Fire Walls

a. In a diked enclosure where more than one tank is located, fire walls of a minimum height of 600 mm should be provided to prevent spills from one tank endangering any other tank in the same enclosure.

b. A group of small tanks, each not exceeding 9 m in diameter and in all not exceeding 5000 m³ in capacity, should be treated as one tank for the provision of a fire wall.

c. For excluded petroleum product storage, a fire wall of a height not less than 300 mm should be provided by limiting the number of tanks to 10 or the capacity of group of tanks to 5000 m³, whichever is lower.

3.9.4 Separation Distances between Tanks and Off-Site Facilities

The following stipulations should apply for the separation distances for above-ground tanks storing petroleum:

a. For a larger installation, minimum separation distances should be as specified in Tables 3.8 and 3.9. The tables are applicable where total storage capacity for class A and class B petroleum products is more than 5000 m³ or the diameter of a class A or class B product tank is more than 9 m.

b. For a smaller installation, minimum separation distances should be as specified in Table 3.10. This table is applicable where total storage capacity of class A and class B is less than 5000 m³ and the diameter of any tank storing class A and class B petroleum product does not exceed 9 m. Table 3.10 is also applicable for the installation storing only class C petroleum.

c. Excluded petroleum should be treated as class C petroleum for the purpose of separation distances and Table 3.10 is applicable for their separation distances.

TABLE 3.8

Proximity of Atmospheric Storage Tanks to Boundaries and Other Facilities

Boundary Lines or Other Facilities	Minimum Distance From:			
	Low-Flash or Crude Stocks in Floating Roof Tanks	Low-Flash Stocks in Fixed Roof Tanks	Crude Stocks in Fixed Roof Tanks	High-Flash Stocks in Any Type of Tank
Property lines adjacent to land that is developed or could be built upon, public highways, mainline railroads, and manifolds located on marine piers	60 m	60 m	60 m	45 m
Buildings of high occupancy (offices, shops, labs, warehouses, etc.)	One-half tank diameter, but not less than 45 m need not exceed 60 m	One and a half tank diameter, but not less than 45 m need not exceed 60 m	60 m	One tank diameter, but not less than 30 m need not exceed 45 m
Nearest process equipment, or utility plant (or nearest unit limits if firm layout not available)	45 m	45 m	60 m	One tank diameter, but not less than 30 m need not exceed 45 m

TABLE 3.9

Proximity of Atmospheric Storage Tanks to Each Other

Types of Stocks and Tankage	Minimum Spacing Between[a]		
	Single or Paired Tanks	Grouped Tanks	Adjacent Rows of Tanks in Separate Groups (Spacing between High-Flash and Low-Flash Tank Groups Shall Be Governed by the Low-Flash Criteria)
Low-flash stocks in floating roof tanks	Three-quarter tank diameter, need not exceed 60 m	One-half tank diameter, need not exceed 60 m	Three-quarter tank diameter, not less than 25 m, need not exceed 60 m
Low-flash stocks in fixed roof tanks	One tank diameter	One-half tank diameter	One tank diameter not less than 30 m
Crude oil stocks in floating roof tanks	Three-quarter tank diameter, need not exceed 60 m	Not permitted	–
Crude oil stocks in fixed roof tanks	One and a half tank diameter (pairing not permitted)	Not permitted	–
High-flash stocks in any type tank	One-half tank diameter, need not exceed 60 m	One-half tank diameter, need not exceed 60 m Finished stocks with a closed-up flash point above 93°C may be spaced at a minimum of 2 m apart provided that all of the following requirements are met: a. The stock is stored at ambient temperature of, if heated, not above 93°C and not within 10°C of its flash point b. The stock is not received directly from a process unit where upset conditions could lower its flash point below the limits of requirement a above	One-half tank diameter not less than 15 m, need not exceed 60 m

(continued)

TABLE 3.9 (Continued)

Proximity of Atmospheric Storage Tanks to Each Other

Types of Stocks and Tankage	Minimum Spacing Between[a]		Adjacent Rows of Tanks in Separate Groups (Spacing between High-Flash and Low-Flash Tank Groups Shall Be Governed by the Low-Flash Criteria)
	Single or Paired Tanks	Grouped Tanks	
		c. There are not tanks storing low-flash stocks within the same group	
		Finished stocks with a close-up flash point of 55°C or higher but less than 93°C may be spaced one-sixth of the aim of their diameters apart, except where the diameter of one tank is less than one-half the diameter of the adjacent tank, the spacing between the tanks shall not be less than one-half the diameter of the smaller tank, provided that all of the following requirements are met:	
		a. The spacing between tanks is not less than 2 m	
		b. The stock is not heated above 93°C and not within 10°C of its flash point	
		c. Corporations do not exceed at total capacity of 15,900 m³ (100,000 bbl) and there are no tanks storing low-flash stocks within the same group	
		d. The stock is not received directly from a process unit where upset conditions could lower its flash point below the limits of requirement b above	

[a] Spacing between high-flash and low-flash tank groups shall be governed by the low-flash criteria. A minimum spacing of 3 m shall be provided between any tank shell and the peripheral dike or toe wall.

TABLE 3.10

Proximity of Nonrefrigerated Pressure Storage Vessels/Drums to the Boundaries of Other Facilities

Boundary Lines and Other Facilities	Minimum Spacing to Spheres, Spheroids, and Drums
Property lines adjacent to land that is developed or could be built upon, public ways, main railroads, and manifolds located on marine piers	60 m (distance from boundary line or facility to centerline of peripheral dike wall surrounding the storage vessel shall not be less than 30 m at any point)
Buildings of high occupancy (offices, shops, labs, warehouses, etc.)	60 m (distance from boundary line or facility to centerline of peripheral dike wall surrounding the storage vessel shall not be less than 30 m at any point)
Nearest process equipment or utilities point (or nearest unit admits if firm layout is not available)	60 m (distance from boundary line or facility to centerline of peripheral dike wall surrounding the storage vessel shall not be less than 30 m at any point)
Refrigerated storage facilities	Three-quarter tank diameter, but not less 30 m, need not exceed 60 m
Atmospheric storage tanks (stock closed-up flash point of 55°C and below)	One tank diameter, but not less than 30 m, need not exceed 60 m
Atmospheric storage tanks (stock closed-up flash point above 55°C)	One-half tank diameter, but not less than 30 m, need not exceed 45 m

3.10 Layout of LPG Facilities

The requirements given below are applicable to above-ground LPG storage facilities:

a. Vessels should be arranged into groups, each with a maximum of six vessels. The capacity of each group should be limited to 15,000 m³. Each group should be provided with a curb wall.

b. Any vessel in one group should be separated from a vessel in another group by a minimum distance of 30 m.

c. Spheres and bullets should be treated as separate groups with a 30-m separation distance between the two groups.

d. Longitudinal axes of horizontal vessels (bullets) should not point toward other vessels, vital process equipments, and control room.

e. Storage vessels should be located downwind of process units, important buildings, and facilities.

f. LPG storage vessels should not be located within the same dikes where other liquid hydrocarbons are stored.

g. Storage vessels should be laid out in a single row in the case of both spheres and bullets. Storage vessels should not be stacked one above the other.

h. A spillage collection shallow sump should be located at a distance where the flames from a sump fire will not impinge on the vessel. This distance should not be less than the diameter of the nearest vessel or 15 m, whichever is higher.

i. The curb wall around a storage tank should have a minimum height of 30 cm. However, it should not exceed 60 cm at the shallow sump position, otherwise evaporation of spilled LPG may become affected.

3.10.1 LPG Bottling Facility

a. LPG bottling facilities should be located at a safe distance from other facilities with minimum ingress to trucking traffic and downwind to storage.

b. There should not be any deep ditches in the surrounding area to avoid LPG settling.

c. Stacking areas for empty and filled cylinders should be located separately. Cylinders shall be stacked vertically. Filling machines and testing facilities shall be organized in sequential manner distinctly in a separate area.

d. Filled LPG cylinders shall not be stored in the vicinity of cylinders containing other gases or hazardous substances.

e. Trucking traffic shall be smooth to avoid blocking/obstruction for loading and unloading of cylinders.

3.10.2 Bulk Handling Facilities

a. The LPG truck loading/unloading gantry should be located in a separate block and should not be grouped with other petroleum products. The maximum number of LPG tank truck bays should be restricted to eight in one group. The bay should be designed in such a way that the driver's cabin faces the exit direction and has no obstruction.

b. The LPG rail loading/unloading gantry should be located on a separate rail spur and should not be grouped with other petroleum products.

c. Rail loading/unloading of LPG should be restricted to a maximum of half rake. Full rake loading/unloading should be done on two separate rail gantries having a minimum distance of 50 m.

Figure 3.9 illustrates a layout considered in and a footprint of the fireproofing zones. Fireproofing zones suggested by API 2218 and API 2510 for the LPG storage vessels are also shown.

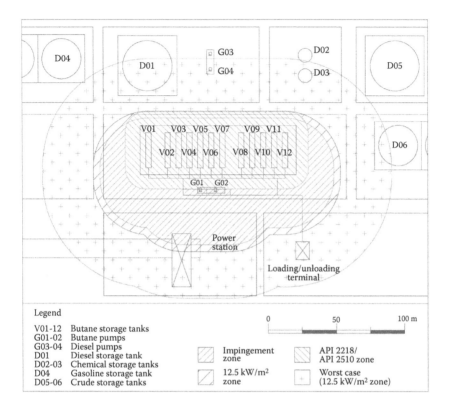

FIGURE 3.9
Layout considered in and footprint of the fireproofing zones. Fireproofing zones suggested by API 2218 and API 2510 for the LPG storage vessels are also shown. (Reprinted from *Journal of Hazardous Materials*, 191, A. Di Padova et al., Identification of fireproofing zones in oil and gas facilities by a risk-based procedure, 83–93, Copyright 2011, with permission from Elsevier.)

The data given below shows typical distances between equipment surface to equipment surface, and clearances that must be adjusted based on the result of the piping layout at the detail planning stage:

a. Distances between individual equipment:

a-1: Column to column, 3 m

a-2: Drum to drum, 2 m

a-3: Exchanger shell to exchanger shell, 1 m

a-4: Pump to pump (foundation) for

Small pumps 3.7 kW and less mount on common foundations with suitable center-to-center distances

Medium pumps 22.5 kW and less, 1 m

Large pumps above 22.5 kW, 1.5 m

b. Distance of equipment to other boundaries and facilities:

 b-1: Exchangers to other equipment, 1-m minimum clear aisle

 b-2: Pipe rack to equipment, 6 m (except flammable material pumps)

 b-3: Pipe rack to structure, 5 m

 b-4: Driver end of pumps to truck, 3-m access (if required)

3.11 Equipment Layout and Spacing

Minimum separation distances between various blocks/facilities described above should be as per Table 3.11. The table should be read in conjunction with the accompanying footnotes. Minimum separation distances between various equipment within process units are given in Table 3.12. The distances recommended should be followed to the extent feasible. Equipment spacing within the process unit may be varied to meet the requirements specified by licensors or engineering consultants. However, the distances specified in the footnotes to Table 3.12 should be met. Table 3.13 lists separation distances between tank/off-site facilities for large installations. Table 3.14 lists separation distances between storage tanks within a dike. Table 3.15 lists separation distances between tank/off-site facilities for small installations. Table 3.16 lists separation distances for LPG facilities. Table 3.17 lists separation distances between LPG storage vessels and boundary/group of buildings not associated with LPG facilities.

3.11.1 Reactors

Adequate space should be provided for handling and storing catalysts (both fresh and spent), chemicals, hydrogen, nitrogen, and so forth, including truck access where appropriate.

 In general, maximum use should be made of mobile equipment for transporting and handling these commodities.

3.11.2 Towers

Towers closely related with processing equipment such as overhead condensers, overhead receivers, or reboilers should be arranged adjacent to each other.

 The location of the tower should be studied with consideration to the transportation route and erection procedure.

 Space should be provided for assembling and disassembling tower internals such as tray demistors.

TABLE 3.11

Separation Distances between Blocks/Facilities

No.	From/To	1	2	3	4	5	6	7	8	9	10	11	12	13	14	15	16
1	Process units	Note 1	Note 3	30	30	30	60	90	45	45	60	45	30	60	60	30	90
2	Process control room (note 2)	Note 3	x	Note 4	Note 5	30	60	90	45	45	30	Note 3	x	30	15	30	30
3	Storage tanks, class A	30	Note 4	Note 6	Note 6	Note 6	30	90	30	30	60	(90)	30	T3	60	30	50
4	Storage tanks, class B	30	Note 5	Note 6	Note 6	Note 6	30	90	30	30	60	(90)	30	T3	30	30	50
5	Storage tank, class C	30	30	Note 6	Note 6	Note 6	30	90	30	30	60	(90)	30	T3	30	30	50
6	Pressurized storage: LPG/C4 and lighter/H2	60	60	30	30	30	T7	90	30	T6	90	(90)	30	T7	45	30	60
7	Flare (note 7)	90	90	90	90	90	90	90	90	90	90	90	90	90	90	90	90
8	Bulk loading Petroleum, Oils and Lubricants (POL) (rail/road)	45	45	30	30	30	30	90	Note 8	Note 9	60	30	Note 10	T3	60	30	50
9	Bulk loading LPG (rail/road)	45	45	30	30	30	T6	90	Note 9	T6	90	(90)	T6	T6	60	30	50
10	Fire station/first aid center	60	30	60	60	60	90	90	60	90	x	30	30	12	12	30	90
11	Boiler house/process unit heaters (note 11)	45	Note 3	(90)	(90)	(90)	(90)	90	30	(90)	30	x	15	50	30	30	Note 12
12	Rail spur	30	x	30	30	30	30	90	Note 10	T6	30	15	x	30	6	15	50

(continued)

TABLE 3.11 (Continued)

Separation Distances between Blocks/Facilities

Sr No.	From/To	1	2	3	4	5	6	7	8	9	10	11	12	13	14	15	16
13	Boundary wall around installation	60	30	T3	T3	T3	T7	90	T3	T6	12	50	30	x	6	30	50
14	Service buildings	60	15	60	30	30	45	90	60	60	12	30	6	6	x	30	50
15	Cooling tower	30	30	30	30	30	30	90	30	30	30	30	15	30	30	x	15
16	API separators/oil sludge pit	90	30	50	50	50	60	90	50	50	90	Note 12	50	50	50	15	x

General notes

a. All distances are in meters. T indicates the table number to be referred, and x means any distance suitable for constructional or operational convenience.

b. All distances shall be measured between the nearest points on the perimeter of each facility except (1) in the case of a tank vehicle loading/unloading area where the distance shall be from the center of nearest bay, and (2) where the distances given in the brackets () are from the shell of the heater/boiler/furnace/still.

Specific notes

Note 1: This shall be 36 m considering the 6-m wide road passing through the center. The edge of the road shall not be less than 15 m away from the edge of the unit.

Note 2: Type of construction shall be as per local authority permission.

Note 3: Process control room to process units/boiler house/heaters, the minimum separation distance shall be 16 m. For gas processing plants, it shall be a minimum of 30 m irrespective of whether it is for one or more units.

Note 4: Shall be 60 m for nonblast construction and 30 m for blast-resistant construction.

Note 5: Shall be 45 m for nonblast construction and 30 m for blast-resistant construction.

Note 6: Separation distances between the nearest tanks located in two dikes shall be equivalent to the diameter of the larger tank or 30 m, whichever is more. For distances within a dike, it shall be as per Tables 3.13 and 3.14.

Note 7: The distances specified are for the elevated flare. For ground flare, these distances shall be 150 m. For exploration and production installations, this shall be in line with oil and mine regulations.

Note 8: Separation distance between a tank truck gantry and a wagon gantry shall be 50 m.

Note 9: The separation distance shall be 50 m. However, for an LPG tank truck bulk loading to a POL tank truck bulk loading it shall be 30 m.

Note 10: Separation distance between a tank truck gantry and a rail spur shall be 50 m.

Note 11: The boiler house or heater of a process unit is to be treated as a separate identity only for the consideration of surrounding blocks/facilities. However, the heater of a process unit remains an integral part of the process unit to which it is attached and in that case the interequipment distances should be in line with Table 3.12.

Note 12: Centralized/common API separators, corrugated plate interceptor (CPI), open oil separators shall be categorized under the same risk and shall be located at a distance of 90 m from heaters/boilers. However, if these are covered from the top and provided with adequate venting to safe location, the minimum separation distance shall be 30 m.

TABLE 3.12

Separation Distances between Equipment within a Process Unit

No.	From/To	1	2	3	4	5	6	7	8	9	10	11	12	13	14	15	16
1	Fired heater/any fired equipment	x	15	15	15	22	15	15	20	15	15	15	x	18	6	30	15
2	Reactors	15	2	2	6	8	7	15	7	7	4	3	15	5	3	15	3
3	Distillation column	15	2	3	4	7	5	15	5	5	2	3	15	3	3	15	3
4	Accumulators, hydrocarbon	15	6	4	2	8	5	15	4	4	2	3	15	3	3	15	3
5	Compressors, hydrocarbon	22	8	7	8	3	7	15	7	7	7	7	15	4	3	20	7
6	Hot oil pump	15	7	5	5	7	1	7	1	1	2	2	15	3	x	15	x
7	Fuel oil/HCS day tank	15	15	15	15	15	7	T-5	15	15	15	15	15	15	x	15	15
8	Pumps for class A and all above autoignition temperatures	20	7	5	4	7	1	15	1	1	2	2	15	3	x	15	x
9	Pumps for all other hydrocarbons	15	7	5	4	7	1	15	1	1	2	2	15	3	x	15	x
10	Heat exchangers	15	4	2	2	7	2	15	2	2	2	2	15	2	2	15	x
11	Air fin coolers for hydrocarbons	15	3	3	3	7	2	15	2	2	2	x	15	2	x	15	2
12	Fired heater, local control panel	x	15	15	15	15	15	15	15	15	15	15	x	10	x	15	5

(continued)

TABLE 3.12 (Continued)

Separation Distances between Equipment within a Process Unit

No.	From/To	1	2	3	4	5	6	7	8	9	10	11	12	13	14	15	16
13	Pressure vessels/drums of hydrocarbons	18	5	3	3	4	3	15	3	3	2	2	10	2	3	15	2
14	Main pipe rack	6	3	3	3	3	x	x	x	x	2	x	x	3	x	15	x
15	Blowdown facility, drum, pump, vent stack	30	15	15	15	20	15	15	15	15	15	15	15	15	15	x	15
16	Structural main, technological platforms	15	3	3	3	7	x	15	x	x	x	2	5	2	x	15	x

General notes

a. All distances are face-to-face clear minimum distances in meters.

b. x indicates suitable distance as per good engineering practices to meet construction, operational, and maintenance requirements.

c. Distances specified in this table are the minimum recommended distances to which the industry should adhere. These could be suitably modified as required to suit space constraints and relevant engineering practices except for the following:

    i. The blowdown facility (open-pit type)/oil catcher shall be located at a distance not less than 30 m from a fired heater/any fired equipment. If the blowdown drum is located underground and the oil catcher is covered and is vented to a safe location, the minimum separation distance shall be 15 m.

    ii. The fuel oil day tank shall be located at a distance of not less than 15 m from equipment except those facilities such as heat exchanger, pump connected directly with the fuel oil system.

d. Firewater hydrant/monitors shall be a minimum of 15 m away from the equipment that is being protected.

e. The water spray deluge valve shall be a minimum of 15 m from equipment handling hydrocarbons.

f. The fuel gas knockout drum shall be located at a minimum separation distance of 15 m from the heater.

g. Separation distances specified in other standards or elsewhere in this book (other than this table) shall be followed as recommended.

TABLE 3.13

Separation Distances between Tank/Off-site Facilities (Large Installations)

	Tanks/Facility	1	2	3	4	5	6	7	8	9
1	Storage tank for petroleum class A/class B	T4	T4	15	15	15	15	8	15	0.5 D Min 20 m
2	Storage tank for petroleum class C	T4	x	15	x	8	x	x	x	0.5 D Min 20 m
3	Storage/filling shed for petroleum class A or class B	15	15	x	8	15	15	8	15	15
4	Storage/filling shed for petroleum class C	15	x	8	x	8	x	x	x	10
5	Tank vehicle loading/unloading for petroleum class A or class B	15	15	15	8	x	x	8	15	20
6	Tank vehicle loading/unloading for class C	15	x	15	x	x	x	x	x	10
7	Flameproof electric motor	8	x	8	x	8	x	x	8	x
8	Nonflameproof electric motor	15	x	15	x	15	x	8	x	x
9	Boundary wall	0.5 D Min 20 m	0.5 D Min 20 m	15	10	20	10	x	x	x

General notes

a. All distances are in meters.

b. x indicates suitable distance as per good engineering practices to meet construction, operational, and maintenance requirements

c. D + d stands for diameter of larger and smaller tanks. Distances given are shell to shell in the same dike.

d. All distances shall be measured between the nearest points on the perimeter of each facility except in the case of a tank vehicle loading/unloading area where the distance shall be measured from the center of each bay.

e. For different combination of storage tanks, the stringent of the applicable formulas shall be considered for minimum separation distance.

f. The distance of storage tanks from a boundary wall is applicable for
   i. Floating roof tanks having protection from exposure
   ii. Tanks with a weak roof-to-shell joint having an approved foam or inerting system and the tank diameter not exceeding 50 m.

g. For the facilities not covered in this table, refer to Table 3.11.

TABLE 3.14

Separation Distances between Storage Tanks within a Dike

Item	Between Floating Roof Tanks (Class A and Class B)	Between Fixed Roof Tanks (Class A and Class B)	Between Class C Petroleum Storage Tanks
1 All tanks with a diameter up to 50 m	(D + d)/4 Min 10 m	(D + d)/4 Min 10 m	(D + d)/6 Min 6 m
2 Tanks with a diameter exceeding 50 m	(D + d)/4	(D + d)/3	(D + d)/4

General notes
a. All distances are in meters.
b. x indicates suitable distance as per good engineering practices to meet construction, operational, and maintenance requirements.
c. D + d stands for the diameter of larger and smaller tanks. Distances given are shell to shell in the same dike.
d. Distances given are shell to shell in the same dike.
e. For different combination of storage tanks, the stringent of the applicable formulas shall be considered for minimum separation distance.
f. The distance of storage tanks from a boundary wall is applicable for
   i. Floating roof tanks having protection for exposure
   ii. Tanks with weak roof-to-shell joint having an approved foam or inerting system and the tank diameter not exceeding 50 m.

Where two or more towers are installed, their center-lines should be aligned parallel with the pipe rack, except those small in diameter, which may be grouped and aligned perpendicular to the pipe rack.

Towers and drums should be lined up on the basis of centerlines.

Self-standing towers exceeding 30 in L/D may require support structures. Hence, in the case of such towers, the equipment design engineer in charge should be contacted in advance of the company for confirmation.

3.11.3 Fired Equipment

Furnace and boilers should be located on the windward side of the plant to avoid contact with inflammable gases (light hydrocarbon) that may leak out.

Where practical and economical, isolating fired equipment should preferably be grouped together. Where a common stack should be employed, isolating dampers or suitable barriers may be provided in individual ducting to the stack.

Heaters should be located near the edge of a process area rather than in the center of the area. This provides more area for maintenance and helps isolate them from other equipment. In many places it is now necessary to have tall stacks. Because of the high cost and the minimum size, a tall stack can be for all heaters and must be located in one area or at least in clusters. This can be a major factor in layout and must be resolved early.

TABLE 3.15

Separation Distances between Tanks/Off-Site Facilities (Small Installations)

		1	2	3	4	5	6	7	8	9	10	11	12	13
1	Storage tank class A	0.5 D	0.5 D	0.5 D/6.0	9	9	9	15	15	15	3	15	15	15
2	Storage tank class B	0.5 D	0.5 D	0.5 D/6.0	9	9	0.5 D	9	4.5	4.5	3	4.5	D Min 4.5	D Min 4.5
3	Storage tank class C	0.5 D/6.0	0.5 D/6.0	x	x	0.5 D	x	9	4.5	x	x	x	0.5 D Min 3.0	0.5 D Min 3.0
4	Storage/filling shed for petroleum class A	9	9	9	x	4.5	6	9	9	9	3	9	9	9
5	Storage/filling shed for petroleum class B	9	0.5 D	0.5 D	4.5	x	1.5	9	4.5	4.5	1.5	4.5	4.5	4.5
6	Storage/filling shed for petroleum class C	9	0.5 D	x	6	1.5	x	9	4.5	x	x	x	3	3
7	Tank vehicle loading/unloading, class A	15	9	9	9	9	9	x	9	9	3	9	9	9
8	Tank vehicle loading/unloading, class B	15	4.5	4.5	9	4.5	4.5	9	x	4.5	1.5	4.5	4.5	4.5
9	Tank vehicle loading/unloading, class C	15	4.5	x	9	4.5	x	9	4.5	x	x	x	3	3
10	Flameproof electric motors	3	3	x	3	1.5	x	3	1.5	x	x	3	x	x
11	Nonflameproof electric motors	15	4.5	x	9	4.5	x	9	4.5	x	3	x	x	x
12	Office building, stores, amenities	15	D Min 4.5	0.5 D Min 3.0	9	4.5	3	9	4.5	3	x	x	x	x
13	Boundary wall	15	D Min 4.5	0.5 D Min 3.0	9	4.5	3	9	4.5	3	x	x	x	x

General notes

a. All distances are in meters and the table specifies the minimum requirement.

b. x indicates suitable distance as per good engineering practices to meet construction, operational, and maintenance requirements.

c. D indicates the diameter of the larger tank.

d. Distances given for the tanks are shell to shell in the same dike.

e. Where alternate distances are specified (e.g., 0.5 D/6.0), the minimum thereof shall be used.

f. All distances shall be measured between the nearest points on the perimeter of each facility except in the case of the tank vehicle loading/unloading area, where the distance shall be from the center of each bay.

g. Pig launcher/receiver at liquid hydrocarbon handling pipeline installations should be located at least 5 m from boundary.

TABLE 3.16

Separation Distances for LPG Facilities

Sr No.	From/To	1	2	3	4	5	6	7	8
1	LPG storage vessels	Note 1	T7	30	30	50	30	15	60
2	Boundary wall/group of buildings not associated with LPG plant	T7	x	30	30	50	30	30	x
3	Shed for filling and storage of LPG, cold repair shed, cylinder evacuation facilities	30	30	15	30	50	30	15	60
4	Tank truck loading/unloading gantry	30	30	30	30	50	50	30	60
5	Tank wagon gantry	50	50	50	50	50	50	30	60
6	Rail spur	30	30	30	50	50	50	30	60
7	Pump house/compressor house (LPG)	15	30	15	30	30	30	x	60
8	Fire water pump house	60	x	60	60	60	60	60	x

General notes

a. Table 3.16 is applicable for total storage of above 100 tonnes.

b. All distances are in meters.

c. x indicates suitable distance as per good engineering practices to meet construction, operational, and maintenance requirements.

d. T7 indicates Table 3.17.

e. Distance of a stabling line shall be as per railway standards.

Specific note

The distance shall be 2 m or one-quarter of the sum of the diameters of the adjacent vessels or half the diameter of the larger of the two adjacent vessels in the same group, whichever is greater.

TABLE 3.17

Separation Distances between LPG Storage Vessels and Boundary Wall/Group of Buildings Not Associated with LPG Facilities

Capacity of each vessel in m³ of water	10–20	21–40	41–350	351–450	451–750	751–3800
Distance in meters	15	20	30	40	60	90

General notes
a. All distances are in meters.
b. Distance of stabling line shall be as per railway standards.

Tube pulling areas should not encroach on any main roadways or other process areas.

Good drainage should be provided around and under fired heaters to direct any liquid spills to a safe location.

Fired equipment may be placed within 15 m from a source of hazard depending on the requirement of the process design.

3.11.4 Heat Exchangers

Heat exchangers should be located close to the related vessels or equipment. Some items such as bottom coolers may be placed away from the vessels.

Horizontal clearance between heat exchanger shells and between heat exchangers and major equipment, for access purposes, should be a minimum of 1.0 m in any direction.

Where there is a plural number of heat exchangers, as a rule the centerlines of their channel nozzles should be aligned.

Piping around heat exchangers and their relevant equipment in high-pressure/temperature service, should be provided with sufficient flexibility against thermal stress.

Clearance around heat exchangers should be adequate to permit safe installation and removal of bolting, and also pulling out bundles.

Where air fins are used extensively they may actually determine the unit length. Air fins should not be located close to a heater.

Where air-cooled heat exchangers are installed on pipe racks or structures, adequate space should be provided around them to perform maintenance work.

Attention should be paid to prevent hot air being taken into air-cooled heat exchangers from other adjacent high-temperature equipment since the atmospheric temperature at the suction side is the design base for the air-cooled heat exchangers.

Air cooled exchangers should all be located at the same level within a unit unless they are so widely separated that one will not be able to suck up the discharged air from another.

Thermosiphon reboilers should be located next to the related vessels.

3.11.5 Vessels and Drums

Vessels and drums should principally be laid out as close as possible to the related equipment.

Where horizontal drums are arranged near a pipe way, the horizontal centerline of the drums should be located at right angles to the pipe way.

The centerlines of vertical drums that are located adjacent to vertical type equipment should be aligned with the centerline of the said vertical equipment.

For spacing of drums see Figure 3.6 and Table 3.8.

3.11.6 Pumps

Pumps should be located as a group where feasible to facilitate their operation and maintenance. Pumps should be located so that the suction lines are short.

Pumps should generally be located in a row or rows under or adjacent to the pipe racks. The drivers should be located toward the center of the pipe rack.

Some pumps, such as vacuum column bottom pumps may be located closer to the equipment they take suction from and do not conform to 9.5.6.3.

Aisles between rows of pumps should be 3 m minimum (clear).

The suggested spacing for pumps requiring a 0.5 m to 1.0 m wide foundation is 2 m center to center (a range of 1.5 to 3.0 m is acceptable).

The location of a small chemical pump and inline and injection pumps is governed by the above minimums.

Pump house spacing where feasible should conform to distances given in Tables 3.3 through 3.5 with respective to oil and gas processing plants.

3.11.7 Compressors

Where there are several large compressors in a unit, it is economical for operation and maintenance to locate them in one area.

Compressor orientation should consider the possibility of major mechanical failure in relation to surrounding equipment.

Access for firefighting must be available from at least two sides of the building.

Associated intercoolers, knockout drums, and so forth may be located in the compressor area provided that they do not restrict access for firefighting and maintenance.

Large capital investments such as major equipment and compressors should be protected from fires involving other equipment.

Compressors should be located adjacent to an access way for ease of maintenance.

Space should be provided, at grade, next to the compressors for manifold piping and compressors auxiliary equipment such as suction drums and intercoolers.

Compressors should be located to minimize the pressure drop at the suction side.

Compressors should be located as close as possible to the control room and substation, since a large amount of electrical and instrument cabling is required for the compressors.

Sufficient space should be provided around compressors to permit maintenance.

3.11.8 Storage Vessels/Tanks

Location, layout, and spacing of storage vessels/tanks should be subject to the following mandatory requirements.

Vessels should be located to permit maximum dissipation of vapors by free circulation of air. Ground contours and other obstacles should be taken into account for their effects on air circulation.

Vessels should be arranged in rows not more than two deep. Every vessel should be adjacent to a road or access way.

A firewater system must be provided, otherwise more stringent requirements apply.

The minimum spacing from vessels to boundaries or between vessels and other facilities should be as given in Table 3.7 and/or in accordance with the latest editions of NFPA 30 Flammable and Combustible Code.

Spill control should be accomplished by dike enclosure:

- Dome roof tanks: one vessel diameter
- Spheres or spheroids: three-quarter vessel diameter
- Dome roof tanks and spheres or spheroids: one vessel diameters

For atmospheric storage tanks designed for 20 kPa or less containing flammable and combustible liquids:

- Tanks containing crude or low-flash stocks should be located in areas remote from process units, property lines, and other areas of high occupancy.
- The minimum spacing from atmospheric tanks to boundaries or other facilities should be as specified in Table 3.8 and/or Tables 3.11 through 3.13 where applicable.
- Tanks for crude or low-flash stocks should be arranged in rows not more than two deep. Every tank should be adjacent to a road or access way.

- Tanks for high flash stocks should be arranged in rows not more than three deep. None of the tanks should be more than one row away from a road or access way.
- Minimum spacing between atmospheric storage tanks should be as specified in Table 3.9 and/or Tables 3.11 through 3.13 where applicable.

Dike enclosures should be accomplished for spill control for nonrefrigerated pressure storage vessels designed for 20 kPa or greater containing flammable liquid or liquefied compressed gases.

Vessels should be located to permit maximum dissipation of vapors by free circulation of air. Ground contours and other obstacles should be taken into account for their effects on air circulation.

Spheres and spheroids should be arranged in rows not more than two deep. At least one side of every vessel should be adjacent to a road or access way.

The minimum spacing between the vessels and boundaries or between the vessels and other facilities should be as specified in Table 3.10 provided that it does not contradict with those specified in Tables 3.11 through 3.13.

Pressure storage LP gas tanks may not be provided with spill dikes. However, a dividing bond of 600 mm must be provided between each vessel. Such bonds must not be totally enclosed.

The location of LP gas storage tanks should be in accordance with the latest edition of NFPA 59 Liquefied Petroleum Gases at Utility Gas Plant.

Spacing between pressure storage vessels should be as follows:

(a) No less than a three-quarter vessel diameter between any two spheres and spheroids
(b) No less than one shell diameter between shells of any two drums

3.12 Layout of Piping

All equipment and piping should be indicated or accounted for on the layout. The layout should be made to scale.

Piping should be routed in accordance with the piping and instrumentation diagrams and project specifications.

Piping should be routed to provide convenience, to provide ease of erection and maintenance, and to provide consistency in appearance. These requirements should be met with consideration given to economy.

Piping should be routed in groups overhead wherever possible. Firewater and sewer systems should be buried. All piping should be arranged to avoid or minimize gas and liquid traps except when noted otherwise on the piping and instrumentation diagrams.

Piping should be routed to permit normal bends and offsets to take thermal expansion. Where this is not sufficient, expansion loops or other means to accommodate thermal expansion should be provided. Do not use trenches unless specified otherwise.

All branch lines off the instrument air, steam, and cooling water supply headers are taken off the top of the header.

Overhead clearance should be provided above access areas of 2.2 m minimum. The clear gap between underground pipes should be 300 mm minimum. Clearance for above ground is normal-flange to bare pipe (or insulation) plus 25 mm.

Operating drains should be so arranged that the discharge is visible from the drain valve.

Uninsulated lines should lie directly on the pipe support member. Heat insulate lines should be set on 100 mm (4-inch) T-bar supports (shoes). Height should be adjusted if insulation is greater than 100 mm thick.

3.13 Utility Layout and Spacing

The utility area should be near the process area. The utility area should be arranged for easy access and adequate working area provided around all equipment for maintenance.

The cooling tower should be located to provide the least possible restriction to the free flow of air, and away from areas where drift or fogging might create a problem.

The circulating fuel oil system that supplies oil for process heaters and boilers is usually located in one corner of the utility area including tanks and circulating pumps. Tanks are to be diked.

All boilers should be grouped together with space provided for at least one future boiler. All boiler auxiliaries including deaerator, feed pumps, flash drums, and chemical feed systems should be located in close proximity to the boiler. Consideration must be given to single stacks for each boiler or one common stack.

Plant and instrument air compressors including dryers should be located in the utility area. Switchgear for the electrical system should be placed in an enclosed building and located within the utility area. Substation serving process units and off-site facilities should usually be located in OGP process areas dependent upon the areas served.

Utility control house should be provided to house all board-mounted instruments used for operation and control of utility equipment.

Raw water storage and fire pumps should be located adjacent to either the boilers or the cooling towers, whichever provides the more economic arrangement.

Critical steam and power facilities feeding major portions of the plant should be protected from possible fire or explosion in equipment handling hydrocarbons.

3.14 Off-Site Facilities

A large number of facilities, including storage facilities and loading and unloading facilities, should be located in an off-site area.

3.14.1 Tank Farm

The tank farm area should be adjacent to the process and utility area.

Product storage tanks should be located on the lee side of, and preferably downslope from, the remainder of the plant.

Horizontal product storage tanks should be located so that their longitudinal axes are not in line with buildings and plant equipment. Exposure of storage tanks to potential sources of fire should be minimized.

The minimum distance between the above-ground storage vessels/tanks containing highly dangerous, flammable, and combustible liquids to the boundaries of other facilities should be in accordance with the requirements of the available standard.

Dikes are not usually required for LPG storage. The storage areas should be graded to drain to a safe area.

Sufficient areas around all diked areas should be provided for firefighting, maintenance, and required pipeways. The number of tanks within a dike, spacing within the dike, and volume of dikes should be in accordance with OIA (Tables 3.4 through 3.6).

Transfer pumps should be grouped in one or more locations, located outside a tank dike, and provided with a minimum of a roof shelter. The number of pump locations should be at a minimum. In general, piping in the tank farm should be run on sleepers located between roadways and tank dikes.

Figure 3.10 shows a typical layout of crude oil storage tanks.

3.14.2 Loading and Unloading Facilities

The main loading and unloading racks for tank trucks and rail tank cars should be consolidated at one location as near to the plant site as practical, and close to an access gate, so that traffic through the plant is minimized and high-risk areas are avoided.

Loading facilities must be provided with adequate space and roadways for safe truck maneuvering and parking. In the case of rail loading and unloading areas, allowance must be made for parking and shunting of tank cars.

FIGURE 3.10
Layout of crude oil storage tanks.

Allow for dispersal of vapors and liquid spills so as to minimize the damage to other equipment in case of fire. Truck and rail loading racks for flammable and combustible liquids should be located at a distance of at least 30 m from a process unit and other facilities to avoid truck traffic near process areas.

An LP gas truck and rail loading rack should be placed at least 75 m from a process unit, 30 m from other types of truck loading racks, and 60 m from atmospheric or pressurized storage tanks.

Wharves handling flammable liquids should be spaced a minimum of 60 m from a process unit and should be spaced at least 75 m from fired heaters or other continuously exposed sources of ignition.

An adequate parking area for trucks waiting to load must be provided out of the path of moving trucks. If weigh scales are required, for truck loading they should be located near the entrance for ease in weighing trucks as they enter and just prior to their leaving.

3.14.3 Flare

For flaring of process units that generate combustible by products (gas or vapor), when required, sufficient space must be left between the flare and the nearest adjacent equipment to keep the radiant heat flux below allowable limits.

The flare stack should be located remotely from off-site and process facilities and preferably downwind from any areas where personnel are required for continuous operation.

The flare stack should be located at least 90 m from other facilities.

There should be a totally clear area surrounding the flare stack. The size of this area is dependent on heat intensity requirements, which depend on the height of the flare and relieving quantity.

The flare knockout drum, pump-out pump, and flare ignition system should be located at the periphery of the clear area.

Spacing of flares from process equipment depends on the flare stack height, flare load in pounds per hour, and the allowable heat intensity at the equipment location. Flare locations should be at lower elevations than process areas, should be curbed to contain hydrocarbon carry-over, and should be at least 60 m from equipment containing hydrocarbons. Areas where personnel may be present and where the public has free access must also be considered.

3.15 Waste Treatment Facilities

The preferred location of the waste treatment area should be at a refinery/plant low point to insure gravity flow from all areas.

The waste treatment area should be remote from the process and utility area and arranged to permit future expansion of the system.

The layout of the area must include vehicle accessibility for maintenance purposes.

3.15.1 Wastewater Separators

Wastewater separators handling hydrocarbons should be spaced at least 30 m from process unit equipment handling flammable liquids and 60 m from heaters or other continuous sources of ignition. Preferably, wastewater separators should be located downgrade of process equipment and tankage.

3.16 Emergency Shutdown System

a. Gas and Product Line Control Valves

High-pressure gas lines should not pass through a process area or run within 30 m of important structures or equipment without shutdown valving to insure that portions of piping within the process area can be isolated from the main gas line and depressurized in the event of an emergency. However, extensive use of shutdown valves may not be needed, since the increased complexity of the system will require a greater degree of preventive maintenance if unwarranted shutdowns are to be avoided. Shutoff valves, sometimes known as station isolation valves, should be provided on all gas and product pipelines into and out of the plant. A bypass line with a normally shut valve may be required between the plant inlet and discharge lines.

All station isolation valves and bypass valves, if any, should be located at a minimum distance of 75 m but not more than 150 m from any part of the plant operations. Care should be taken in locating these valves so that they will not be exposed to damage by plant equipment or vehicular traffic.

b. Emergency Shutdown Stations

At least two remote emergency shutdown stations, located at a minimum distance of 75 m apart, should be provided. Locate actuating points at least 30 m from compressor buildings and high-pressure gas lines. More than two shutdown stations may be required depending on the size and complexity of a given plant. One of the actuating stations should be located in the control room. It should be distinctively marked and equipped with signs stating the proper method of actuation in the event of an emergency.

3.17 Blowdown Drums

Blowdown drums are used for liquid relief in emergencies and are not usually installed when a suitable pressure-relieving system and flare are provided. When used, blowdown drums should be 30 m minimum from the process unit battery limits and 60 m from storage tanks and other refinery facilities.

3.18 Fire Training Areas

Fire training areas are ignition sources when in use. Because of the smoke produced, they can also create a nuisance for the refinery and neighboring facilities. Fire training areas should be 60 m from process unit battery limits, main control rooms, fired steam generators, fire pumps, cooling towers, and all types of storage tanks. They should also be 75 m from property boundaries, administration, shops, and similar buildings and from the main substation.

3.19 Tetraethyl Lead Blending Plants

Tetraethyl lead (TEL) blending plants should be spaced 30 m from process equipment handling flammable liquids and 45 m from fired heaters or other continuously exposed sources of ignition. Arrange to reduce any possibility of flammable liquids draining near the TEL plant.

References

API (American Petroleum Institute) RP-500, *A Recommended Practice for Classification of Location for Electrical Installation in Petroleum Refineries*, Fourth Edition, January 1982. API Std. 620, On Large, Welded, Low Pressure Storage Tanks; API Std. 650: On Welded Steel Tanks for Oil Storage.

ASCE (American Society of Civil Engineers) Minimum Design Loads for Structures.

ASME (American Society of Mechanical Engineers) Boilers and Pressure Vessel Codes: Section I, Power Boilers; Section VIII, Pressure Vessels.

Di Padova, A., Tugnoli, A., Cozzani, V. and Barbaresit, T.F. 2011. Identification of fireproofing zones in oil and gas facilities by a risk-based procedure, *Journal of Hazardous Materials* 191, 83–93.

NFPA 30. 2000. *Flammable and Combustible Liquids Code*, National Fire Protection Association, Quincy, MA.

Tugnoli, A., Cozzani, V., Di Padova, A., Barbaresi, T. and Tallone, F. 2012. Mitigation of fire damage and escalation by fireproofing: A risk-based strategy, *Reliability Engineering and System Safety* 105, 25–35.

Further Readings

American Insurance Association. 1968. Hazard Survey of the Chemical and Allied Industries, Technical Survey No. 3, 1968, New York.

API RP 521. 1982. *Guide for Pressure-Relieving and Depressurizing Systems*, American Petroleum Institute, Washington, DC.

API RP 752. 1995. Management of Hazards Associated with Location of Process Plant Buildings, American Petroleum Institute, Washington, DC.

Chiu, C.-H. 2007. Design considerations for offshore liquefaction process facility, 2007 AIChE Spring National Meeting. Houston, TX, April 22–27, 2007.

Dreux, M.S. 2004. Defending OSHA facility siting citations, Center for Chemical Process Safety, 19th Annual International Conference—Emergency Planning Preparedness, Prevention and Response, pp. 385–388.

Elgamel, M.A. and Bayoumi, M.A. 2003. An efficient minimum area spacing algorithm for noise reduction, Proceedings of the IEEE International Conference on Electronics, Circuits, and Systems 2, No. 301923, pp. 862–865.

House, F.F. 1969. *An Engineer's Guide To Process-Plant Layout, Chemical Engineering*, McGraw Hill, New York.

Jalnapurkar, K.M. and Amale, P.D. 2001. Plant layout doing it economically, *Chemical Engineering World* 36 (9), 39–41.

NFPA 58. 2001. *Liquefied Natural Gas*, National Fire Protection Association, Quincy, MA.

NFPA 496. 1998. *Purged and Pressurized Enclosures for Electrical Equipment in Hazardous (Classified) Locations*, National Fire Protection Association, Quincy, MA.

Onodera, H., Sakamoto, M., Kurihara, T. and Tamaru, K. 1989. Step by step placement strategies for building block layout, Proceedings—IEEE International Symposium on Circuits and Systems 2, pp. 921–926.

Simon, A., Pret, J.-C. and Johnson, A.P. 1997. A fast algorithm for bottom-up document layout analysis, *IEEE Transactions on Pattern Analysis and Machine Intelligence* 19 (3), 273–277.

Taylor, D.W. 2007. The role of consequence modeling in LNG facility siting, *Journal of Hazardous Materials* 142 (3), 776–785.

Yamakoshi, K., Watanabe, T., Takei, Y. and Sawada, H. 1997. Layout technique, *NTT R and D* 46 (10), 1071–1077.

4

Fire Alarm Systems and Automatic Detectors

4.1 Introduction

A key aspect of fire protection is to identify a developing fire emergency in a timely manner and to alert the building's occupants and fire emergency organizations. This is the role of fire detection and alarm systems.

Depending on the anticipated fire scenario, building and use type, number and type of occupants, and criticality of contents and mission, these systems can provide several main functions.

First, they provide a means to identify a developing fire through either manual or automatic methods, and second, they alert building occupants to a fire condition and the need to evacuate.

Another common function is the transmission of an alarm notification signal to the fire department or other emergency response organization. They may also shut down electrical and air handling equipment or special process operations, and they may be used to initiate automatic suppression systems. This section will describe the basic aspects of fire detection and alarm systems.

The early detection of a developing fire and an early warning to operational and firefighting personnel form an important aspect in the basic concept of fire protection. Automatic detection is of the utmost importance when immediate action is required (i.e., where products are handled that may ignite spontaneously on leaking into the atmosphere).

Typical examples are within the enclosed compartments of gas turbines, at the seal areas of floating roof tanks, in computer and control rooms, and material stores. This chapter provides general guidelines for the design and engineering of fire and gas detection on oil, gas, and chemical industry installations.

4.2 Basic Principles

The fire and gas detection systems (FGDSs) on an installation are provided to enable the detection of escaped hydrocarbon gas as well as the development of any fire at the earliest possible stage, so that protective measures can be taken before the situation gets out of control.

The system consists of strategically positioned sensors connected to a central panel. The centralized alarm and control systems should be installed in the designated control center for the specific area of product movement, jetty control room, building entrances, and so forth. Slave displays should be installed for other locations where the occurrence of a fire needs to be known, such as the general control center, the fire station, or the gatehouse entrance to the plant or refinery.

Slave displays should be installed only in centers that are manned 24 hours a day.

The display system should include all alarm detection and fire protection systems for flammable gas, fire, and smoke.

Alarm siren(s) should be installed that will sound automatically upon activation by push buttons and fire detection systems. Sirens should have a range of at least 1.5 km in still air.

At minimum, all gas detectors and those components of the detection system that are located outdoors and in an enclosed hazardous area should satisfy the requirements for division II area as specified in Chapter 2.

In gas plants and off-shore installations where the fire and gas detection systems are part of the emergency support system, its components should satisfy the requirements for division II area as specified in Chapter 2.

The detection system should also be fed from the emergency power system to enable operation for a period of 24 hours after loss of main power.

The fire alarm and control system for accommodation and service spaces should be located in the fire station if this is permanently manned; where this is not the case, it should be located in the main control center.

4.3 Alarms and Status Indication for Plant Units

Although the FGDSs with associated safety measures are designed to be fully automatic, the presentation of system status to the personnel in the central control room (CCR) is very important. It is vital that operators are immediately made aware of where a detection has been made. If a fault has occurred in the system it should be detected where in the system it has occurred. Status indication should be provided by means of a "mimic," a simplified layout of the installation indicating, for example, detection (fire or gas) and fault for the various areas.

Presentation in the CCR should also enable the operator to supervise that the correct chain of events takes place in case of detection. This can be provided with a functional matrix where, for each area, the different detections and manual calls and the operation of the different safety measures are indicated.

There should be a possibility for manual activation of these safety measures from the CCR.

The fire alarm system should be initiated by

a. Manual switches (call points) located at strategic points in the plant area, at roadsides, jetties, loading stations, tank farms, etc.
b. Automatic switches, such as
    1. On sprinkler systems
    2. Smoke and fire detectors in buildings

4.4 Buildings, Warehouses' Fire Detection, and Control Panel

All buildings, including warehouses, accommodation spaces, and service spaces should be provided with fire detection and alarm systems. The alarm panel should be located in fire station or in a permanently manned location if there is no fire station existing in the plant. The system should consist of suitably positioned manual call points and fire and smoke detectors in conjunction with the multizone alarm panel.

4.4.1 Manual Fire Detection

Manual fire detection is the oldest method of detection. In the simplest form, a person yelling can provide fire warning. In buildings, however, a person's voice may not always transmit throughout the structure. For this reason, manual alarm stations are installed. The general design philosophy is to place stations within reach along paths of escape. It is for this reason that they can usually be found near exit doors in corridors and large rooms.

The advantage of manual alarm stations is that, upon discovering the fire, they provide occupants with a readily identifiable means to activate the building fire alarm system. The alarm system can then serve in lieu of the shouting person's voice. They are simple devices and can be highly reliable when the building is occupied. The key disadvantage of manual stations is that they will not work when the building is unoccupied. They may also be used for malicious alarm activations. Nonetheless, they are an important component in any fire alarm system.

The manual call points should be located on exit routes, floor landings, exits to open air, and possibly other areas depending on the layout of the building. The distance that a person must travel to raise an alarm should not exceed 30 m. Obviously the layout of the building could make this considerably less. The standard height for a manual call point should be 1.4 m above the finished ground. The points should, in general, be surface-mounted for ease of viewing. If they are semiflush then they must have a clearly visible side profile.

4.4.2 Heat Detectors

Heat or thermal detectors are the oldest type of automatic detection device, originating in the mid 1800s, with several styles still in production today. The most common units are fixed temperature devices that operate when the room reaches a predetermined temperature (usually in the 57°C–74°C range). The second most common type of thermal sensor is the rate-of-rise detector, which identifies an abnormally fast temperature climb over a short time period. Both of these units are spot type detectors, which means that they are periodically spaced along a ceiling or high on a wall. The third detector type is the fixed temperature line type detector, which consists of two cables and an insulated sheathing that is designed to break down when exposed to heat. The advantage of line type over spot detection is that thermal sensing density can be increased at lower cost.

Thermal detectors are highly reliable and have good resistance to operation from nonhostile sources. They are also very easy and inexpensive to maintain. On the down side, they do not function until room temperatures have reached a substantial temperature, at which point the fire is well underway and damage is growing exponentially. Subsequently, thermal detectors are usually not permitted in life safety applications. They are also not recommended in locations where there is a desire to identify a fire before substantial flames occur, such as spaces where high-value, thermal-sensitive contents are housed.

The maximum ceiling height for heat detector applications should be considered as 7.5 m. The maximum area coverage by a single detector should not be more than 50 m². Requirements of BS-5839 Part 1 (1988) should be considered in the design of heat detector layouts.

4.4.3 Smoke Detectors

Smoke detectors are a much newer technology, having gained wide usage during the 1970s and 1980s in residential and life safety applications. As the name implies, these devices are designed to identify a fire while in its smoldering or early flame stages, replicating the human sense of smell. The most common smoke detectors are spot type units that are placed along ceilings or high on walls in a manner similar to spot thermal units.

They operate on either an ionization or photoelectric principle, with each type having advantages in different applications. For large open spaces such as galleries and atria, a frequently used smoke detector is a projected beam unit.

This detector consists of two components, a light transmitter and a receiver, that are mounted at some distance (up to 100 m) apart. As smoke migrates between the two components, the transmitted light beam becomes obstructed and the receiver is no longer able to see the full beam intensity. This is interpreted as a smoke condition, and the alarm activation signal is transmitted to the fire alarm panel.

A third type of smoke detector, which has become widely used in extremely sensitive applications, is the air aspirating system. This device consists of two main components: a control unit that houses the detection chamber, an aspiration fan, and operation circuitry, and a network of sampling tubes or pipes. Along the pipes are a series of ports that are designed to permit air to enter the tubes and be transported to the detector.

Under normal conditions, the detector constantly draws an air samples into the detection chamber via the pipe network. The sample is analyzed for the existence of smoke, and then returned to the atmosphere. If smoke becomes present in the sample, it is detected and an alarm signal is transmitted to the main fire alarm control panel.

Air aspirating detectors are extremely sensitive and are typically the fastest responding automatic detection method. Many high-technology organizations, such as telephone companies, have standardized on aspiration systems. In cultural properties they are used for areas such as collections storage vaults and highly valuable rooms. These are also frequently used in aesthetically sensitive applications since components are often easier to conceal when compared with other detection methods.

The key advantage of smoke detectors is their ability to identify a fire while it is still in its incipient stages. As such, they provide added opportunity for emergency personnel to respond and control the developing fire before severe damage occurs.

They are usually the preferred detection method in life safety and high-content-value applications. The disadvantage of smoke detectors is that they are usually more expensive to install when compared to thermal sensors and are more resistant to inadvertent alarms. However, when properly selected and designed, they can be highly reliable with a very low probability of false alarm.

The maximum ceiling height for smoke detectors should be considered as 10.5 m. The maximum area coverage by a single smoke detector should not be considered more than 100 m^2 approximately. Requirements of BS-5839 (1988) should be considered in preparation of the layout of detectors.

Video smoke detection is also similar to flame detection. In this case, the similarity is in the idea of viewing an open area for evidence of a fire. In the case of video smoke detection, the system's ability to detect a fire is based on computer analysis of the visual data. The processor looks for specific motion patterns of smoke and fire (while ignoring other onscreen movement patterns). Applications include large open areas where traditional smoke detection may be impractical or inefficient, buildings with high ceilings, areas with high air movement, and so forth. Figure 4.1 shows a typical video smoke detection system.

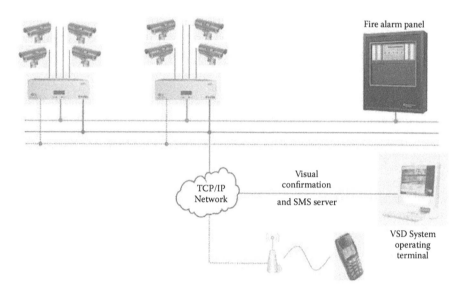

FIGURE 4.1
Video smoke detection. (Reproduced from Henan Inte Electrical Equipment Co. With permission.)

4.4.4 Audible and Visual Alarms

When considering the siting and selection of audible and visual alarms, the following points should be considered:

1. In accommodation places where people are asleep, the sound level at behead should be minimally 75 dBA with doors closed. The maximum sound level should not be in excess of 100 dBA.
2. There should be at least two sounders on a system. At least one audible sounder should be considered for each zone in a manner to provide a sound level at least 5 dB above the surrounding noise.
3. It is normal that a master sounder should be sited in the immediacy of the control/indicating equipment. The number, location, and type of sounder should be easily distinguished from background and other noise levels, and also should be clearly audible throughout the premises. In the designation of the areas for multizone panels, the following points should be considered:
   a. The number of floors in the building.
   b. The compartmentation of the floors.
   c. Rooms containing high-value equipment such as computers. These

areas could also have additional protection such as fixed extinguishing systems and in some cases can have their own separate detection systems with a simple link to the central fire alarm panel.

d. Stairways should be considered as a separate zone on multistory buildings.

e. A single zone on the alarm panel should not cover an area exceeding 2000 m².

In plant units, where noise level does not permit application of sounders, consideration should be given to a visual beacon alarm. A visual beacon alarm should be considered for analyzer houses to raise the alarm when a purge system has failed. This alarm should also be activated by fire alarm panel to indicate a fire hazard condition.

Different type of hazards, such as toxic/flammable gas concentration or fire, should be indicated in suitable plant area locations by different colors of the beacons installed.

4.4.5 Ultraviolet/Infrared Flame Detectors

Flame detectors represent the third major type of automatic detection method; this method imitates the human sense of sight. They are line-of-sight devices that operate on an infrared (IR), ultraviolet (UV), or combination principle. As radiant energy in the approximate 4000- to 7700-angstrom range occurs, as indicative of a flaming condition, their sensing equipment recognizes the fire signature and sends a signal to the fire alarm panel.

The advantage of flame detection is that it is extremely reliable in a hostile environment. Flame detectors are usually used in high-value energy and transportation applications where other detectors would be subject to spurious activation. Common uses include locomotive and aircraft maintenance facilities, refineries and fuel loading platforms, and mines.

A disadvantage is that they can be very expensive and labor-intensive to maintain. Flame detectors must be looking directly at the fire source, unlike thermal and smoke detectors that can identify migrating fire signatures. Their use in cultural properties is extremely limited.

UV or combined UV/IR flame detectors should be used in a general area where flames are expected to be one of the prime indications of fire, such as open outdoor areas, hydrocarbon areas, and fuel areas.

IR flame detectors should be used in enclosed areas where the smoke and heat detector limitations do not permit their application. The number and location of flame detectors should be based on their coverage pattern in a manner that there is no blind corner left undetected.

Figure 4.2 represents the leading-edge technology UV optical flame detectors. This flame detector senses radiant energy in the UV spectrum. The radiant emissions of flaming fires will alert the detector to their presence.

FIGURE 4.2
UV optical flame detectors. (Reproduced from Haider. With permission.)

4.5 Fire Detection System

Fire detectors should, except for fusible plugs, be of a resettable type such that after activation they can be restored to normal surveillance without the renewal of any components.

Fire detectors connected in loops should further have a visible indication to show that they have been operated. The indication should continue until the loop has been manually reset. If the loop-connected detector that has been operated can be identified from the central control room, there is no need for identification on the detector itself. The electrical interconnections should be monitored for faults (i.e., an alarm should be given in case of short circuit, earth fault, and open circuit).

Several initiating devices can be wired to one input to give a group alarm in case any of these devices is actuated (e.g., push-button stations for a group of oil tanks).

Only the first incoming alarm in a group should give an audible alarm; subsequent alarms in the same group should be ignored until the alarms have been reset.

Alarms in other groups should be accepted as first-incoming alarms and give an audible alarm.

The alarm system should have an input memory so that momentary alarms are held until acknowledged manually.

The incoming alarms should automatically

- Operate indicating lamps on the alarm panel
- Operate a claxon in the fire station, control center, and any other location specified in the project
- Operate the siren
- Start the fire-fighting pumps (if necessary)

4.5.1 Selection of Detectors

The selection of detectors for the various areas on an installation should be based on evaluation of the prime fire indications and ambient conditions.

A fire development in a process or wellhead area is likely to be associated with flammable fluids and have a fast development. The prime indications of a fire in such areas will be flames and a high heat output.

A fire development in a switchgear or control room will typically be caused by overheating of insulation in electrical components and give rise to a very slow fire development initially characterized by invisible products of combustion. The choice of fire detectors for an area must reflect the anticipated prime indications of a fire under development.

The second aspect that should be considered is the ambient conditions in the various areas under which the detectors will have to operate. This will involve evaluation of natural environmental conditions such as wind, temperature, solar radiation, salinity, and humidity, as well as industrial conditions such as dust, oily atmosphere, and vibration.

These conditions will impose restrictions in the choice and effectiveness of the detectors and also on the operation and maintenance requirements.

Table 4.1 lists some commonly used detector types with their characteristics, application, and environmental resistance.

4.5.2 Detector Layout

When it is decided which types of detectors are to be employed in the various areas, the next task should be the location of individual detectors. Final positioning should be decided on-site after equipment, pipes, ventilation ducts, and so forth have been installed. The number of detectors and their layout should be decided during the design stage.

Performance of heat and smoke detectors should be in accordance with European Standards EN 54-5 and EN 54-7, respectively. Tables 4.2 through 4.4 should be used as guidance in the spacing layout of the detectors.

All detectors for process area applications should be considered to be explosion-proof for class I, division 1, and groups A, B, C, and D.

The siting should also take account of intensity and pattern of ventilation to ensure that the fire signature from a developing fire will reach the detector. The detectors should be located clear of beams and other features likely to shield the detectors. Smoke tests with equipment and ventilation running should be carried out at commissioning stage to verify adequate location.

Fire detectors should be suitably protected against physical damage caused by normal activity in an area. The accessibility of the detectors should be satisfactory with respect to maintenance.

Where detectors are located behind panels, in false ceilings/floor, voids, or in other invisible locations, a remote fire indication showing the operation of

TABLE 4.1

Fire Detector Parameters and Applications

Detector Type	Advantages	Disadvantages	Application	Environmental Resistance
UV flame detectors	• Fast response • Large coverage • Unaffected by wind • Infrared radiation not absorbed by smoke	• Need a straight line of sight • Radiation from sun and hot vibratory • Machinery may give rise to false alarms • Ultraviolet radiation absorbed by heavy smoke	• In general areas where flames are expected to be one of the prime indications of fire • Hydrocarbon areas and fuel areas • Open outdoor areas	• Very good • Unaffected by rain, wind, etc. • Solar blind if sensitivity below 2800°C
Heat detectors	• Reliable	• Relatively slow response	• General areas where ambient conditions are too rough for smoke detectors • As backup for flame detectors in high-hazard areas	• Good, although response affected by wind makes them less suitable for open outdoor areas
Smoke detectors	• Very sensitive • Detect smoldering fires at early stage	• Require a relatively clean atmosphere	• General clean areas not associated with flammables control room • Switchgear room • Void spaces behind false floors and ceilings • Accommodations	• Not suitable for open outdoor areas or naturally ventilated areas

TABLE 4.2

Limits for Siting Point-Type Heat Detectors in Naturally Ventilated Open Areas

Maximum Floor Area to Be Covered by One Detector (m²)	Maximum Distance between Detector Centers (m)	Maximum Distance Away from Any Bulkhead (m)[a]	Ceiling Heights, Higher Limit (m)[b]
25	7	3.5	4–7

[a] Detectors should not be mounted less than 0.5 m away from any outside wall or dividing partition. This should be applied for detectors mounted adjacent to bulkheads.
[b] For fast response detectors, maximum ceiling height limit is 7 m. For slow response detectors, maximum ceiling height limit is 4 m.

TABLE 4.3

Limits for Siting Point-Type Heat Detectors in Mechanically Ventilated Enclosed Areas

Maximum Floor Area to Be Covered by One Detector (m²)	Maximum Distance between Detector Centers (m)	Maximum Distance Away from Any Bulkhead (m)[a]	Ceiling Heights, Higher Limit (m)[b]
37	9	4.5	5.5–8.5

[a] Detectors should not be mounted less than 0.5 m away from any outside wall or dividing partition. This should be applied for detectors mounted adjacent to bulkheads.
[b] For fast response detector, ceiling height limit is 8.5 m. For slow response detector, ceiling height limit is 5.5 m.

TABLE 4.4

Limits for Siting Point-Type Smoke Detectors in Enclosed Areas

Maximum Floor Area to Be Covered by One Detector (m²)	Maximum Distance between Detector Centers (m)	Maximum Distance Away from Any Bulkhead (m)[a]	Ceiling Heights, Higher Limit (m)[b]
50	10	5	7.5

[a] Detectors should not be mounted less than 0.5 m away from any outside wall or dividing partition. This should be applied for detectors mounted adjacent to bulkheads.
[b] For fast response detector, ceiling height limit is 8.5 m. For slow response detector, ceiling height limit is 5.5 m.

the detectors should be arranged in an adjacent area normally occupied (i.e., corridor, hall, general area).

4.5.3 System Configuration

Certain safety measures are often initiated based on fire detection. Keeping in mind that fire detectors do not respond to fire as such, but rather to certain characteristics commonly associated with fire (i.e., smoke, heat, radiation), it is unavoidable that some nuisance activation of sensors will take place. This because the above-mentioned characteristics are also present during normal operating conditions with varying frequency and intensity due to events other than fire.

To safeguard against initiation of safety measures on a false basis, it is common to require two fire detectors to operate before this is initiated (e.g., 2-out-of-3, which implies that with a group of three detectors a signal from any two will initiate a safety function. A signal from any one detector should always give alarm. It is important with such systems that the detectors involved are located such that each will sense the fire development sufficiently fast to enable the safety measures to be initiated before the fire has developed to a critical level. A 2-out-of-2 voting system should not be considered as this implies reduced availability.

The safety actions to be initiated depend on the type of area where the detection has been made. The following are examples of typical actions that should be initiated as applicable to the particular area:

- Fire alarm to be activated in the central control room and areas influenced by the fire
- The flow of hydrocarbon to and from the area where the detection has been made to be shut down
- Mechanical ventilation to the area where the fire has been detected to be shut down and fire dampers to be closed
- Fixed fire extinguishing system to be activated
- Fuel supply to fired units to be shut down (except prime movers for emergency equipment)

4.5.4 Alarm Panel

The control panel is the "brain" of the fire detection and alarm system. It is responsible for monitoring the various alarm input devices, such as manual and automatic detection components, and then activating alarm output devices, such as horns, bells, warning lights, emergency telephone dialers, and building controls.

Control panels may range from simple units with a single input and output zone to complex computer-driven systems that monitor several buildings over an entire campus. There are two main control panel arrangements, conventional and addressable, which will be discussed below.

Conventional or point-wired fire detection and alarm systems were for many years the standard method for providing emergency signaling. In a conventional system one or more circuits are routed through the protected space or building. One or more detection devices are placed along each circuit. Selection and placement of these detectors is dependent on a variety of factors including the need for automatic or manual initiation, ambient temperature and environmental conditions, the anticipated type of fire, and the desired speed of response. One or more device types are commonly located along a circuit to address a variety of needs and concerns.

Upon fire occurrence, one or more detectors will operate. This action closes the circuit, which the fire control panel recognizes as an emergency condition. The panel will then activate one or more signaling circuits to sound building alarms and summon emergency help. The panel may also send the signal to another alarm panel so that it can be monitored from a remote point.

In order to help ensure that the system is functioning properly, these systems monitor the condition of each circuit by sending a small current through the wires. Should a fault occur, such as due to a wiring break, this current cannot proceed and is registered as a "trouble" condition. The indication is a need for service somewhere along the respective circuit.

In a conventional alarm system, all alarm initiating and signaling is accomplished by the system's hardware, which includes multiple sets of wire, various closing and opening relays, and assorted diodes. Because of this arrangement, these systems are actually monitoring and controlling circuits and not individual devices.

The alarm panel should have indicating lamps for

- Individual alarms
- Group alarms
- Power "on"
- System faults such as electricity supply failure, overcurrent, low battery voltage, and system failure

The panel should have the following operating controls:

- On/off (per group)
- Test (simulating alarm condition)
- Cancel the audible alarm
- Reset the system (clear input memory)

When the alarm and control panels are located in the fire station, consideration should be given to the use of the typical fire alarm systems available commercially. The control panel should match the size and appearance of the alarm panel. The complete alarm and pertinent control system, together with logic circuitry and electricity supply, should preferably be arranged in one cubicle or cabinet.

When the alarm and control panels are located in the control room, preference should be given to a design matching other alarm/control systems in the control room. The logic circuitry and electricity supply should then be accommodated in the auxiliary room.

4.5.5 Electricity Supply

For the control system and for the alarm system when mounted in the control room, an electricity supply should be provided with rectifier and batteries with a minimum voltage of 24 V DC for 24 hours minimally to be suitable for voltage variations of ±10%. This electricity supply should also operate lamps, claxons, and so forth, and should be completely stand-alone from other systems (including those for process safeguarding, telecommunication, etc.)

The siren should be connected to the AC main power supply.

4.6 Gas Detection System

The principal use for gas detection on fixed off-shore installations and gas processing plants is normally that of catalytic combustion. This principle has particular advantages and disadvantages that should be considered in the design and installation of the system.

An advantage is that this type of measurement is very direct; it measures flammability directly by an exotherm oxidation on the sensor element, the heat of oxidation being directly proportional to the percentage of the LEL existing at the sensor head. This measuring principle does not easily give rise to nuisance alarms as only the presence of flammable gas on the sensor element normally causes the detector to be activated. A disadvantage with this measuring principle is the inherent non-fail-safe mode (i.e., that on loss of sensitivity, which is the most experienced and probable failure for these detectors, no alarm is given). The only way to demonstrate satisfactory performance of such sensors is to expose them to a concentration of gas (e.g., 50% LEL) and read off the response.

The electrical connections should be monitored as for fire detection systems (i.e., a fault alarm should be given in case of open circuit, short circuit and earth fault).

At least two adjustable alarm levels (e.g., set to 20% and 60% LEL) should be provided with independent, voltage-free contact outputs.

It should be possible to test the alarm levels from the control unit by simple means, such as a local miniature switch.

Figure 4.3 shows a typical gas detection and control system.

4.6.1 Detector Layout

The location of detectors should reflect a combination of two philosophies. First, the gas should be detected near the probable sources of leakage. This implies that gas detectors should be installed in hazardous areas and ventilation outlets from mechanically hazardous areas.

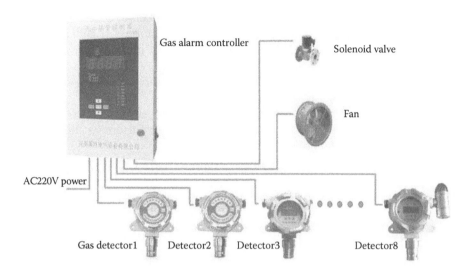

FIGURE 4.3
Typical gas detection and control system. (Reproduced from Henan Inte Electrical Equipment Co. With permission.)

Second, any gas approaching areas where possible ignition sources are located should be detected. Such areas are utility areas where, for example, nonexplosion-protected electrical equipment or combustion engines are located. This implies that gas detectors should be located in ventilation inlets to nonhazardous areas and in combustion air inlets.

The layout of gas detectors should not be precisely determined before major equipment, pipes, and ducts are installed. The full effect of the ventilation pattern and the normal ambient conditions at the various locations should be known beforehand.

Consideration should be given to the molecular weight of the gas mixture in the various parts of the process, and whether a lighter-than-air or heavier-than-air gas leak would result in case of a leakage.

Weather protection covers should be employed in areas as recommended by the manufacturer. An evaluation of the influence on the response time caused by such accessories should be considered.

The access for maintenance is particularly important for gas detectors as they are instrument systems heavily dependent on proper testing and maintenance.

4.6.2 Poisoning of Catalytic Gas Detectors

A problem to be aware of with catalytic gas detection system is the phenomenon of catalyst poisoning. This implies that the sensing element becomes desensitized in the presence of small quantities of certain chemical substances in the atmosphere. Permanent loss of sensitivity will generally be caused by compounds containing silicones or heavy metals such as lead, copper, and zinc.

The following considerations should be related to the poisoning problem:

- Be aware of the compounds that are liable catalyst poisons (consult the manufacturer) and try to limit their use. Nonpoisonous alternatives are often available
- Make use of protective filters (e.g., carbon or ceramic filters) in problem areas (i.e., areas where a poisonous atmosphere is unavoidable)

4.6.3 System Configuration

The use of a voting principle for gas detectors should be carefully evaluated due to the non-fail-safe mode of gas detectors. If a 2-out-of-3 voting principle is used, it should be ensured that all the detectors within a group can sense the same gas leakage. Normally, the distance between any two detectors working concurrently should not be more than 3 m. A 2-out-of-2 voting system should be avoided. The safety actions to be initiated upon gas detection in an area should be as follows:

- All flow of hydrocarbons to and from the actual area should be shut down
- All potential ignition sources in the area should be eliminated

In case of maximum 50% LEL gas detection in a ventilation intake, the following actions should take place:

- The ventilation fan should be stopped
- The fire damper should be closed
- The heating element should be shut off
- All ignition sources within the space being ventilated should be eliminated

In case of maximum 50% LEL gas detection in combustion air intakes, the machinery should be stopped. Prime movers for fire pumps are exempted from this.

For turbines, shutdown should be effected at considerably lower gas concentration at the combustion air intake. Shutdown at 15%–25% LEL is often recommended. Turbine manufacturers should be consulted. Table 4.5 lists some typical applications for fire, smoke, and flammable gas detections.

TABLE 4.5

Typical Applications for Fire, Smoke, and Flammable Gas Detection

Areas of Application	Detection Type						Remarks
	Flame	Linear Heat	Spot Heat	Spot Smoke	Area Smoke	Gas	
Floating roof tank rim area		✓	✓				Heat-sensitive tubing/quart-zoid bulbs
Selected hydrocarbon pumps		✓					Heat-sensitive tubing
Selected areas or equipment holding hydrocarbons		✓					Heat-sensitive tubing
Analyzer house	✓					✓	—
Gas turbine/gas compressor in enclosures			✓			✓	Rate of rise heat detection
Plant laboratory				✓ Spot smoke detectors should be of the integral heat detection type			—
Main laboratory				✓ Spot smoke detectors should be of the integral heat detection type			—
Instrument auxiliary room, cabinets, floor cavity, cable routes					✓		Combined HCL smoke detection
Computer auxiliary room					✓		Combined HCL smoke detection

(continued)

TABLE 4.5 (Continued)

Typical Applications for Fire, Smoke, and Flammable Gas Detection

Areas of Application	Detection Type						Remarks
	Flame	Linear Heat	Spot Heat	Spot Smoke	Area Smoke	Gas	
Computer operation room			✓Spot smoke detectors should be of the integral heat detection type	✓			
Battery room							
LPG bottle-filling plant			✓			✓	Plant/area
Workshop, general			✓Spot smoke detectors should be of the integral heat detection type	✓			
Workhouse process analyzers				✓		✓	
Warehouse, general			✓Spot smoke detectors should be of the integral heat detection type	✓			
Warehouse, LPG storage		✓				✓	
Warehouse, hydrocarbons			✓			✓	

Location		Notes
Administration buildings	✓	Spot smoke detectors should be of the integral heat detection type
Telephone exchange/radio room	✓	Spot smoke detectors should be of the integral heat detection type
Canteen	✓	✓
Kitchen area	✓	
Training center	✓	
Fire station	✓	Spot smoke detectors should be of the integral heat detection type
Garage	✓	Spot smoke detectors should be of the integral heat detection type
LNG containment area	✓	Also with low-temperature detection

4.7 Gas Detectors: Materials

This section gives details of construction, safety performance, and testing requirements of portable and transportable gas detecting instruments. These instruments can be used for detection of

- Combustible gases
- Toxic gases
- Oxygen deficiency and excess

These instruments are constructed to sense the presence of combustible, toxic, and concentrations of gas or vapor reading in parts per million (ppm) (for toxic gases) and LEL (for explosive gases and vapors) and are in the following groups:

1. *Monitoring system:* Provides early warning of the incipient accumulation of combustible and/or toxic gases. Gas monitoring systems are multichannel gas detecting with remotely installed sensors that provide automatic alarm and control systems. Monitoring systems can also provide oxygen deficiency or excess detections.
2. *Transportable and/or portable:* Monitoring of combustible or hazardous gases that can be used for leak seeking or continuous monitoring of areas in which maintenance work such as hot works are in progress in gas-free areas.
3. *Chemical-sensing detector tubes:* For measurement of atmospheric contaminants. Portable gas-detecting instruments are also used to protect employees who are working in dangerous locations such as process units, storage vessels, sewer systems, and any plant handling oil, gas or chemicals.

(**NOTE:** Contaminants at concentration of occupational exposure limit [OEL]).

This section covers the minimum requirements for material and equipment of three types of portable or transportable gas-detecting instruments for sensing the presence of combustible and common toxic gas and vapor concentration, and does not cover gas detection instruments of the laboratory used for analysis of measurement or continuous gas-monitoring fixed systems.

The section gives also guidance on the use, care, and maintenance of

1. Battery-operated transportable or portable apparatus that indicate the presence of a combustible or potentially explosive mixture of gas

or vapor with air by using an electrical signal from a gas sensor to produce a meter reading to activate a visual or audible preset alarms.

2. Portable combination gas (hydrogen sulfide or hydrocarbon oxygen deficiency or excess).

3. Toxic gas testing detector by chemical sensing tubes.

Gas detection instruments that are intended to detect gas or vapor present in the area should be suitable for use in zone 1 hazardous locations. Gas detection instruments specifically intended for use in the presence of corrosive vapors or gases that may produce corrosive by-products as a result of a catalytic oxidation or other chemical process should be constructed of materials known to be resistant to corrosion by such substances. All instruments should be provided with the means for facilitating regular accuracy checks.

Electrical assemblies and components should comply with the construction and test requirements; in addition, all parts of combustible gas detection instrument should employ material and construction suitable for continuous operation within an ambient temperature range or as specified in the following clauses:

a. Flameproof enclosures shall comply with the requirements of BS 5501: Part 5 or BS 4683: Parts 1 and 2

b. Intrinsically safe and associated apparatus shall comply with the requirements of BS 5501: Part 7

c. Increased safety electrical apparatus shall comply with the requirements of BS 5501 Part 6 or BS 4683: Parts 1 and 4

The design of a combustible gas detection instrument should be such that all material used in the construction and components including electrical and electronic parts should be used within the manufacturer's rating or the limits specified by them.

The instrument should show, in a location on or adjacent to the meter or other indicator, the gas with which it has been calibrated.

- Handheld spot reading (portable) instruments should not exceed 1 kg in mass
- An indication should be provided to show that the instrument is switched on
- If individual indicating lights are fitted they should be colored as follows:

  The alarm indicating the presence of gas in potentially dangerous concentrations should be colored red

  The equipment fault indicator should be colored yellow

  The power supply indicator light should be colored green

In addition to the color requirements, the indicator lights should be adequately labeled to show their function.

Alarm devices provided as part of portable gas detection instrument reading up to 100% LEL should only be set to operate at a gas concentration not higher than 60% of the LEL.

Alarm levels for a multigas detector should be set to 20% LEL, 19% and 24% oxygen, 10 ppm hydrogen sulfide and 50 ppm carbon monoxide.

Fault signals in transportable gas-detection instruments should provide a fault signal in the event of failure of power to the instrument, short circuit in one or more of the wires, or loss of continuity of any gas-sensing system. This fault signal should be distinguishable from any other alarm. Instruments powered with integral batteries should be provided with an indication of low battery condition and the nature and purpose of this indication should be clearly explained in the instruction manual.

4.7.1 Batteries

- *Portable continuous duty instruments:* Instruments with fresh or fully charged batteries should be capable of continuous nonalarm operation for a period of at least 10 hours without replacement of recharged batteries.

- *Portable spot reading instruments:* Instruments with fresh or fully charged batteries should be capable of operation at a duty cycle of 10 minutes on, 10 minutes off for a period of 8 hours (total accumulative "on" time of 4 hours) without replacing or recharging the batteries.

4.7.2 Carrying Case

A portable instrument should be outfitted with a carrying case to protect it against severe shock and to store the detector probe when not in use.

4.7.3 Panel

The panel should have the following controls:

- Zero control
- Alarm set potentiometer, on-off battery check range switch

4.7.4 Selection

Transportable apparatus should normally be selected for such purposes as monitoring work areas (hot work) and areas where flammable liquid, vapor, or gas is present in process units.

Portable apparatus should normally be selected for such purposes as leak checking, verification of gas-free conditions, safety checks, and similar applications.

Factors that are important in selecting portable or transportable apparatus include its size, mass, robustness, power supply requirement, the type of indication required, and the visibility or audibility of the alarm.

Portable and transportable gas-detecting instruments are used for short periods of time in almost any environment, outdoors or indoors. They may be subject to contamination, dirt, wind, rain, dust, and handling.

Adequate robustness and weatherproofness should be considered in the selection of the apparatus. Weather conditions should be kept in mind if the apparatus is to be subject to rapid temperature changes and rapid air flow, particularly where accuracy is important. Gas-detecting instruments should be sealed to IP 65 to ensure that the electronic systems are safe from corrosion and dust particles and also protected from suffering accidental dropping onto hard surfaces.

4.7.5 Labeling and Instructions

4.7.5.1 Identification

Each instrument should carry a label stating the manufacturer's name and the instrument type identification and serial number together with the number of the appropriate part of the official accepted standard.

The word "caution" should be attached onto a label in capital letters at least 3 mm high and other wording should be in capital letters 2.5 mm high.

The label should be visible, legible, and permanently attached on each portable or transportable gas-detection instrument.

The labeling required should appear on a surface of the instrument and/or carrying case and be exposed when the instrument is in use.

4.7.5.2 Calibration Gas

A portable and transportable instrument should carry a label indicating the gas with which the instrument has been calibrated by the manufacturer. This label may be on the meter, if provided, or on the surface of the instrument (adjacent to the meter).

4.7.5.3 Instruction Manual

Each gas-detection instrument or group of instruments should be provided with a suitable instruction manual. The instruction manual should contain complete, clear, and accurate instructions for safe and proper operation and servicing of the instrument. It should include the following information:

a. Initial start-up operation.

b. Operating instructions and adjustments.

c. Instructions for checking and/or calibration on a routine bases.

d. Detail of operational limitations including the following items where applicable:

   1. Range of gases for which the instrument is suitable

   2. Ambient temperature limits

   3. Humidity range

   4. Battery life

   5. Maximum and minimum storage temperature limits

   6. Sample velocity limits

e. Applications, information on the adverse effects of contaminating gases or substances, and oxygen-enriched or -deficient atmospheres on the proper performance of the instrument. In the case of an oxygen-enriched atmosphere, the safety of electrical components of the instrument.

f. For instruments of the aspirated type, wording to indicate the minimum and/or maximum flow rate or range of flow rates, pressure, and tubing type and size of proper operation.

g. Clear statements of the nature and significance of all alarms and fault signals, the duration of such alarms and signals (if they are time-limited or self-restoring), and any provision that may be made for silencing or resetting of such alarms and signals as applicable.

h. Detail of any method for determination of the possible sources of malfunctioning and any corrective procedures.

i. A statement that alarm devices, outputs, or contacts are of the non-latching types, where applicable.

j. For portable continuous-duty instruments of the aspirated type that are provided with an integral flow indicating device, detected instructions regarding one or more suitable techniques that the user may employ to ensure that sample lines are intact and a proper flow is established.

k. For battery-operated instruments, installation and maintenance instructions for the batteries.

l. A recommended replacement part list.

m. The storage life and recommended storage conditions for critical replacement parts.

n. Where the special nature of the instrument requires additional instructions and/or special information that are alternative to, or in addition to the requirements, the instructions/information should be provided.

o. Detail of certification to the relevant accepted standard and any special condition of service.

4.7.6 Test Requirement

The manufacturer should supply test certification for each instrument dispatched. The instruments excluding all optional or accessory parts should be subjected to the entire test applicable to that type of instrument in accordance with BS EN 50059 or any other tests methods accepted by the company.

4.8 Gas Detector Tubes

This section specifies requirements for short-term detector tubes and associated aspirating pumps used at normal atmospheric pressure and temperature for evaluating atmospheric contaminants at concentration in the range of OEL values. It covers the length of stain tubes and color-match tubes, which are designed to give an indication of concentration over a short period of time.

4.8.1 Performance Requirements

The accuracy of an aspirating pump used with a given detector tube should be certified by the manufacturer of the tube.

As supplied, the pump should aspirate a volume of air within ±5% of the volume stated by the manufacturer. Means of indicating the completion of each aspiration should be provided. (Note that aspirating pumps should be checked for leakage in accordance with the manufacturer's instructions before each use and checked for accuracy at least every three months. Pumps should be maintained in accordance with the manufacturer's instructions and should not be used if their accuracy falls outside 10% of the nominal capacity.)

4.8.1.1 Length-of-Stain Short-Term Gas Detector Tubes

4.8.1.1.1 Graduation

Tubes should either be graduated in volume or mass per unit volume of air or should be accompanied by a calibration graph in the same units. The graduation of the tubes should permit the measurement of concentrations of half the OEL and twice the OEL.

4.8.1.1.2 *Length of End Point of Stain*

The length of a stain shall not be less than 15 mm after aspiration of the recommended volume of the gas mixture at the concentration of the OEL of the gas being detected.

The maximum variation of stain length around the circumference of the tube at the interface between the stained and unstained indicator layer should not exceed 20% of the stain length when measured at its points of maximum length at concentrations of half the OEL and above of the gas being detected.

4.8.1.1.3 *Boundary between Indicator Layer and Packing*

The boundary between the indicator layer and any inert packing or cleansing layers should be such that the difference between the longest and shortest length of packing or cleansing layers around the circumference of the tube does not exceed 1 mm.

4.8.1.1.4 *Direction of Gas Flow*

Tubes should be marked to indicate the direction of gas flow.

4.8.2 Color-Match Short-Term Gas Detector Tubes

4.8.2.1 Accuracy

Tubes should be calibrated against standard atmospheres in accordance with the manufacturer's instructions. There should be a distinct difference in color produced by concentrations 20% below and 20% above that of the OEL when operated according to the manufacturer's instructions.

The tubes should meet these requirements at all temperatures between 5°C and 35°C. (Note that color-match tubes are generally less accurate than length-of-stain tubes because of the subjective nature of color comparison.)

4.8.2.2 Assessment of Results

Tubes should be accompanied by a color chart, a table, or a mathematical formula for evaluating the concentration in units of volume or mass per unit volume of air. Tubes should permit the measurement of concentrations of half the OEL and twice the OEL.

4.8.3 Life of Tubes

Short-term detector tubes should be used for a period of not more than two years from the date of batch testing provided they are stored in accordance with the manufacturer's recommended storage temperatures.

4.8.4 Recommendations for Use

This section should be read in conjunction with the accepted international OELs.

Particular attention is drawn to the definitions of control limits and recommended limits in the publication and to the fact that for many substances two limits are set, one relating to an 8-hour time-weighted exposure and the other relating to a 10-minute short-term exposure period.

Gas detector tubes may be used for a number of purposes. When short-term detector tubes are used to assess exposure against an OEL it is important to ensure that sufficient information is available on exposure throughout the relevant period for a valid time-weighted average to be estimated. This may require sequential testing throughout the exposure period. OELs relate to personal exposure and tests should be carried out in the breathing zone, taken to be within 300 mm of the mouth or nose.

4.8.5 Instructions

4.8.5.1 Aspirating Pumps

Manufacturer's instructions should be supplied with each aspirating pump and should include the following:

a. A warning stating that because pumps made by different manufacturers may not operate at the same rate even when they draw the same volume, they should not be used interchangeably
b. Instructions for testing for leakage before each use
c. Maintenance instructions

4.8.5.2 Short-Term Detector Tubes

Manufacturer's instructions should be supplied with each box of detector tubes and should include the following:

a. Directions stating that aspiration pumps shall be tested for leakage before each use.
b. Information on the general reactions involved in the system and the levels at which other gases and vapors, including water vapor, are likely to interfere to the extent of reducing the accuracy below the level.
c. Where the contents of tubes are hazardous, a warning to this effect together with disposal instructions.
d. Where special requirements for lighting are needed to ensure reliable reading of color-match and length-of-stain tubes, and details of such requirements.

e. For color-match tubes, instructions for evaluating the color change.

f. The time required for the completion of one aspiration and the limits on this time.

g. A statement about the limitations of reuse.

4.9 Marking

4.9.1 Boxes

Each box of short-term detector tubes should be marked with the following information:

a. The manufacturer's name, trademark, or other means of identification

b. The number and date of the standard used

c. The names of the gases or vapors for which the short-term detector tubes may be used and the concentration ranges

d. The manufacturer's batch numbers and date of batch testing

e. The expiry date

f. Recommended storage temperatures

4.9.2 Tubes

Each tube should be marked with the following information:

a. The manufacturer's name, trademark, or other means of identification

b. An indication of the gases and vapors for which the tubes are intended to be used

4.10 Instruction Manual

Each gas-detection instrument should be provided with an instruction manual containing complete, clear, and accurate instructions for safe and proper operation. It should include the following information:

a. Operating instructions and adjustment

b. Instructions for checking and/or calibration

c. Detail of operational limitations, including, where applicable, the following:

1. Range of gases for which the instrument is suitable
2. Ambient temperature limits
3. Humidity range
4. Battery life
5. Maximum and minimum storage temperature limit
6. Sample velocity limit

d. Information on the adverse effects of contaminating gases and substances and oxygen-enriched or -deficient atmospheres on the proper performance of the instrument

4.11 Principle of Operation, Combustible Gas Detectors

4.11.1 Catalytic Sensor

The principle of operation of the catalytic sensor depends on the oxidation of flammable gas at the surface of an electrically heated catalytic element (i.e., filament or bead). This oxidation causes the temperature of the sensing element to change as a function of the concentration of gas detected. The change of electrical resistance is determined in a bridge circuit and the apparatus is calibrated to provide indication of gas concentrations and alarms.

Since oxidation depends on the presence of oxygen, detection apparatus using this type of sensor should only be used for gas/air concentrations up to the LEL.

By their very nature, catalytic sensors will positively detect the presence of any combustible gas mixture. The other types of sensors infer the presence of combustible gases by relating the response to the gas detected to the calibration of the detection apparatus.

The catalytic sensor is the most widely used type of gas detector, used in either diffusion mode or in aspirating or sampling systems.

4.11.2 Thermal Conductivity Sensor

The principle of operation of the thermal conductivity sensor depends on the heat loss by conduction of an electrically heated resistance element (i.e., filament or bead, located in a gas sample stream of fixed velocity).

The resulting change of electrical resistance is compared with that of a similar sensing element located in a reference cell, both electrical elements forming part of an electrical bridge or other measuring circuit and the apparatus are calibrated in any suitable range up to 100% gas to provide indications of gas concentrations and alarms.

This type of sensor is best used for the detection of specified single gases of relatively high thermal conductivity with respect to air (hydrogen, methane, etc.) at a concentration above the LEL.

By their very nature, thermal conductivity detectors normally function within some form of sampling systems and not by diffusion.

4.11.3 Infrared Sensor

The principle of operation of the infrared sensor depends on the absorption of a beam of infrared radiation by the gas being detected. Detection apparatus with infrared sensors may take various forms but may be categorized as one of the following:

1. Specifically adapted analyzers with sampling systems
2. Single-point, self-contained infrared detection apparatus suitable for installation in potentially explosive atmospheres
3. Open-path detection apparatus that protect an infrared beam along an open path through the area being monitored

In cases 1 and 2, the absorption of infrared radiation by the gas is detected by photoelectric means and produces an electrical signal to provide an indication of gas concentration and alarms.

In case 3, the absorption of infrared radiation by gas anywhere along the open path is detected by photoelectric means and also produces an electrical signal to provide alarms and indications of the concentration of gas integrated along the path.

Open-path infrared detection apparatus differs from the other types mentioned in that it does not measure the concentration of gas at a particular point location, but rather measures the path integral of gas concentration along an investigative beam. It is therefore capable of detecting the presence of gas over a wider area than other types. However, it is inherently not capable of distinguishing between a high concentration of gas occupying a short section of the open path and a low concentration of gas occupying a long section of the path.

As in the case of thermal conductivity sensors, infrared sensors may be used for the detection of specified combustible gas(es) in any specified range of concentrations up to 100% gas.

4.11.4 Semiconductor Sensor

The principle of operation of the semiconductor sensor depends on changes of electrical conductance that occur by chemisorption when the heated semiconductor sensing element is exposed to gas. The changes of conduction are then determined in an appropriate electrical circuit and the apparatus is calibrated in any suitable range to provide indications of gas concentrations and alarms.

This type of sensor is normally only used for the detection of a specified gas in a nominated range of concentrations.

Semiconductor sensors may be used in either diffusion mode or in a sampling system.

4.11.5 Intended Application of Apparatuses

4.11.5.1 Transportable Apparatuses

Transportable apparatuses should normally be selected for such purposes as monitoring temporary work area hot work and areas where combustible liquids, vapors, or gases may be transferred.

4.11.5.2 Portable Apparatuses

Portable apparatuses should normally be selected for such purposes as leak seeking, verification of gas-free conditions, safety checks, and similar applications.

A portable apparatus is normally used in diffusion mode, but where leak seeking is involved or where the apparatus is used for the detection of gas in confined spaces beyond the normal reach of the user, either a static sample probe or a hand- or mechanically aspirated sample probe will be necessary.

Where a portable apparatus is, from time to time, likely to be exposed to gas concentrations greater than the LEL, care should be taken to select an apparatus suitable for that purpose.

4.11.5.3 Portability of the Apparatus

Factors that are important in selecting a portable or transportable apparatus include its size, mass, and robustness, its power supply requirements, the type of indication required, and the visibility or audibility of any alarms.

4.12 Use of Portable and Transportable Combustible Gas-Detection Apparatuses

The various types of portable and transportable gas-detection apparatuses may be used in a variety of ways according to their particular design and specification.

Small, handheld apparatuses may be used for leak seeking or spot checks, while larger portable devices, some with visual and/or audible alarms, may be used in multirole modes so as to include leak seeking, spot checking, and local area monitoring functions according to the particular needs of the user.

Transportable apparatuses are intended for use as temporary area monitors in locations where the operations involved are potentially hazardous and are of a temporary nature (e.g., during the loading or unloading of fuel or chemical tankers or where temporary hot work (in connection with maintenance activities) may be in progress in classified hazardous areas under the authority of a gas-free certificate).

Transportable apparatuses are not intended to be hand-carried for long periods of time, but are intended to be in place for hours, days, or weeks.

While both the portable and transportable types of apparatuses are likely to be exposed to adverse climatic and handling conditions from time to time, transportable apparatuses are more likely to be exposed to these conditions due to the nature of their use, whether indoors or outdoors, and particular attention should be paid to protecting them from climatic or handling damage.

Anyone required to use portable or transportable gas-detection apparatuses should be properly trained in their use and have ready access to the operating instructions.

4.12.1 Guidance on the Use of Portable and Transportable Apparatuses

In areas being surveyed with portable apparatuses where gases or vapors may form a layer rather than be uniformly mixed, spot checks may be made at different levels using an extension probe.

- When sampling vapor above a liquid, care should be taken to avoid the sample line or sensor from coming into contact with the liquid, since this may block the gas entry to the apparatus. (Note: only sample lines recommended by the manufacturer should be used.)
- When taking a portable apparatus from a cool environment to a warm environment, it is important that time is taken to allow the apparatus temperature to rise sufficiently to avoid vapor condensation that may otherwise interfere with its correct operation.
- Combustible gas-detection apparatuses are not normally designed to detect the presence of combustible materials that are not in a volatile state under the conditions that measurements are made.
- Combustible gas-detection apparatuses are not intended to be used to indicate the presence of either combustible dusts or fibers.
- Many types of combustible gas-detection apparatuses are not sensitive to a specific gas. The presence of other gases than that for which the apparatus is calibrated may adversely influence its indications.

- Erratic indications may indicate apparatus malfunction or some atmospheric disturbance. Where doubt exists, a check should be made with a second apparatus of the same type and/or the apparatus should be checked under controlled conditions before its continued use.
- The presence of very low concentrations of combustible gas can produce indications that may be mistaken for zero drifts. When in doubt, the apparatus should be removed to a clean air environment and rechecked.
- Saturated steam may physically block the flame arrestors of certain types of gas sensor, making them inoperative, and care should be exercised accordingly.
- Where an apparatus is to be used to detect the presence of more than one gas it should be calibrated accordingly. The calibration may be with several gases or with the gas to which the apparatus is least sensitive. However, a gas-detection apparatus of this kind should not be regarded as being suitable for gas analysis.
- Where off-scale indications occur (in either direction), this may indicate the presence of a potentially explosive atmosphere. It may then be necessary to flush the sensor with clean air and to cross-check for the presence of gas by taking the reading again or by using another type of gas-detection apparatus. In any event, in such cases, the presence of a potentially explosive atmosphere should be assumed until proven otherwise.
- Care should be taken to ensure that the materials from which the apparatus has been constructed are compatible with the gas or vapor to be detected. For example, copper should not be contained in any apparatus likely to be used for the detection of acetylene or its derivatives because of the possibility of the formation of potentially dangerous acetylides.
- When using a portable gas-detection apparatus, it is necessary to be aware that some combustible gases and vapors are also toxic and may at least cause serious discomfort and at worst cause death.
- Any portable or transportable gas-detection apparatus that is used infrequently should be regularly inspected, maintained, and calibrated, so that it may be available for immediate use when required.
- If a portable or transportable gas-detection apparatus is dropped or otherwise damaged, it should immediately be taken out of service for inspection, repair, and recalibration as necessary before reuse.

4.13 Maintenance Routine Procedures and General Administrative Control

Portable and transportable gas-detecting apparatuses are units taken to sites where a mixture of gas/air atmospheres is expected to be present. Inadequate maintenance, incorrect zero adjustments and deteriorated batteries are all causes of gas-detection errors. Errors and failures in an apparatus may not be self-evident, therefore gas detectors should not be relied on unless they are regularly and properly checked and maintained.

The accuracy of gas-detection apparatus is particularly dependent on the accuracy of the concentration of the test gas used for calibration. All types of gas detectors should be checked at intervals with the test gas recommended by the manufacturer.

When it is necessary to detect the presence of several gases mixed with air, the sensitivities to these gases should be checked periodically with appropriate test gases.

Any repair or maintenance should not invalidate the certification of the apparatus for use in potentially explosive atmospheres.

4.13.1 Repair

Portable and transportable gas-detection apparatuses should be removed to a nonhazardous location for repair and testing under close supervision of a competent person.

4.13.2 Routine Test and Recalibration Procedures

All detection apparatuses, whether portable or transportable, should be subjected to routine tests and recalibration procedures. These should be carried out in accordance with the instruction manual and using the recommended field test kit.

In general, gas-detection apparatuses should be

- Calibrated in accordance with the manufacturer's instructions, using the recommended test kit/equipment.
- Regularly inspected for possible malfunction, damage or other deterioration, and if necessary, recalibrated immediately before each occasion of use.

4.13.3 Accuracy

When calibrated against atmospheres whose concentrations of appropriate gas or vapor in air have been standardized to within ±5% of the desired concentrations (standard atmosphere), tubes should, when aspirated in accordance with the manufacturer's instructions at all temperatures between 5°C and 35°C, be accurate with a 95% confidence limit between

- +30% and –20% at a concentration equal to the OEL
- +50% and –20% at a concentration of twice the OEL and half the OEL

4.13.4 General Administrative Control

Portable and transportable apparatuses should be marked with an indication of the calibration date, and according to the control system employed, the length of time to the next recalibration.

Apparatuses awaiting recalibration or maintenance should be kept separately from apparatuses awaiting return to service after such operations.

Spares may deteriorate in storage owing to mishandling or age and should always be tested before use.

4.14 Combination Explosive, Toxic, and Oxygen Deficiency or Excess Detectors (Portable and Transportable)

The gas detector unit can be used in the industrial environment from oil rigs to sewer systems and designed for use in explosive or toxic atmospheres and powered from rechargeable nickel-cadmium batteries.

The unit can be used for leak seeking or for a long period of time for static site monitoring provided that the batteries are charged. The instrument can be safely used for 10 to 12 hours by fully charged batteries.

4.14.1 Design

4.14.1.1 Sensors

a. The sensor for flammable gases is usually of a pellistor type calibrated for methane, pentane, or other saturated hydrocarbons
b. A toxic gas sensor for hydrogen sulfide is of electrochemical type
c. Detection of oxygen deficiency or excess is of oxygen sensor type

4.14.1.2 Alarms

- The instrument should be provided with a control knob in the test mode to test that the battery is charged and three alarm lamps and siren are operative. The alarm levels selected for gas monitoring should be indicated as ready check.

- To indicate normal operation, a green indicator light to pulse every 10 seconds followed by an audible click should be provided.
- If the gas is escaped, the instrument should react regardless of the control knob. The alarm latches cannot be reset until the gas concentration drops back to the safe level.
- The instrument should be provided with a steady continuous note to sound a warning when the battery needs recharging.

4.14.2 Safety Features

- The instrument should be made intrinsically safe and certified to be used in maximum safety in all potentially hazardous areas of zones 0, 1, or 2.
- The detector, in addition to its function to monitor in open areas, should also test the air in a confined area by using a manual aspirator to suck the air into the unit for testing and checking where the gas may become trapped. The length of aspirator hose should be 2 m.
- The instrument should contain a calibration adjustment for course and fine level. The preset control should be covered and locked and temper resistance to reduce the risk of accidental adjustment.
- The case should be built to work in a tough environment and should be of sealed stainless steel. A tailor-made case should not be provided with neck and waist straps and for maximum convenience should allow the operator to work in comfort while using the unit.

4.14.3 Calibration

Regular functional checking of the instrument should be undertaken using a gas test in accordance with the manufacturer's instruction manual.

Gas detectors should be supplied with operation and maintenance instructions that should be followed. Repair and maintenance should be carried out by a fully qualified engineer.

4.15 Chemical Sensing Detectors and Tubes

This section specifies operating instructions of short-term and long-term detector tubes and associated aspiratory pumps used at normal atmosphere for evaluating atmospheric contaminants at concentrations of OELs. It also covers color-match tubes that are designed to give an indication of concentration over a short period of time.

4.15.1 Detector Tubes

Detector tubes with direct reading colorimetric indication have many applications. Approximately 350 different substances can be measured with detector tubes.

Tubes are usually capable of only being used once; repeated measurement of the same substance can be performed with electrochemical sensor, which is more practical and economical.

When complex mixtures are present, only a laboratory analysis will suffice. Contaminant air is trapped in a sorbent sampling tube like silica gel or activated charcoal and analysis is performed in the laboratory. Since many OELs are in the range of a 1-ppm measurement, the flexibility of the detector tube method lends itself well to the diverse measurement applications encountered in the workplace.

Before each measurement, an assessment of the situation should be made as to what contaminants are in question, at what location, at what time, and then action is taken according to established safety requirements.

4.15.2 Classification

Detector tubes have different applications and are classified as follows:

a. Air investigation at workplace, measurement in the range of the OELs

b. Technical gas analysis-detector tube measurement in the area of emission concentration

c. Compressed air for breathing apparatuses

Specially calibrated detector tubes should be used to determine the quality of compressed breathing air. The typical contaminants are cobalt (CO), carbon dioxide (CO_2), water, and oil.

4.15.2.1 Short-Term Detector Tubes

Short-term detector tubes are designed for on-the-spot measurement at a particular location over a relatively short time period from 10 seconds to 15 minutes or so, depending on the particular detector tube and sampling pump. Some applications for short-term tubes are the evaluation of concentration fluctuation in the workplace, the measurement of contaminants in the workers' breathing zone, the investigation of confined spaces (e.g., chemical, hydrocarbon tanks, sewers) prior to entry and additionally to check for gas leaks in process areas.

4.15.2.2 Detector Pump

Detector pumps and short-term tubes are usually designed and calibrated as a unit and the use of other short-term detector tubes is not recommended. The difference in the flow characteristics of the pump and tube can result in considerable measurement errors.

4.15.2.3 Long-Term Tubes

Long-term tubes provide integrated measurement that presents the average concentration during the sampling time. The long-term tubes are used for measurements between one and eight hours. Long-term tubes can be used as personal monitors or area monitors to determine the time-weighted average concentration. A constant flow pump is used for long-term tubes.

In addition to the long-term tubes, there are also direct-reading diffusion tubes and badges for measurement.

4.15.3 Measurement

An assessment of the measurement for choosing the best detector tubes needs the following considerations:

- The ambient condition
- Limits of use

Although the detector tube is an easily operated gas measurement method, it belongs in the hands of specialists or trained employees.

NOTES

1. Threshold limit value, time-weighted average (TLV-TWA).
2. The TWA concentration for a normal 8-hour workday and 40-hour work week to which nearly all workers may be repeatedly exposed day after day, without adverse effect.
3. TLV, short-term exposure limit (STEL).
4. The concentration to which workers can be exposed continuously for a short period of time without suffering from
   a. Irritation
   b. Chronic or irreversible tissue damage
   c. Narcosis of a sufficient degree to increase the likelihood of accidental injury, impair self-rescue, or materially reduce work efficiency should be considered
5. Threshold limit value, ceiling (C).

6. The concentration that should not be exceeded during any part of the working exposure. In conventional industrial hygiene practice, if instantaneous monitoring is not feasible, then the TLV-C can be assessed by sampling over a 15-minute period except for those substances that may cause immediate irritation when exposure is short.

4.15.4 Detector Tube Measurement System

A detector tube is a vial that contains a chemical component that reacts with the measured substance by changing color. The normal shelf life of a detector tube is two years and the tube tips are fused on both ends.

a. The system consists of a detector tube and a gas-detector pump. The pump precisely matches the reaction of the reagent system in the tube. Therefore the gas-detector pump delivering the correct volume must be used.

b. Short-term and long-term measurement is distinguished between two different types of pumps: short-term pumps and the long-term pumps.

c. There are four types of detector tube pumps:
   1. Hand-operated bellows pumps for leak detection.
   2. Battery-operated pumps, where, after choosing the number of strokes necessary for measurement and pressing the stroke and stop button, the pump is automatically actuated. The pump must be Ex approved for dangerous areas.
   3. Microprocessor-controlled automatic gas-detector pumps, which can be programmed with number of strokes and feature displays indicating the number of preselected and the actual number of strokes. This type also must be Ex approved for dangerous areas.
   4. Long-term pumps, which are a constant-flow type that drives a preset flow rate and the battery capacity ensures continuous operation for more than 8 hours. This type must also be Ex approved for dangerous areas as specified.

4.15.5 Maintenance of Gas-Detector Pumps

To ensure precise results, it is particularly important to confirm that the pump is operating properly. Short-term pumps should be checked before each measurement for leaks and suction capacity according to the manufacturer's operating instructions. In addition, after a measurement, short-term pumps should be flushed with clean air by performing several strokes without a detector tube in the pump. This purges the pump of reaction products that enter the bellows due to a reaction in the tube. The long-term pumps should be checked for flow volume according to the operating instructions.

4.15.6 Tubes for Short-Term Measurement

The design of short-term tubes depends on the measurement task, particularly on the substances to be measured.

There are several types of short-term tubes:

- Tubes with an indication layer
- Tubes with one or more prelayers plus an indication layer
- A combination of two tubes
- Tubes with a connecting tube
- Tubes with a built-in reagent ampoule
- Tubes for simultaneous measurement

4.15.7 Evaluation of Tubes

The evaluation of the indication on the detector tube is another important factor to be considered. The following are guidelines for interpreting the indication.

a. Continuously observe the tube during measurement
b. Evaluate the indication immediately following the measurement according to the instruction of use
c. Use sufficient lighting
d. Use a light-colored background
e. Compare with an unused tube as follows:
   1. Observing the tube during the measurement is particularly important to make sure that a complete discoloration of the tube has not occurred without being realized. This complete discoloration can sometimes occur abruptly with high concentrations even during the course of a first stroke.
   2. A sufficient lighting source is necessary. Direct sunlight should be avoided because UV radiation from the sun may cause a change in discoloration.
   3. The reading of the tube should be done immediately following the measurement; keeping the used tube as proof is not useful.

4.15.8 Expiration Date, Storage, and Disposal

Detector tubes contain reagent systems designed to undergo a chemical reaction with the particular substance. Since chemicals and chemical reagents are not stable indefinitely, each box of detector tubes is stamped with an expiration date. Tubes used beyond the expiration date cannot be relied on to give accurate results.

Tubes should be stored in their original packages at a room temperature of 25°C. Avoid temperatures that are excessively low (less than 2°C) or higher than 25°C, and do not subject the tubes to light for prolonged periods.

Do not dispose of the tubes in domestic waste.

Further Readings

American National Standards Institute (ANSI).

ANSI/UL-827. 1988. Central Station for Watchman Fire Alarm and Supervisory Services.

ANSI/UL-268. 1988. Smoke Detectors for Fire Protection Signaling Systems.

ANSI/UL-217. 1985. Single and Multiple Station Smoke Detectors.

ANSI/NFPA-92A. 1988. Smoke Control System.

ANSI/NFPA-72E. 1987. Automatic Fire Detectors.

ANSI/NFPA-72B. 1986. Auxiliary Protective Signaling Systems for Fire Alarm Services.

ANSI/NFPA-325M. 1989. Fire Hazards Properties of Flammable Liquids, Gases, Volatile Solids.

British Standards Institution (BSI).

BSI-5343. 1986. Specification for Short Term Gas Detector Tubes-Part 1.

BSI-5445. 1984. Automatic Fire Detection System Parts 1, 5, 7, 8 and 9.

BSI-5839. 1988. Fire Detection of Alarm Systems for Buildings Parts 1, 2, 3, 4 and 5.

Chen, S.-J., Hovde, D.C., Peterson, K.A. and Marshall, A. 2007. Fire detection using smoke and gas sensors, *Fire Safety Journal* 42 (8), 507–515.

Fowler, J. 1995. Raising standards in the fire detection and alarm industry. *Fire Prevention* (281), 16–19.

Gupta, Y. and Dharmadhikari, A. 1985. Analysis of false alarms given by automatic fire detection systems. *Reliability Engineering* 13 (3), 163–174.

He, Z., Pu, J. and Cai, Y. 2002. Multi-sensor fire detection algorithm for ship fire alarm system using neural fuzzy network. *Process in Safety Science and Technology Part A* 3, 142–146.

Holt, M. 2006. Fire alarm system facts, *EC and M: Electrical Construction and Maintenance* 105 (11), 40–44.

International Organization for Standardization (ISO).

ISO-7240-1. 1988. Fire Detection and Alarm Systems General and Definitions.

ISO-7731. 1986. Danger Signals for Work Places.

Klose, J. 1991. Analysis, synthesis and simulation of fire signals as a tool for the test of automatic fire detection systems. *Fire Safety Journal* 17 (6), 499–518.

Minster, K. 2008. Multi-criteria fire detection—An intelligent response. *Building Engineer* 83 (9), 24–25.

Morgan, R.B. 1991. Designing fire detection and alarm systems. *EC and M: Electrical Construction and Maintenance* 90 (4), 33–43.

Ollero, A., Arrue, B.C., Martinez, J.R. and Murillo, J.J. 1999. Techniques for reducing false alarms in infrared forest-fire automatic detection systems. *Control Engineering Practice* 7 (1), 123–131.

Spearpoint, M.J. and Smithies, J.N. 1997. Experimental analysis of the performance of a multi-sensor two-stage fire detection and water discharge algorithm. *Fire Safety Journal* 29 (2–3), 141–157.

Sun, T., Grattan, K.T.V., Sun, W.M., Wade, S.A. and Powell, B.D. 2003. Rare-earth doped optical fiber approach to an alarm system for fire and heat detection. *Review of Scientific Instruments* 74 (1), 250–255.

Tong, D. and Canter, D. 1985. Informative warnings: In situ evaluations of fire alarms. *Fire Safety Journal* 9 (3), 267–279.

Tung, T.X. and Kim, J.-M. 2011. An effective four-stage smoke-detection algorithm using video images for early fire-alarm systems. *Fire Safety Journal* 46 (5), 276–282.

Young, N. 2004. Fire detection and alarm systems. *Fire Prevention and Fire Engineers Journal* 64 (243), 53–55.

Zhang, Y., Xiang, H. and Tang, G. 2006. Application of optical fiber Bragg grating fire detection and alarm system in crude oil tank farm. *Petroleum Refinery Engineering* 36 (11), 56–57.

5

Firefighting Sprinkler Systems

5.1 Introduction

For most fires, water represents the ideal extinguishing agent. Fire sprinklers utilize water by direct application onto flames and heat, which causes cooling of the combustion process and prevents ignition of adjacent combustibles. They are most effective during the fire's initial flame growth stage, while the fire is relatively easy to control. A properly selected sprinkler will detect the fire's heat, initiate alarm, and begin suppression within moments after flames appear. In most instances sprinklers will control fire advancement within a few minutes of their activation, which will in turn result in significantly less damage than otherwise would happen without sprinklers.

Among the potential benefits of sprinklers are the following:

- *Immediate identification and control of a developing fire.* Sprinkler systems respond at all times, including periods of low occupancy. Control is generally instantaneous.
- *Immediate alert.* In conjunction with the building fire alarm system, automatic sprinkler systems will notify occupants and emergency response personnel of the developing fire.
- *Reduced heat and smoke damage.* Significantly less heat and smoke will be generated when the fire is extinguished at an early stage.
- *Enhanced life safety.* Staff, visitors, and firefighters will be subject to less danger when fire growth is checked.
- *Design flexibility.* Egress route and fire/smoke barrier placement becomes less restrictive since early fire control minimizes demand on these systems. Many fire and building codes will permit design and operations flexibility based on the presence of a fire sprinkler system.
- *Enhanced security.* A sprinkler-controlled fire can reduce demand on security forces by minimizing intrusion and theft opportunities.
- *Decreased insurance expenditure.* Sprinkler-controlled fires are less damaging than fires in buildings without sprinklers. Insurance underwriters may offer reduced premiums for sprinkler-protected properties.

These benefits should be considered when deciding on the selection of automatic fire sprinkler protection.

This chapter provides the minimum requirements for the design of fire sprinkler systems in buildings and industrial plants and open-head, deluge-type foam-water sprinklers and foam-water spray systems. It covers the classification of hazards, provision of water supplies, and components to be used.

A sprinkler system is a specialized fire protection system and requires knowledgeable and experienced design and installation. A sprinkler system consists of a water supply (or supplies) and one or more sprinkler installations; each installation consists of a set of installation main control valves and a pipe array fitted with sprinkler heads. The sprinkler heads are fitted at specified locations at the roof or ceiling, and where necessary, between racks, below shelves, and in ovens or stoves. The main elements of a typical installation are shown in Figure 5.1.

The sprinklers operate at predetermined temperatures to discharge water over the affected part of the area below the flow of water through the alarm

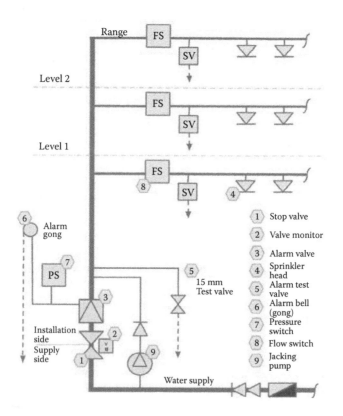

FIGURE 5.1
Main elements of a sprinkler installation.

valve, initiating a fire alarm. The operating temperature is generally selected to suit ambient temperature conditions.

Only sprinklers in the vicinity of the fire (i.e., those that become sufficiently heated) operate.

In some life safety applications an authority should require sprinkler protection only in certain designated areas and solely to maintain safe conditions for the evacuation of persons from the sprinkler-protected areas. Such a system will not provide protection against a fire that starts in a nonsprinklered part of the premises and develops to some size before spreading to the sprinkled parts, and for more complete protection the sprinkler system is extended throughout the premises with only limited exceptions.

It should not be assumed that the provision of a sprinkler system entirely obviates the need for other means of fighting fires and it is important to consider the fire precautions in the premises as a whole.

Structural fire resistance, escape routes, fire alarm systems, particular hazards needing other fire protection methods and provision of hose reels, fire hydrants, portable fire extinguishers, and so forth, safe working and goods handling methods, management supervision, and good housekeeping all need consideration.

5.2 Outline Design

Consideration should be given to any benefits that will be gained by changes in building design, work procedures and so forth when preparing the outline design. In planning the site layout and building design, particular consideration should be given to the following:

- The occupancy hazard class and goods category that determine the water discharge density and water supply pressure and flow.
- The siting of any town main water supply connection(s).
- The siting of any water supply tank(s) or reservoir.
- The siting of any pump house.
- The maximum quantity of water available and maximum rate of supply (based on site tests at periods of maximum demand) from the supply source compared with the system requirements.
- The location of sprinkler installation control valves, together with the access thereto, indication of their position, and the disposal of drainage and water supply test water.
- The source and means of supply of electric power and so forth.
- The protection of valve sets, pipework, and sprinklers against accidental damage. It is important to consider building design in the context

of fire protection, for example, in the choice of materials, support of sprinkler pipework with regard to the load imposed on structure by the weight of sprinkler pipework and the contained water, building heating, need for inbuilt drainage (which is strongly advised for computer areas), raising a base of stacked goods above the floor where water damage may be severe, and so forth. When storage of goods is involved it should be appropriate to consider the height of the building and of material stacks, and the height and type of any storage racks that will have a considerable bearing on fire protection costs.

The design of double-entry storage racks should be influenced by the need to mount sprinklers therein. Where sprinklers are fitted in racks, additional rack structural members may be needed to prevent impact damage to the sprinkler head and pipework.

5.3 Interaction with Other Fire Protection Measures

Account should be taken of possible interaction between sprinkler systems and other fire protection measures. Examples of possible adverse interaction between sprinkler protection and other fire protection measures are

- Water damage to an inadequately shielded fire alarm control panel in a sprinkler-protected area with consequent possible failure of the fire alarm system.
- Operation or failure of smoke detectors in zones adjacent to one in which water discharge is taking place because of the water spray mist traveling into adjacent zones.

Such possible interactions need particularly careful consideration in the case of systems that are part of life safety measures.

5.4 Buildings to Be Sprinkler-Protected

The sprinkler system should provide protection to all parts not specified as exceptions in Section 5.4.1 of the following:

- The building under consideration
- Any building communicating directly or indirectly with the building under consideration

5.4.1 Exceptions (Buildings and Parts of Buildings Not Sprinkler-Protected)

5.4.1.1 Obligatory Exceptions

Sprinkler protection should not be provided in the following parts of a building or plant:

- Areas, rooms, or places where the water discharged from a sprinkler may present a hazard.

5.4.1.2 Optional Exceptions

Sprinkler protection should be considered for, but need not be provided in, the following parts of a building or plant:

a. Stairs, spaces below stair headings (but not rooms above a stair), and lift wells. Any part not provided with sprinkler protection should be enclosed by walls, ceilings, and floors with a fire resistance of not less than 2 hours, in which all doors are of not less than 1-hour fire resistance, and in which all glazed areas either are of not less than 1-hour fire resistance, or in the case of stairs, are protected by cutoff sprinklers. The area of glazing in any part not provided with sprinkler protection should not exceed 1.5 m² in each story.

b. Washrooms, toilets, and water closets (WCs) (but not cloakrooms).

c. Rooms or compartments containing electric power distribution apparatus, such as switchgear and transformers, and used for no other purpose(s). Any part not provided with sprinkler protection should be enclosed by walls, ceilings, and floors of not less than 2-hour fire resistance in which all doors are of not less than 1-hour fire resistance.

d. Areas containing oil or similar flammable liquids.

5.4.1.3 Communicating Buildings

Sprinkler protection should be considered for, but need not be provided in, the following communicating buildings or structures:

- A building or story separated from the sprinklered building by walls of not less than 6-hour fire resistance in which each opening is protected by two fire doors or fire shutters (arranged in series), each of not less than 2-hour fire resistance.
- Canopies of noncombustible construction not extending beyond 2.3 m from the building wall. Any such canopy not provided with

sprinkler protection should be fitted with cutoff sprinklers under the canopy over each opening between it and the sprinklered building. Any opening 2.5 m or less in width should be provided with a cutoff sprinkler positioned centrally over the opening. Openings exceeding 2.5 m in width should be provided with cutoff sprinklers over the opening and not more than 2.5 m apart and with a sprinkler not more than 1.25 m from each side.

- Exterior loading docks and platforms either of noncombustible construction or with the space beneath closed off against accumulation of debris.

- Buildings used solely as offices and/or private dwelling(s). Any part not provided with sprinkler protection should be separated from the sprinkler-protected building by a wall of not less than 6-hour fire resistance in which any glazed areas are of not less than 1-hour fire resistance and are provided with cutoff sprinklers, and in which all door openings are protected by either

  Single fire doors or single fire shutters of not less than 2-hour fire resistance

  Fire doors of not less than 1-hour fire resistance and cutoff sprinklers

- Buildings, storys, or rooms of noncombustible construction used mainly for wet processes.

- Stairs, washrooms, and WCs external to the sprinkler-protected building in which all openings to the sprinkler-protected building are protected by doors of not less than 1-hour fire resistance.

- Staircases, washrooms, toilets, and WCs external or internal to the sprinkler-protected building that form a means of communication between the sprinklered building and a nonsprinklered building. In any such part not provided with sprinkler protection, all openings into the communicating area from the sprinklered building and from the nonsprinklered building should be protected by fire doors of not less than 1-hour fire resistance.

5.5 Classification of Occupancies

Occupancy classifications for this standard relate to sprinkler installations and their water supplies only. They are not intended to be a general classification of occupancy hazards.

Light Hazard

Occupancies or portions of other occupancies where the quantity and/or combustibility of contents is low and fires with relatively low rates of heat release are expected.

Ordinary Hazard (Group 1)

Occupancies or portions of other occupancies where combustibility is low, quantity of combustibles is moderate, stockpiles of combustibles do not exceed 2.4 m, and fires with moderate rates of heat release are expected.

Ordinary Hazard (Group 2)

Occupancies or portions of other occupancies where quantity and combustibility of contents is moderate, stockpiles do not exceed 3.7 m, and fires with moderate rate of heat release are expected.

Ordinary Hazard (Group 3)

Occupancies or portions of other occupancies where quantity and/or combustibility of contents is high and fires of a high rate of heat release are expected.

5.5.1 Extra Hazard Occupancies

Extra hazard occupancies are portions of other occupancies where quantity and combustibility of contents is very high, and flammable and combustible liquids, dust, lint, or other materials are present, introducing the probability of rapidly developing fires with high rates of heat release.

Extra hazard occupancies involve a wide range of variables that may produce severe fires. The following should be used to evaluate the severity of extra hazard occupancies:

- *Extra hazard (group 1)* includes occupancies described above with little or no flammable or combustible liquids
- *Extra hazard (group 2)* includes occupancies described above with moderate to substantial amounts of flammable or combustible liquids or where shielding of combustibles is extensive

5.6 Types

A sprinkler installation should be based on one of the following main types:

- Wet pipe
- Alternate (wet and dry pipe)
- Dry pipe
- Preaction
- Recycling
- Tail-end alternate
- Tail-end dry pipe
- Deluge

Wet pipe installations are preferred. However, if the temperature of the premises cannot be guaranteed to remain above freezing at all times, an alternate installation should be fitted. Where only part of the premises falls below 5°C during the winter, a tail-end alternate extension should be installed in that part as an extension to the wet installation.

Where freezing or elevated temperatures are experienced either frequently or continuously, a dry pipe installation should be installed, or only in small areas should tail-end dry pipe extensions be installed as extensions to the main installation.

Sprinkler installations should incorporate deluge systems to cover small areas of flammable liquid hazards such as oil-fired boiler rooms.

Figure 5.2 shows a typical sprinkler installation of wet and dry systems.

5.7 Wet Pipe Installations

A wet pipe sprinkler system (Figures 5.3 and 5.4) is a sprinkler system employing automatic sprinkler heads attached to a piping system containing water and connected to a water supply so that water discharges immediately from sprinklers opened by heat from a fire.

Wet pipe installations should only be installed where there is no danger at any time of the water in the pipes freezing, and where the temperature will not exceed 70°C. Antifreeze should not be employed as a means of preventing the water freezing in the pipes.

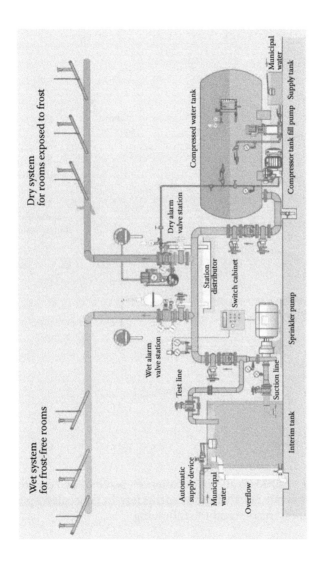

FIGURE 5.2
Sprinkler installation of wet and dry systems.

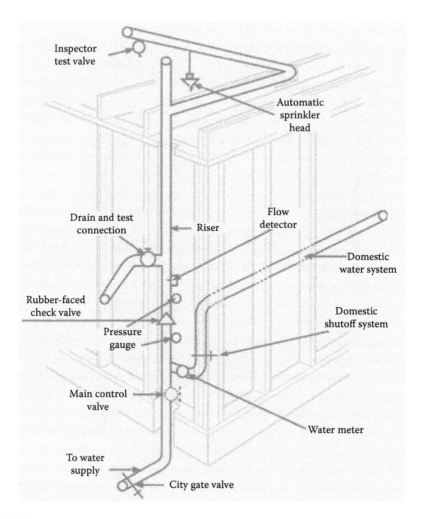

FIGURE 5.3
A wet pipe sprinkler system.

The number of sprinklers in an installation, section, or zone (including tail-end extensions, but not including sprinklers in concealed spaces or in machines, etc.) should not exceed the following:

- Light hazard: 500 per installation
- Ordinary hazard: 1000 per installation
- High hazard: 1000 per installation

Each sprinkler is activated individually when it is heated to its design temperature. Most sprinklers discharge approximately 20–25 gallons per minute

Typical wet pipe system

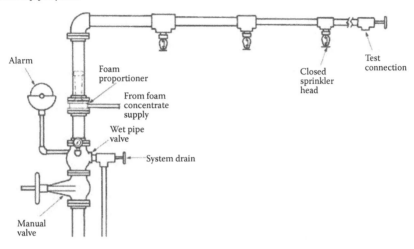

FIGURE 5.4
Schematic of a wet pipe sprinkler system.

(gpm), depending on the system design. Sprinklers for special applications are designed to discharge up to 100 gpm.

Figure 5.4 shows a schematic of a wet pipe sprinkler system.

5.8 Alternate (Wet and Dry Pipe) Installations

Alternate installations should only be installed where there is an intermittent danger of the water in the pipes freezing (e.g., during the winter months), and where the ambient temperature does not exceed 70°C.

TABLE 5.1

Number of Sprinklers, Including Any in Tail-End Extensions for Alternate Installations

Light hazard	250 per installation
Light hazard without accelerator or exhauster	125 per installation
Ordinary and high hazard with accelerator or exhauster	500 per installation
Ordinary and high hazard without accelerator or exhauster	250 per installation
Combined light and ordinary hazard	500 per installation
Combined light and high hazard	250 per installation

Alternate installations should not be used for the protection of high-hazard storage. Areas where freezing conditions are experienced at times when the installation is in the wet mode can be protected by a tail-end dry or alternate system, or with dry upright or dry pendent sprinklers projecting into the low-temperature area.

The number of sprinklers, including any in tail-end extensions, should not exceed the data that is presented in Table 5.1.

5.9 Dry Pipe Installations

A *dry pipe sprinkler system* (Figures 5.5 and 5.6) is a system with automatic sprinkler heads attached to a piping system containing air or nitrogen under

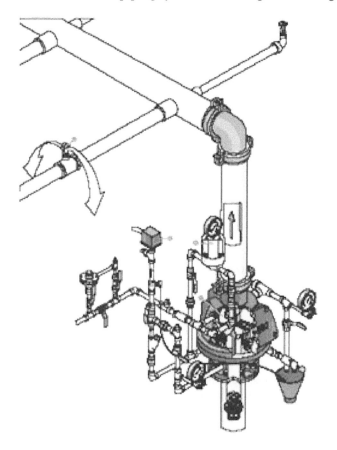

FIGURE 5.5
Dry pipe installations.

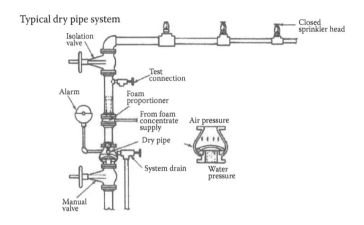

FIGURE 5.6
Schematic of a dry pipe installation.

TABLE 5.2

Maximum Number of Sprinklers for Dry Pipe Installations

Light hazard with accelerator or exhauster	250 per installation
Light hazard without accelerator or exhauster	125 per installation
Ordinary and/or high hazard with accelerator or exhauster	500 per installation
Ordinary and/or high hazard without accelerator or exhauster	250 per installation
Combined light and ordinary hazard	500 per installation
Combined light and high hazard	250 per installation

pressure. The release of this pressure (as from the opening of a sprinkler) permits the water pressure to open a valve known as a dry pipe valve and the water then flows into the piping system and out of the open sprinkler head.

Dry pipe sprinkler systems are installed in areas where wet pipe systems may be inappropriate such as areas where freezing temperatures might be expected.

Dry pipe installations should only be installed where the conditions are such that a wet pipe system or alternate installation cannot be used.

The number of sprinklers should not exceed that listed in Table 5.2. Figure 5.6 shows a schematic of dry pipe installations.

5.10 Preaction Installations

A preaction sprinkler system (Figures 5.7 and 5.8) is similar to a deluge sprinkler system except the sprinklers are closed. This type of system is typically

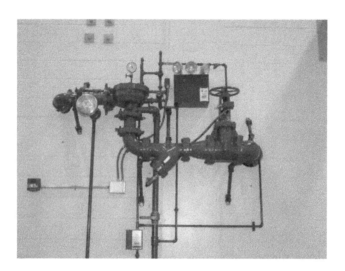

FIGURE 5.7
A preaction sprinkler system.

used in areas containing high-value equipment or contents and spaces that are highly sensitive to the effects of accidental sprinkler water discharge. The preaction valve is normally closed and is operated by a separate detection system.

Activation of a fire detector will open the preaction valve, allowing water to enter the system piping. Water will not flow from the sprinklers until heat activates the operating element in individual sprinklers. Opening of the preaction valve effectively converts the system to a wet pipe sprinkler system.

In a preaction system the piping is pressurized with air or nitrogen; monitoring of this air pressure provides a means of supervising the system piping. Loss of the supervisory air pressure in the system piping results in a trouble signal at the alarm panel.

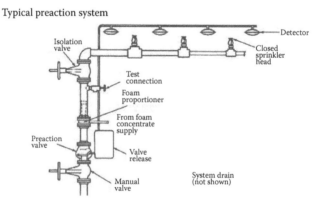

FIGURE 5.8
Schematic of a preaction sprinkler system.

There are two types of preaction installation:

1. *Type 1,* which should be installed only to prevent a premature discharge of water from pipework or sprinklers that have suffered mechanical damage
2. *Type 2,* which should be installed only to facilitate an early discharge of water from a dry pipe or alternate installation by opening the installation main control valve, thus filling the installation pipework with water upon operation of a fire detection system

The number of sprinklers should not exceed the following:

- Light hazard: 500 per installation
- Ordinary hazard: 1000 per installation
- High hazard: 1000 per installation

Figure 5.8 illustrates the schematic of a preaction sprinkler system.

5.11 Recycling Installations

Recycling installations should only be installed where it is necessary for the following reasons:

- To restrict water damage after a fire is extinguished.
- To avoid closure of the main installation stop valve if modifications are made to the installation pipework or if sprinkler heads are to be replaced.
- To prevent water damage caused by accidental mechanical damage of the installation pipework or sprinklers. The heat detectors and control equipment should be suitable for use in recycling preaction sprinkler installations. The number of sprinklers should not exceed 1000 per installation.

5.12 Tail-End Alternate Pipe and Tail-End Dry Pipe Extensions

Tail-end alternate extensions should be installed only in comparatively small areas where there is a possible frost danger in an otherwise adequately heated building as extensions to wet pipe installations. Tail-end dry extensions should be installed only as the following:

- Extensions to wet or alternate installations in high-temperature ovens or stoves
- Extensions to wet, dry, or alternate installations in buildings where freezing conditions may occur and with an air/gas pressure not less than the air/gas pressure between the main installation control valve and the tail-end valve

The number of sprinklers on any tail-end extension should not exceed 100. Where more than two tail-end extensions are controlled by one installation control valve set, the total number of sprinklers in the tail-end extensions should not exceed 250.

5.13 Deluge Installations

The arrangement of *deluge fire sprinkler system* piping (Figures 5.9 and 5.10) is similar to a wet or dry pipe system with two major differences:

1. Standard sprinklers are used, but they are all open. The activating elements have been removed so that when the control valve is opened water will flow from all of the sprinklers simultaneously and deluge the area with water.
2. The deluge valve is normally closed. The valve is opened by the activation of a separate fire detection system.

Deluge systems are used where large quantities of water are needed quickly to control a fast-developing fire. Deluge valves can be electrically, pneumatically, or hydraulically operated. Deluge installations should be installed only where it is necessary to apply water over an entire area in which a fire may originate.

Figure 5.10 shows the schematic of a deluge fire sprinkler system.

5.14 Water Supplies

5.14.1 Reliability

All practical steps should be taken to ensure the continuity and reliability of water supplies. The flow from town mains to the sprinkler system should be reduced by fire brigade operations.

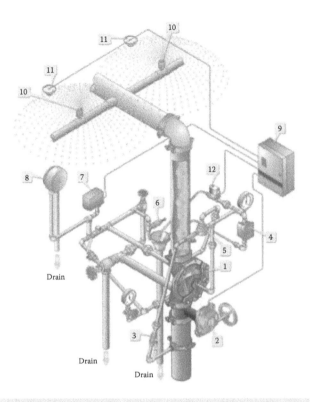

FIGURE 5.9
A deluge fire sprinkler system.

Legend:
1	Deluge valve (DV-5)	9	Releasing panel
2	Isolation valve	10	Sprinkler
3	Diaphragm supply valve	11	Smoke/heat detector
4	Manual control station	12	Pneumatic actuator
5	FSV (fail-safe valve)	13	Air supply inlet
6	Automatic drain valve	14	Pressure switch (air)
7	Pressure switch water	15	Check valve
8	Water motor gong	16	Solenoid valve

Typical deluge system

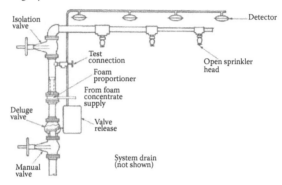

FIGURE 5.10
Schematic of a deluge fire sprinkler system.

Water supplies should preferably be under the control of the user, or guaranteed by the organization having control. The pressure and flow capacity of town mains are not guaranteed.

5.14.2 Frost Protection

The installation main control valve set and the feed pipe should be maintained at a minimum temperature of 4°C.

5.14.3 Quality

Water supplies for sprinkler installations should be free from suspended fibrous or other matter that accumulates in the system pipework.

Salt or brackish water should not be retained in sprinkler installation pipework.

Where there is no suitable freshwater source available, a salt or brackish water supply should be used provided the installation is normally charged with freshwater.

5.14.4 Ring Mains

Where sprinkler systems are fed by a ring main supply pipe arrangement on the premises, any isolating valves on the ring main should be of the interlocking key type. Isolating valves positioned at various suitable points on a ring main enable the supply to be maintained to the maximum possible extent in the event of fracture or other need to close down part of the ring main.

5.14.5 Housing of Equipment for Water Supplies

Equipment such as pumps, pressure tanks, and gravity tanks should not be housed in buildings or sections of premises in which there are hazardous processes or explosion hazards.

5.14.6 Types of Water Supply

Below is a low-rise system selection of suitable supplies.

Systems for light- and ordinary-hazard occupancies should be provided with the following:

a. A single supply
b. A superior supply
c. Duplicate supplies

Wherever practical, a superior supply or duplicate supplies should be provided. Systems for high-hazard occupancies should be provided with the following:

a. A superior supply
b. Duplicate supplies

5.14.6.1 Provision of Fire Brigade Inlet

Systems supplied only from a pressure tank and/or a gravity tank and/or a pump suction tank should, if possible, be fitted with a fire brigade inlet. It is strongly emphasized that a fire brigade inlet should be fitted to all systems to allow the brigade to pump water into the system using their own equipment. The water authority will not normally allow a fire brigade inlet on systems with town main supplies, because water from the inlet could enter the town main.

5.14.6.2 Single Supplies

A single supply should be one of the following:

a. A town main
b. A single automatic suction pump, drawing water from a source
c. A single automatic booster pump drawing water from a town main

5.14.6.3 Superior Supplies

A superior supply should be one of the following:

a. A town main

b. Two automatic suction pumps drawing water from a suction tank

c. Two automatic booster pumps

d. An elevated private reservoir

e. A gravity tank

f. A pressure tank for light-hazard and/or ordinary-hazard group 1 occupancies only

5.14.6.4 Duplicate Supplies

Duplicate supplies should include at least one of the suitable combinations given in Table 5.3. The supply pipes from each source should be joined into a common trunk main at a point as close as possible to the protected premises.

The common trunk main should neither traverse ground not under the control of the user nor be under a public roadway. The common trunk main should serve more than one installation in a system.

5.14.6.5 High-Rise Systems

The water supply for a high-rise system should be either

a. A gravity tank

b. An automatic suction pump arrangement in which each installation is served by either a separate pump or a separate stage of a multi-stage pump

---

5.15 Design Density and Assumed Maximum Area of Operation for Fully Hydraulically Calculated Installations

For fully hydraulically calculated installations, the density of discharge calculated should not be less than the appropriate value when all the ceiling or roof sprinklers in the room concerned, or in the assumed maximum area of operation (AMAO), whichever is fewer, plus any supplementary sprinklers, and/or sprayers, and/or medium-velocity sprayers, and/or high-velocity sprayers, installed below the roof or ceiling sprinklers considered to be in operation, are in operation.

The basis for full hydraulic calculation for all fire hazard classes is the specification of a minimum design density from a group of sprinklers (four in number if the number in open communication is four or more) in a specified geometric pattern. This group is the most hydraulically remote from

TABLE 5.3

Combinations Suitable for Duplicate Water Supplies

Supply	Town main with or without pump	Booster pump from elevated private reservoir	Suction pump	Gravity tank	Elevated private reservoir	Pressure tank
Pressure Tank	Light Ordinary	Light Ordinary	Light Ordinary	Suitable only with a third supply given as suitable	Light Ordinary	Suitable only with a third supply given as suitable
Elevated Private Reservoir	Not suitable	Light Ordinary	Light Ordinary High	Light Ordinary High	Light Ordinary High	
Gravity Tank	Not suitable	Light Ordinary	Light Ordinary High	Light Ordinary High (a divided tank or two separate tanks may be used)		
Suction Pump	Not usually permitted	Light Ordinary High	Light Ordinary High			
Booster Pump from Elevated Private Reservoir	Not usually permitted	Light Ordinary High				
Town Main with or without Booster Pump	Light Ordinary High					

the water supply and is part of a larger group of sprinklers assumed to be discharging simultaneously. The larger group is the AMAO and is specified for each hazard class. The hydraulically most unfavorably AMAO is used to calculate design density.

5.15.1 Light Hazard

The design density and the AMAO of roof or ceiling sprinklers should not be less than as given in Table 5.4.

Not more than six sprinklers should be installed in a room, except in either a corridor where there is a single line of sprinklers or a concealed space protected.

5.15.2 Ordinary Hazard

The design density and AMAO of roof or ceiling sprinklers for each of the various groups should be not less than as given in Table 5.4.

TABLE 5.4

Minimum Design Density and AMAO for Light-, Ordinary-, and High-Hazard (Processes) Roof or Ceiling Sprinklers

Hazard	Minimum Design Density (mm/min)	AMAO (m²)
Light	2.25	84
Ordinary		
Group I	5	72
Group II	5	144
Group III	5	216
High (Processes)		
Type 1	7.5	260
Type 2[a]	10	260
Type 3	12.5	260
Type 4	10	Complete deluge protection for each building

[a] Type 2 category III in Table 4 of BS 5306 Pt 2 for process of high hazards includes flammable liquids in combustible containers, rubber goods, and wooden pallets and wooden flats (idle).

5.15.3 High Hazard (Process Hazard)

The minimum design density and AMAO should not be less than as given in Table 5.4.

5.15.4 High-Piled Storage Hazards (Goods), Storage Classes S1 and S4

Where the storage height exceeds that for ordinary hazard, the design density and AMAO of roof or ceiling sprinklers should be not less than the appropriate value given in Table 5.5.

Note that Classes S1 thru S8 are the types of storage methods as given hereunder:

- S1: Free-standing or block stacking
- S2: Post or box pallets in single rows
- S3: Post or box pallets in multiple rows
- S4: Open-bottom post pallets
- S5: Palletized rack (beam pallet racking)
- S6: Solid or slatted shelves 1 m or less wide
- S7: Solid or slatted shelves over 1 m and not more than 6 m wide
- S8: Solid or slatted shelves over 1 m wide where intermediate sprinklers cannot be installed

5.15.5 High-Piled Storage Hazards (Goods), Storage Classes S2 and S5

Where the storage height exceeds that for ordinary hazard, the design density and AMAO should not be less than the appropriate value given in Table 5.6.

5.15.6 High-Piled Storage Hazards (Goods), Storage Classes S3, S6, S7, and S8

For goods of classes S3 and S6 stored to heights above 5.7 m and for goods of classes S7 and S8 the design density and AMAO should not be less than as given in Table 5.7.

High-piled storage hazards (goods) and roof or ceiling protection where intermediate sprinklers are fitted in racks or under shelves.

Where roof or ceiling sprinklers are more than 3 m above the top of the goods, the roof or ceiling sprinklers should have a design density of not less than 7.5 mm minimum and an AMAO of not less than 260 m^2 and intermediate sprinklers should be provided at each tier, including the top tier, of storage.

TABLE 5.5

Minimum Design Density and AMAO for High-Piled Storage Hazards (Goods), Storage Types S1 and S4 Roof or Ceiling Sprinklers

Category I S1 Only Stack Height		Category II S1 Only Stack Height		Category III S1 and S4 (Class S4 Includes Only Category III Rubber Types) Stack Height		Category IV S1 Only Stack Height		Minimum Design Density (mm/min)	AMAO	
More Than (m)	Not More Than (m)	More Than (m)	Not More Than (m)	More Than (m)	Not More Than (m)	More Than (m)	Not More Than (m)		Wet Pipe, Preaction, and Recycling Systems (m²)	Dry Pipe and Alternate Systems (m²)
0	5.3	0	4.1	0	2.9	0	1.6	7.5	260	325
5.3	6.5	4.1	5	2.9	3.5	1.6	2	10	260	325
6.5	7.6	5	5.9	3.5	4.1	2	2.3	12.5	260	325
—	—	5.9	6.7	4.1	4.7	2.3	2.7	15	260	325
—	—	6.7	7.5	4.7	5.2	2.7	3	17.5	260	325
—	—	—	—	5.2	5.7	3	3.3	20	300	375
—	—	—	—	5.7	6.3	3.3	3.6	22.5	300	375
—	—	—	—	6.3	6.7	3.6	3.8	25	300	375
—	—	—	—	6.7	7.2	3.8	4.1	27.5	300	375
—	—	—	—	—	—	4.1	4.4	30	300	375

TABLE 5.6

Minimum Design Density and AMAO for High-Piled Storage Hazards (Goods), Storage Types S2, and S5 Roof or Ceiling Sprinklers

Category I Stack Height		Category II Stack Height		Category III Stack Height		Category IV Stack Height		Minimum Design Density (mm/min)	AMAO Wet Pipe, Preaction, and Recycling Systems (m²)	AMAO Dry Pipe and Alternate Systems (m²)
More Than (m)	Not More Than (m)	More Than (m)	Not More Than (m)	More Than (m)	Not More Than (m)	More Than (m)	Not More Than (m)			
0	4.7	0	3.4	0	2.2	–	–	7.5	260	325
4.7	5.7	3.4	4.2	2.2	2.6	1.6	2	10	260	325
5.7	6.8	4.2	5	2.6	3.2	2	2.3	12.5	260	325
–		5	5.6	3.2	3.7	2.3	2.7	15	260	325
–		5.6	6	3.7	4.1	2.7	3	17.5	260	325
–				4.1	4.4	3	3.3	20	300	375
–				4.4	5.3	3.3	3.8	25	300	375
–				5.3	6	3.8	4.4	30	300	375

TABLE 5.7

Minimum Design Density and AMAO for High-Piled Storage Hazards (Goods), Storage Types S3, S6, S7, and S8 Roof or Ceiling Sprinklers

Category I Stack Height		Category II Stack Height		Category III Stack Height		Category IV Stack Height		Minimum Design Density (mm/min)	AMAO Wet Pipe, Preaction, and Recycling Systems (m²)	Dry Pipe and Alternate Systems (m²)
More Than (m)	Not More Than (m)	More Than (m)	Not More Than (m)	More Than (m)	Not More Than (m)	More Than (m)	Not More Than (m)			
0	4.7	0	3.4	0	2.2	0	1.6	7.5	260	325
4.7	5.7	3.4	4.2	2.2	2.6	1.6	2	10	260	325
–	–	4.2	5	2.6	3.2	2	2.3	12.5	260	325
–	–	–	–	–	–	2.3	2.7	15	260	325
–	–	–	–	–	–	2.7	3	17.5	260	325

TABLE 5.8

Minimum Design Density and AMAO for High-Piled Storage Hazards (Goods), Top-Tier Protection by Roof or Ceiling Sprinklers Only

| Category I | Category II | Category III | | Category IV | | | | AMAO, Wet Pipe, |
| | | Stack Height | | Stack Height | | | | |
Stack Height Not More Than (m)	Stack Height Not More Than (m)	More Than (m)	Not More Than (m)	More Than (m)	Not More Than (m)	Minimum Design Density	Preaction, and Recycling Systems (m²)
4.7	3.5	–	2.2	–	1.6	7.5	260
–	–	2.2	2.6	1.6	2	10	260
–	–	2.6	3.2	2	2.3	12.5	260
–	–	3.2	3.5	–	–	15	260

Where roof or ceiling sprinklers are not more than 3 m above the top of the goods, the roof or ceiling sprinklers should have a design density and an AMAO not less than as given in Table 5.8 and intermediate sprinklers should be provided at each tier, except the top tier, of storage.

5.16 Water Supply Pressure-Flow Characteristics and Velocity

- *Application.* For the purposes of this section, requirements applicable to wet pipe installations also apply to preaction and recycling instal-lations and requirements applicable to alternate installations apply also to dry pipe, tail-end dry pipe, and tail-end alternate installations.
- *High hazards, extra sprinklers.* Where additional sprinklers are installed because of obstructions within racks or to protect columns within storage areas, the water supply for the extra sprinklers within the AMAO should be added to that for the normal installation.

5.16.1 Precalculated Pipe Size Installations

5.16.1.1 Light Hazard

The water supply running pressure at the C gage should be not less than 2.2 bar plus the static pressure equivalent of the height of the highest sprinkler in the installation above the C gage when a water flow rate of 225 L/min is established through the drain and test valve.

5.16.1.2 Ordinary Hazard

The water supply running pressure at each section control in a high-rise installation, or at the C gage in a low-rise installation, should not be less than

TABLE 5.9

Pressure and Flow Requirements for Ordinary-Hazard Installations

	Lower Flow Rate		Higher Flow Rate	
Hazard Group	Pressure at C Gage or Section Stop Valve (bar)	Flow Rate through Installation Test Valve (L/min)	Pressure at C Gage or Section Stop Valve (bar)	Flow Rate through Installation Test Valve (L/min)
I	1+S*	375	0.7+S*	540
II	1.4+S*	725	1+S*	1000
III	1.7+S*	1100	1.4+S*	1350

Note: S* is the static pressure difference between the C gage and the highest sprinkler in the installation.

that specified in Table 5.9 when the higher and lower water flow rates are established through the drain and test valve.

5.16.1.3 High-Rise Installations

Each installation rise pipe should be provided with a jockey pump to maintain the static pressure at any check or alarm valve at not less than 1.25 times the static head difference between the valve and the highest sprinkler in the installation. The jockey pump should not be so large as to prevent the operation of suction or booster pumps when a single sprinkler operates.

5.16.2 High Hazard

The water supply running pressure at the control valve gage should not be less than either

a. Where the AMAO is not larger than the area protected Pr + Pf + Ps
b. Where the AMAO is larger than the area protected Pred + Pf + Ps

where

Pred is the running pressure specified in Tables 5.10, 5.11, 5.12, and 5.13 as appropriate at a flow rate equal to area protected times the flow rate specified in the AMAO table (in bar)

Pr is the running pressure at the design point and flow rate specified in Tables 5.10, 5.11, 5.12, and 5.13 as appropriate (in bar)

Pf is the calculated pipe friction loss between the control valve gage and the most hydraulically remote design point (in bar)

TABLE 5.10

Pressure and Flow Requirements for High-Hazard Installations, with 15-mm Sprinklers (Precalculated) and Pipe Sizes from Tables 5.16 and 5.18

| | Flow Rate through Installation Test Valve | | Running Pressure at the Design Point at the Level of the Highest Sprinkler in the High-Hazard Area | | | | | | |
| | | | Floor Area per Sprinkler (m²) | | | | | | |
Minimum Design Density (mm/min)	Wet Pipe, Preaction, and Recycling Installations (L/min)	Alternate and Dry (Including Tail and) Installations (L/min)	6 bar	7 bar	8 bar	9 bar	10 bar	11 bar	12 bar
7.5	2300	2875	–	–	1.8	2.25	2.8	3.35	3.95
10	3050	3825	1.8	2.4	3.15	3.9	4.8	5.75	6.8
12.5	3800	4750	2.7	3.65	4.75	6	7.3	–	–
15	4550	5700	3.8	5.2	6.75	–	–	–	–

TABLE 5.11

Pressure and Flow Requirements for High-Hazard Installation, with 15-mm Sprinklers (Precalculated) and Pipe Sizes from Tables 5.16 and 5.19

Minimum Design Density (mm/min)	Flow Rate through Installation Test Valve		Running Pressure at the Design Point at the Level of the Highest Sprinkler in the High-Hazard Area						
	Wet Pipe, Preaction, and Recycling Installations (L/min)	Alternate and Dry (Including Tail and) Installations (L/min)	Floor Area per Sprinkler (m²)						
			6 bar	7 bar	8 bar	9 bar	10 bar	11 bar	12 bar
7.5	2300	2875	–	–	1.35	1.75	2.15	2.65	3.15
10	3050	3825	1.3	1.8	2.35	3	3.75	4.55	5.45
12.5	3800	4750	2	2.75	3.6	4.6	5.7	7	8.35
15	4550	5700	2.8	3.85	5.1	6.5	–	–	–

TABLE 5.12

Pressure and Flow Requirements for High-Hazard Installations, with 15-mm Sprinklers (Precalculated) and Pipe Sizes from Table 5.16

Minimum Design Density (mm/min)	Flow Rate through Installation Test Valve		Running Pressure at the Design Point at the Level of the Highest Sprinkler in the High-Hazard Area						
	Wet Pipe, Preaction, and Recycling Installations (L/min)	Alternate and Dry (Including Tail and) Installations (L/min)	Floor Area per Sprinkler (m²)						
			6 bar	7 bar	8 bar	9 bar	10 bar	11 bar	12 bar
7.5	2300	2875	–	–	0.7	0.9	1.1	1.35	1.6
10	3050	3825	0.7	0.95	1.25	1.6	1.95	2.35	2.8
12.5	3800	4750	1.1	1.5	1.95	2.45	3.05	3.7	4.35
15	4550	5700	1.6	2.15	2.8	3.55	4.35	5.25	6.25
17.5	4850	6075	2.15	2.9	3.8	4.8	5.9	7.15	–
20	6400	8000	2.8	3.8	5	6.3	7.75	–	–
22.5	7200	9000	3.5	4.8	6.3	7.95	–	–	–
25	8000	10,000	4.35	5.9	7.75	–	–	–	–
27.5	8800	11,000	5.25	7.15	–	–	–	–	–
30	9650	12,100	6.2	–	–	–	–	–	–

TABLE 5.13

Pressure and Flow Requirements for High-Hazard Installations, with 20-mm Sprinklers (Precalculated) and Pipe Sizes from Table 5.17

Minimum Design Density (mm/min)	Flow Rate through Installation Test Valve		Running Pressure at the Design Point at the Level of the Highest Sprinkler in the High-Hazard Area							
	Wet Pipe, Preaction, and Recycling Installations (L/min)	Alternate and Dry (Including Tail and) Installations (L/min)	Floor Area per Sprinkler (m²)							
			6 bar	7 bar	8 bar	9 bar	10 bar	11 bar	12 bar
7.5	2300	2875	–	–	–	–	–	0.8	0.95
10	3050	3825	–	–	–	0.95	1.15	1.4	1.65
12.5	3800	4750	–	0.9	1.15	1.45	1.8	2.15	2.55
15	4550	5700	0.95	1.25	1.65	2.1	2.55	3.1	3.65
17.5	4850	6075	1.25	1.7	2.25	2.8	3.45	4.2	4.95
20	6400	8000	1.65	2.25	2.95	3.7	4.6	5.55	6.55
22.5	7200	9000	2.05	2.85	3.7	4.7	5.75	6.95	–
25	8000	10,000	2.55	3.5	4.55	5.75	7.1	–	–
27.5	8800	11,000	3.05	4.2	5.5	6.9	–	–	–
30	9650	12,100	3.6	4.95	6.5	–	–	–	–

Ps is the static pressure difference between the highest sprinkler downstream of the design point and the control valve gage (in bar)

c. Where AMAO is fed by more than one distribution pipe, the pipe friction loss should be calculated on the basis that the flow rates in the distribution pipes are in proportion to the fraction of the design area fed by each distribution pipe

5.16.3 Mixed High/Ordinary Hazard

Where the area of the high-hazard protection is less than the AMAO but there is an adjacent area of ordinary hazard in the same room (i.e., an area in which sprinklers are liable to operate simultaneously), the high-hazard area flow rate required should be reduced by the ratio of the actual area to the AMAO and to this flow rate should be added the flow rate for the ordinary hazard area taken as 5.0× the excess of the specified high-hazard AMAO over the actual high-hazard area (L/min).

The water supply running pressure should be based on the level of the highest sprinkler in the high-hazard area.

The high-hazard distribution pipe feeding both high-and ordinary-hazard sprinklers should be of a bore not less than as specified in the ordinary-hazard pipe tables.

5.16.4 Fully Hydraulically Calculated Pipe Arrays

The requirements of this section apply to pipe arrays sized by full hydraulic calculation.

5.16.4.1 Minimum Pipe Sizes

The nominal bore of main and other distribution pipes and range pipes should not be less than as follows:

  a. In a light-hazard installation, (20 mm [steel], or 22 mm [copper], or as given in Table 5.14 for single sprinklers)
  b. In an ordinary or high-hazard installation (25 mm)

5.16.4.2 Maximum Range Pipe Size

The nominal bore of range pipes should not be more than 65 mm, except where individual sprinklers are connected to pipes exceeding 65 mm nominal bore when the arrangement should comply with 24.1.2 of BS 5306/Part 2.

5.16.4.3 Installation Pipework

All pipework downstream of a main distribution and range pipes should be designed in accordance with the relevant requirements of 20.2.

TABLE 5.14

Range Pipe Nominal Sizes for Various Pipe Layouts in Ordinary-Hazard Installation

Range Pipe Layout	Pipe Nominal Size (mm)	Maximum Number of Sprinklers to be Fed by Pipe of Size Listed
(a) Ranges at Remote End of Each Distribution Pipe Spur		
(1) Last two ranges in two end-side layout	25	1
	32	2
(2) Last three ranges in three end-side layout	25	2
	32	3
(3) Last range in all other layouts	25	2
	32	3
	40	4
	50	9
(b) All other ranges	25	3
	32	4
	40	6
	50	9

Range pipe sizes and the maximum number of sprinklers fed by each size of pipe in the range should be neither more nor less than as specified hereunder:

a. Light hazard (Table 5.15)

b. Ordinary hazard (Table 5.14)

c. High hazard (Tables 5.16 through 5.19)

TABLE 5.15

Light-Hazard Range Pipe and Terminal Distribution Pipe Sizes

Pipe Material	Nominal Size (mm)	Maximum Length[a]	Maximum Number of Sprinklers Allowed on Pipe of Size Stated
Copper	15	1 (No elbows fitted; 500 mm if one elbow fitted)	1
	22	8	1
	28	–	3 (The limit of three sprinklers does not preclude the use of 25-nominal bore steel or 28-mm copper pipe between the 2/3 sprinkler design point and the installation control valves if hydraulic calculation shows this to be possible, nor does it follow that 25-mm steel or 22-mm copper pipe may be used between the third and fourth sprinklers where the two-sprinkler point is the design point)
Steel	20	8	1
	25	–	3 (The limit of three sprinklers does not preclude the use of 25-mm nominal bore steel or 28-mm copper pipe between the 2/3 sprinkler design point and the installation control valves if hydraulic calculation shows this to be possible, nor does it follow that 25-mm steel or 22-mm copper pipe may be used between the third and fourth sprinklers where the two-sprinkler point is the design point)

[a] Including allowance for changes of direction (m).

TABLE 5.16

Range Pipe Nominal Sizes for Various Pipe Layouts, for High-Hazard Installations with Sprinklers of 15-mm Nominal Size and Pressure-Flow Characteristics as Given in Table 5.10 or Table 5.11

Range Pipe Layout	Pipe Nominal Size (mm)	Maximum Number of Sprinklers to Be Fed by Pipe of Size Listed
(a) Ranges at Remote End of Each Distribution Pipe Spur		
(1) Two end-side, last two ranges	25	1
	32	2
(2) Two end-side, last three ranges	25	2
	32	3
(3) All other layouts, last range	25	2
	32	3
	40	4
(b) All other ranges	25	3
	32	4

Note: This table gives pipe sizes referred to in Tables 5.10, 5.11, and 5.12.

TABLE 5.17

Range Pipe Nominal Sizes for Various Pipe Layouts, in High-Hazard Installations with Sprinklers of 15-mm Nominal Size and Pressure-Flow Characteristics as Given in Table 5.10, or of 20-mm Nominal Size and Pressure-Flow Characteristics as Given in Table 5.13

Range Pipe Layout	Pipe Nominal Size (mm)	Maximum Number of Sprinklers to Be Fed by Pipe of Size Listed
(a) End-side arrangements		
(1) Last three ranges	40	1
	50	3
	65	6
(2) Other ranges	32	1
	40	2
	50	4
	65	6
(b) End-center arrangements		
(1) Two end-center layout		
(i) Last three ranges	32	1
	40	2
(ii) Other ranges	32	2 (32 mm feed to each)
(2) Three and four end-center	32	1
layouts, all ranges	40	2
	50	4

Note: This table gives pipe sizes referred to in Table 5.13.

TABLE 5.18

Distribution Pipe Nominal Sizes Feeding Various Numbers of Sprinklers Downstream of the Design Point, in High-Hazard Installations with Sprinklers of 15-mm Nominal Size and Pressure Flow Characteristics as Given in Table 5.10

Distribution Pipe, Nominal Size (mm)	Maximum Number of Sprinklers to Be Fed by Pipe of Size Listed
32	2
40	4
50	8
65	12
80	18
100	48[a]

Note: This table gives pipe sizes referred to in Table 5.10.

[a] This does not preclude the use of 100-mm nominal size pipe between the design point and the installation main control valves if it complies with the hydraulic calculation requirements.

TABLE 5.19

Distribution Pipe Nominal Sizes Feeding Various Numbers of Sprinklers Downstream of the Design Point in High-Hazard Installations with Sprinklers of 15-mm Nominal Size and Pressure Flow Characteristics as Given in Table 5.9 or Table 5.10 or Sprinklers of 20-mm Nominal Size and Pressure-Flow Characteristics as Given in Table 5.11

Range Pipe Layout	Distribution Pipe Nominal Size (mm)	Maximum Number of Sprinklers to Be Fed by Pipe of Size Listed
Four end-side	50	4
All other layouts	65	8
	80	12
	100	16
	150	48[a]

Note: This table gives pipe sizes referred to in Table 5.11.

[a] This does not preclude the use of 150-mm nominal size pipe between the design point and the installation main control valves if it complies with hydraulic calculation requirements.

5.16.5 Fully Hydraulically Calculated Pipe Size Installation

5.16.5.1 Velocity

The equilibrium water velocity should not exceed 6 m/s at any valve or flow monitoring device, or 10 m/s at any other point in the system for the stabilized flow condition at the demand point involving an AMAO, or, where the system includes intermediate sprinklers, the total number of sprinklers assumed to be in simultaneous operation.

5.16.6 Calculation of Pipework Losses

5.16.6.1 *Static Pressure Difference*

The static pressure difference between two interconnecting points in a system should be calculated from

$$\text{Static pressure difference, } p = 0.1\,h \text{ (bar)}$$

where
  h is the vertical distance between the points (in m).

5.16.6.2 *Pipe Friction Loss*

Frictional pressure loss in pipes should be calculated from the Hazen-Williams formula:

$$p = 6.05\left(\frac{Q^{1.85}}{C^{1.85}\times d^{4.87}}\right)10^5$$

where
  p = pressure loss in bar per meter
  Q = flow through the pipe in L/min
  C = friction loss coefficient
  d = internal diameter of the pipe in millimeters

The valves of C are listed in Table 5.20 and should be used in sprinkler installation and town main calculations.

TABLE 5.20

Friction Loss Coefficient

Pipe Type	Friction Loss Coefficient
Cast iron	100
Ductile iron	110
Mild steel	120
Galvanized steel	120
Spun cement	130
Copper	140
Unplasticized PVC	140
Asbestos cement	140

5.17 Temperature Ratings, Classifications, and Color Coding

The standard temperature ratings of automatic sprinklers are shown in Table 5.21. Automatic sprinklers should have their frame arms colored in accordance with the color code designated in Table 5.21, with the following exceptions:

1. The color identification for corrosion-resistant sprinklers may be a dot on the top of the deflector, the color of the coating material, or colored frame arms
2. Color identification is not required for ornamental sprinklers such as factory-plated or factory-painted sprinklers, or for recessed, flush, or concealed sprinklers

Ordinary temperature-rated sprinklers should be used throughout buildings, with the following exceptions:

1. Where maximum ceiling temperatures exceed (38°C), sprinklers with temperature ratings in accordance with the maximum ceiling temperatures of Table 5.21 should be used
2. Intermediate- and high-temperature sprinklers should be used throughout ordinary and extra-hazard occupancies
3. When permitted or required by other NFPA standards

The following practices should be observed to provide sprinklers of other than ordinary-temperature classification unless maximum expected temperatures are otherwise determined, or unless high-temperature sprinklers are used throughout:

a. *Sprinklers near unit heaters:* Sprinklers in the heater zone should be in the high-temperature classification and sprinklers in the danger zone should be in the intermediate-temperature classification

TABLE 5.21

Temperature Ratings, Classifications, and Color Coding

Maximum Ceiling Temperature (°C)	Temperature Rating (°C)	Temperature Classification	Color Code Frame Arm	Glass Bulb Colors
38	57–77	Ordinary	Uncolored or black	Orange or red
66	79–107	Intermediate	White	Yellow or green
107	121–149	High	Blue	Blue
149	163–191	Extra high	Red	Purple
191	204–246	Very extra high	Green	Black
246	260–302	Ultra high	Orange	Black
329	343	Ultra high		Black

b. Sprinklers located within 305 mm to one side or 762 mm above an uncovered steam main, heating coil, or radiator should be in the intermediate-temperature classification

c. Sprinklers within 2.1 m of a low-pressure blowoff valve that discharges freely in a large room should be in the high-temperature classification

d. Sprinklers under glass or plastic skylights exposed to the direct rays of the sun should be in the intermediate-temperature classification

---

5.18 Hazard to Personnel

5.18.1 High-Temperature Liquids

Sprinklers should not be installed in locations where water discharge from sprinkler heads or sprayers or water leakage from installation pipework may come into contact with high-temperature liquid baths such as salt baths, metal melt pans, frying ranges, and hot-dip bitumen baths.

5.18.2 Water-Reactive Chemicals

The consequences of sprinkler water discharge onto water-reactive chemicals below should be considered and safe storage and usage practices adopted.

Water discharge may ignite certain chemicals or cause a violent reaction and/or emission of poisonous or noxious fumes.

Where water runoff from sprinkler discharge is dangerous owing to contact with water-soluble chemicals or by waterborne dispersion of hazardous materials, construction of suitable drains, sumps, bunds, and so forth should be considered at the planning stage.

5.18.3 Electrical Earthing

All exposed metalwork in systems should be efficiently earthed to prevent the metalwork from becoming electrically charged.

Sprinkler pipework should not be used as a means of earthing electrical equipment.

5.18.4 During Maintenance

Where sprinkler or deluge installation pipework is normally unpressurized, work should not be undertaken involving removal or fitting of sprinkler heads or water sprayers when operation of the installation is possible unless measures are taken to ensure the safety of the erection personnel concerned.

5.19 Outside Sprinklers for Protection against Exposure Fires

Sprinklers installed for protection against exposure fires should be supplied from a standard water supply.

The water supply should be capable of furnishing the total demand for all exposure sprinklers operating simultaneously for protection against the exposure fire under consideration for a duration of not less than 60 minutes.

When automatic systems of sprinklers are installed, water supplies should be from an automatic source.

When fire department connections are used for water supply, they should be located where they will not be affected by the exposing fire.

5.19.1 Control

Each system of outside sprinklers should have an independent control valve. When more than one system is required, the division between systems should be vertical and not horizontal, with the following exceptions:

- When more than six lines are installed, the systems should be divided horizontally with independent risers
- Manually controlled open sprinklers should be used only where constant supervision is present
- Automatic systems of open sprinklers should be controlled by the operation of fire detection devices designed for the specific application

5.19.2 Type

Small orifice sprinklers will normally be used where exposure is light or moderate, the area of coverage is small, or where one horizontal line of window sprinklers is installed at each floor level.

Large orifice sprinklers will normally be used where exposure is severe, or where one horizontal line of window sprinklers is used to protect windows at more than one floor level.

5.19.3 Window Sprinklers

When the exposure hazard is light or moderate, and only one horizontal line of sprinklers is installed, the sprinklers should have 9.5-mm orifices. Where conditions require more than one line of sprinklers, the sprinklers should have orifices as shown in Table 5.22.

Where there are more than six horizontal rows of windows, sprinklers over the first story should be omitted. Sprinklers should also be omitted over the second story windows if a field test indicates wetting of all surfaces.

TABLE 5.22

Orifice Sizes for Sprinklers

	Two Lines (mm)	Three Lines (mm)	Four Lines (mm)	Five Lines (mm)	Six Lines (mm)
Top Line	9.5	9.5	9.5	9.5	9.5
Next below	8	8	9.5	9.5	9.5
Next below	–	6.5	8	8	8
Next below	–	–	6.5	8	8
Next below	–	–	–	6.5	6.5
Next below	–	–	–	–	6.5

Large orifice sprinklers should be used for protecting windows in two or three stories from one line of sprinklers. This will be determined by window and wall construction, such that all parts of the windows and frames will be thoroughly wetted by a single line of sprinklers.

For buildings not over three stories in height, one line of sprinklers will often be sufficient, located at the top-story windows. For buildings more than three stories in height, a line of sprinklers should be used in every other story beginning at the top.

With an odd number of stories, the lowest line can protect the first three stories. When several lines are used, the orifice should be decreased one size for each successive line below the top. In no case should an orifice less than 13 mm be used.

For windows not exceeding 1.5 m wide protected by small orifice sprinklers, one sprinkler should be placed at the center near the top, and located so that water discharged therefrom will wet the upper part of the window, and by running down over the frame and glass, wet the entire window. This may ordinarily be accomplished by placing one sprinkler in the center with the deflector about on a line with the top of the upper sash and 178, 203, and 229 mm in front of the glass, with windows 0.9, 1.2, and 1.5 m wide, respectively. When windows are over 1.5 m wide, or where mullions interfere, two or more sprinklers should be used.

When windows are 0.9 m or less in width, an orifice a size smaller than that required by 28.5 should be used, but in no case should the orifice be smaller than 6.4 mm.

For windows up to 1.5 m wide protected by large orifice sprinklers, use one 13-mm sprinkler at the center of each window. For windows from 1.5 to 2.1 m wide, use a 16-mm sprinkler at the center of each window. For windows from 2.1 to 2.9 m wide, use one 19-mm sprinkler at the center of each window. For windows from 2.9 to 3.7 m wide, use two 13-mm sprinklers at each window.

Large orifice wide-deflector sprinklers should be placed with deflectors 51 mm below the top of the frame and 305 to 380 mm out from the glass.

TABLE 5.23

Space for Window Sprinklers

9.5-mm and 13-mm sprinklers	Not more than 1.5 m apart
16-mm sprinklers	Not more than 2.1 m apart
19-mm sprinklers	Not more than 2.7 m apart

When the face of the glass is close to the exterior wall, cantilever brackets or similar type hangers should be used to maintain the window sprinklers 305 to 380 mm out from the glass.

5.19.4 Cornice Sprinklers

The discharge orifice should be at least 9.5 mm in diameter except when the exposure is severe; then 13-mm or 16-mm cornice sprinklers should be installed.

Sprinklers should not be more than 2.4 m apart, except as noted in 22.1.8.3, and projecting beams or other obstructions should make additional sprinklers necessary.

For cornices with bays up to 2.4 m wide, sprinklers should be placed in the center of each bay. For cornices with bays from 2.4 to 3.0 m wide, sprinkler orifices should be increased one size.

Cornice sprinklers should be located with deflectors approximately 203 mm below the roof plank.

When wood cornices are 762 mm or less above the windows, cornice sprinklers should be supplied by the same pipe used for window sprinklers.

Where the overhang of the cornice is not over 0.3 m, window sprinklers should be used and be spaced as listed in Table 5.23.

The window sprinklers should be placed above the pipe near the outer edge of the cornice with deflectors not more than 76 mm down from the cornice and at such an angle as to throw the water upward and inward. With an overhang of more than 0.3 m, cornice sprinklers should be used.

5.20 Deluge Foam-Water Sprinkler and Foam-Water Spray Systems

This book covers the minimum requirements for open-head deluge-type foam-water sprinkler systems and foam-water spray systems, each of which combines in a single system provision for the alternate discharge of foam or water.

Accordingly, systems can be designed with the required density for either foam or water application as the controlling factor, depending on the design purpose of the protection.

The devices covered herein are intended primarily for use in foam-water deluge sprinkler systems, or foam-water spray systems. This standard is not applicable where separate foam, water sprinkler, or water-spray fixed systems are to be installed.

The purpose of this section is to provide a reasonable degree of protection for life and property from fire through installation requirements for foam-water deluge sprinkler systems and foam-water spray systems based on sound engineering principles, test data, and field experience.

5.20.1 System Design

Automatic operation should be provided and supplemented by auxiliary manual tripping means, with the following exceptions. Manual operation only can be provided when acceptable to the authorities concerned.

- Systems should deliver foam for a definite period at given densities ([L/min]/m²) to the hazards they protect, either prior to water discharge or following water discharge, depending on system-design purpose.
- Following completion of discharge of foam to the hazards protected, these special systems should discharge water until manually shut off.
- Authorities should be consulted as to the means by which a reserve supply of foam concentrate should be made available. The purpose of a reserve supply of concentrate is to have available the means for returning systems to service ready condition following system operation. The reserve supply should be listed for use with system components.

5.20.2 Applicability

Systems of this type should discharge foam or water from the same discharge devices. In view of this dual extinguishing agent discharge characteristic, these systems are selectively applicable to combination class A and class B hazards.

(NOTE: Caution must be exercised when auxiliary extinguishing equipment is used with these systems. Some extinguishing agents will be incompatible with some foams).

Foam-water deluge systems are especially applicable to the protection of most flammable-liquid hazards. They can be used for any of the following purposes or combinations thereof:

a. *Extinguishment.* The primary purpose of such systems is the extinguishment of fire in the protected hazard. For this purpose, suitable foam-solution discharge densities ([L/min]/m^2) should be provided by system design and use of selected discharge devices and by provision of adequate supplies of air-water at suitable pressures to accomplish the system design. Foam-discharge rates should be suitable for the design period and following depletion of foam concentrate supplies, to provide similar rates of water discharge from the system until shut off.

b. *Prevention.* Prevention of fire in the protected hazard is a supplemental feature of such systems. Manual operation of a system to selectively discharge foam or water from the discharge devices in case of accumulations of hazardous materials from spills in such occupancies as garages, aircraft hangars, petrochemical plants, or from other causes in the protected area, will afford protection against ignition pending cleanup measures. In such cases, a manual system operation can provide for foam coverage in the area with water discharge manually available.

c. *Control and exposure protection.* Control of fire to permit controlled burning of combustible materials where extinguishment is not practicable and exposure protection to reduce heat transfer from an exposure fire may be accomplished by water spray and/or foam from these special systems, the degree of accomplishment being related largely to the fixed discharge densities provided by the system design.

Foam of any type is not considered a suitable extinguishing agent on fires involving liquefied or compressed gases (e.g., butane, butadiene, propane) or on materials that will react violently with water (e.g., metallic sodium) or that produce hazardous materials by reacting with water, or on fires involving electrical equipment where the electrical nonconductivity of the extinguishing agent is of first importance.

Ordinary foam concentrates should not be used on fires in water-soluble solvents and polar solvents. Special alcohol-type concentrates are available for production of foams for protection of such hazards.

The design discharge rates for water or foam solution should provide densities of not less than (6.5 [L/min]/m^2) of protected area. This minimum density is required because this is a dual-agent foam-water system.

The foam discharge should continue for a period of 10 minutes at the design rate. If the system discharges at a rate above the minimum, then the operating time may be reduced proportionately but should not be less than 7 minutes.

5.20.3 System Components

5.20.3.1 Approved Devices and Materials

All component parts, including foam concentrates of foam-water sprinkler and foam-water spray systems, should be listed for the intended application.

5.20.3.2 Discharge Devices

Discharge devices should be air aspirating such as foam-water sprinkler and foam-water spray nozzles, or they can be nonair-aspirating, such as standard sprinklers.

Discharge devices and foam concentrates should be listed for use together.

Nonair-aspirating devices should be used only with concentrates such as AFFF that have been tested and listed for use in these devices.

5.20.4 Foam Concentrates

- Foam concentrates should be listed for use with the concentrate pro-portioning equipment and with the discharge device to be used.
- The quantities of foam concentrates to be provided for foam-water sprinkler and spray systems should be sufficient to maintain the dis-charge densities for the application time period used as a base in system design.
- There should be a readily available supply of foam concentrate suf-ficient to meet the design requirements of the system to put the sys-tem back in service after operation. This supply can be in separate tanks or compartments, in drums or cans on the premises, or avail-able from an outside source within 24 hours.
- Replacement supplies of concentrates should be checked by appro-priate tests to determine acceptability.

5.20.5 Foam Concentrate Proportioning Means

Positive pressure-injection is the preferred method for introduction of foam concentrates into the water flowing through the supply piping to the system.

Positive pressure-injection methods will mean one of the following:

a. Foam concentrate pump discharging through a metering orifice into the protection-system riser, with the foam pressure at the upstream side of the orifice exceeding the water pressure in the system riser by a specific design value

b. A balanced-pressure proportioning system (demand-type propor-tioner) utilizing a foam concentrate pump discharging through a metering orifice into a proportioning controller (venturi) or orifice in the protection system riser, with the foam, liquid, and water pressures automatically maintained equal by the use of a pressure control valve

c. Pressure-proportioning tanks with or without a diaphragm to separate the water and foam concentrate

Orifice plates should have "tell-tale" indicators giving orifice diameters and indicating flow direction if flow characteristics vary with flow direction.

Where special conditions warrant, other proportioning methods can be used, such as around-the-pump proportioners and in-line inductors.

5.20.6 Pumps

Foam concentrate pumps and water pumps should have adequate capacities to meet the maximum needs of the system on which they are used. To ensure positive injection of concentrates, the discharge pressure ratings of pumps at the design discharge capacity should be suitably in excess of the maximum water pressure available under any condition at the point of concentrate injection. The excess pressure of 0.7 to 4 bar is recommended.

Foam concentrate pumps should be carefully chosen and have adequate capacity for this special service and special attention should be paid to the type of seals used with regard to the type of concentrate being pumped.

Provision should be made to shut off the foam concentrate pump after the foam supply is exhausted.

5.20.7 Power Supply

Power supply for the drivers of foam concentrate pumps and water pumps should be of maximum reliability. Compliance with the applicable requirements of NFPA 20, Standard for the Installation of Centrifugal Fire Pumps, covering the reliability of power supply for fire pump drivers, is considered to meet the intent of this standard.

Controllers governing the starting of foam concentrate pumps should be of approved types. Control equipment should comply with local standards for the Installation of centrifugal fire pumps.

5.20.8 Air Foam Concentrate Storage Tanks

Storage tanks for foam concentrates should be of construction suitable for the liquid, solidly mounted, and permanently located.

Storage temperatures of foam concentrates should be considered in locating storage tanks.

Storage tanks should have capacities to accommodate only the needed quantities of foam concentrate plus adequate space for thermal expansion, the latter preferably to be accomplished by means of a vertical riser or expansion dome. Tanks meeting this requirement will have minimum surface areas in contact with air and liquid concentrates at the liquid level and thus minimize the possibility of interior corrosion of tanks. Foam concentrate outlets from tanks should be raised above the bottoms of the tanks to provide adequate sediment pockets.

In determining the quantity of foam concentrates, the volume of the sediment pocket should be added to the quantity needed for system operation.

Tanks should be located to furnish a positive head on the pump suction.

5.20.9 Pressure on Foam Concentrate Lines

Where foam concentrate lines to the protective-system injection points are run underground or where they run aboveground for more than 15 m, foam concentrate in these lines should be maintained under pressure to assure prompt foam application and to provide a means of checking on the tightness of the system. Pressure can be maintained by a small auxiliary pump or by other suitable means.

5.20.10 Temperature of Foam Concentrate Lines and Components

Temperature of foam concentrate lines and components should be maintained within the storage temperature limits specified for the foam concentrate.

5.20.11 Location of System-Control Equipment

Equipment items such as storage tanks and proportioners for foam concentrates; pumps for water and foam concentrates, and control valves for water, concentrates, and foam solution should be installed where they will be accessible, especially during a fire emergency in the protected area and where there will be no exposure from the protected hazard.

Automatically controlled valves should be as close to the hazard protected as accessibility permits so that a minimum of piping is required between the automatic control valve and the discharge devices.

5.20.12 Alarms

A local alarm, actuated independently of water flow to indicate operation of the automatic detection equipment, should be provided on each system. An alarm is not required on manually operated systems.

When an alarm is installed, the IP authorities should be consulted regarding the alarm service to be provided and regarding the need for electrical fittings designed for use in hazardous locations in electric alarm installations.

A suitable trouble alarm should be provided for each system to indicate failure of automatic detection equipment (including electric supervisory circuits) or other such devices or equipment upon which the system operation is dependent.

5.20.13 Strainers for Water and Foam Concentrates

Strainers should be listed for fire protection service and should be capable of removing from the water all solids of sufficient size to obstruct the discharge devices. Strainers should be installed so as to be accessible for cleaning during an emergency. Space should be provided for basket removal.

Strainers should be installed in the main water supply lines feeding orifices (or water passage) smaller than 3/8 inch (9.6 mm). Strainers should be installed on systems having larger orifices where water supply conditions warrant. Normally 1/8-inch (3.2 mm) perforations are suitable.

Strainers should be installed in liquid concentrate lines upstream of metering orifices or proportioning devices. Where listed strainers of the proper size are not available, strainers having a ratio of open-basket area to inlet pipe size of at least 10 to 1 should be used.

5.20.14 Water Supplies

5.20.14.1 Types of Water

Water supplied to deluge foam-water sprinkler systems and foam-water spray systems shall be free of constituents not compatible with air foam concentrates.

5.20.14.2 Water Supply Capacity and Pressure

Water supplies for deluge foam-water sprinkler systems and foam-water spray systems should be of capacity and pressure capable of maintaining foam discharge and/or water discharge at the design rate for the required period of discharge over the entire area protected by systems expected to operate simultaneously.

Water supplies should be capable of supplying the systems at the design discharge capacity for at least 60 minutes.

5.20.15 Pipe Fittings

All fittings should be of a type specifically approved for fire protection systems and of a design suitable for the working pressures involved, but not less than 1207 kPa cold water pressures.

5.20.16 Automatic Detection

In automatic systems the detecting equipment should be connected to means for tripping water deluge valves and other system-control equipment. Supplemental manual means for accomplishment of this purpose should also be provided.

In automatic systems foam concentrate injection should be activated automatically by, or concurrently with, activation of the main water supply control valve. Manual operating means should be designed to accomplish this same purpose.

Automatic detection equipment, whether pneumatic, hydraulic, or electric, should be provided with complete supervision so arranged that failure of equipment, loss of supervising air pressure, or loss of electric energy will result in positive notification of the abnormal condition.

Where used in a corrosive atmosphere, the devices should be of materials not subject to corrosion or protected to resist corrosion.

5.20.17 Hydraulic Calculations

System piping should be hydraulically calculated and sized in order to obtain reasonably uniform foam and water distribution and to allow for loss-of-head in water supply piping. The adjustment in pipe sizes should be based on a maximum variation of 15% above the specified discharge rate per sprinkler or nozzle.

Pipe sizes should be adjusted according to detailed friction-loss calculations. These calculations should show the relation between the water supply and demand.

The friction losses in piping carrying foam concentrate should be calculated using the Darcy formula (also known as the Fanning formula). Friction factors for use with this formula should be selected from the chart Friction Factors for Commercial Steel and Cast-Iron Pipe.

In calculating the Reynolds number for selecting friction factors from the charts, the actual density (or specific gravity) of the foam concentrate to be used in the system should be used. The viscosity used should be the actual viscosity of the foam concentrate at the lowest anticipated storage temperature.

For purposes of computing friction loss in piping, the C factors in Table 5.24 should be used for the Williams and Hazen formula.

TABLE 5.24

More C Factors for the Williams and Hazen Formula

Black or galvanized-steel pipe	120
Unlined cast-iron pipe	100
Asbestos-cement or cement-lined cast iron	140

Further Readings

British Standards Institution (BSI). BS 5306 Part 2 Specification for Sprinkler System. 2009.

Chow, W.K. and Fong, N.K. 1991. Numerical simulation on cooling of the fire-induced air flow by sprinkler water sprays. *Fire Safety Journal* 17 (4), 263–290.

Crocker, J.P., Rangwala, A.S., Dembsey, N.A. and LeBlanc, D.J. 2010. Investigation of sprinkler sprays on fire induced doorway flows. *Fire Technology* 46 (2), 347–362.

Davies, G.F. and Nolan, P.F. 2004a. Experimental investigation of the parameters affecting the water coverage of a pressure vessel protected by a deluge system. *Journal of Loss Prevention in the Process Industries* 17 (2), 127–139.

Davies, G.F. and Nolan, P.F. 2004b. Characterisation of two industrial deluge systems designed for the protection of large horizontal, cylindrical LPG vessels. *Journal of Loss Prevention in the Process Industries* 17 (2), 141–150.

Figueroa, M. 2012. Fire sprinklers: Protecting lives, property and assisting water conservation efforts. *Journal—American Water Works Association* 104 (10), 1–28.

Garber, R.I. 2009. Fire sprinkler heads, design, and failure. *Journal of Failure Analysis and Prevention* 9 (6), 495–498.

Gritzo, L.A., Bill, R.G., Jr., Wieczorek, C.J. and Ditch, B. 2011. Environmental impact of automatic fire sprinklers: Part 1. Residential sprinklers revisited in the age of sustainability. *Fire Technology* 47 (3), 751–763.

Hoffman, N., Galea, E.R. and Markatos, N.C. 1989. Mathematical modelling of fire sprinkler systems. *Applied Mathematical Modelling* 13 (5), 298–306.

Jackman, L.A., Lavelle, S.P. and Nolan, P.F. 1991. Characteristics of sprinklers and water spray mists for fire safety. *Proceedings of SPIE—The International Society for Optical Engineering* 1358 (pt 2), 831–842.

Kung, H.-C., Song, B., Li, Y., Liu, X., Tian, L. and Yang, B. 2012. Sprinkler protection of non-storage occupancies with high ceiling clearance. *Fire Safety Journal* 54, 49–56.

Marshall, A.W. and Di Marzo, M. 2004. Modelling aspects of sprinkler spray dynamics in fires. *Process Safety and Environmental Protection* 82 (2 B), 97–104.

Megri, A.C. 2009. Primer on fire sprinkler installation for structural engineers. *Practice Periodical on Structural Design and Construction* 14 (2), 54–62.

Melinek, S.J. 1993. Potential value of sprinklers in reducing fire casualties. *Fire Safety Journal* 20 (3), 275–287.

Morgan, H.P. and Hansell, G.O. 1985. Fire sizes and sprinkler effectiveness in offices—Implications for smoke control design. *Fire Safety Journal* 8 (3), 187–198.

Nam, S. 2005. Fire tests to evaluate CPVC pipe sprinkler systems without fire resistance barriers. *Fire Safety Journal* 40 (7), 595–609.

National Fire Codes (NFC) (NFPA). NFC Volume 1 Installation of Sprinkler Systems, Sections 13 and 16. 1999.

Novozhilov, V., Harvie, D.J.E., Green, A.R. and Kent, J.H. 1997. A computational fluid dynamic model of fire burning rate and extinction by water sprinkler. *Combustion Science and Technology* 123 (1–6), 227–245.

O'Grady, N. and Novozhilov, V. 2009. Large eddy simulation of sprinkler interaction with a fire ceiling jet. *Combustion Science and Technology* 181 (7), 984–1006.

Ren, N., Blum, A.F., Zheng, Y.-H., Do, C. and Marshall, A.W. 2008. Quantifying the initial spray from fire sprinklers. *Fire Safety Science*, 503–514.

Ruffino, P. and DiMarzo, M. 2004. The simulation of fire sprinklers thermal response in presence of water droplets. *Fire Safety Journal* 39 (8), 721–736.

Schwille, J.A., Kung, H.-C., Hjohlman, M., Laverick, G.E. and Gardell, G.W. 2005. Actual delivered density fire test apparatus for sprinklers protecting high commodity storage. *Fire Safety Science*, 823–833.

Sheppard, D.T. and Lueptow, R.M. 2005. Characterization of fire sprinkler sprays using particle image velocimetry. *Atomization and Sprays* 15 (3), 341–362.

Stephens, J. 1993. Improving sprinkler fire control with lower water pressures. *Fire Prevention* 257, 32–36.

Su, P. and Doerr, W.W. 2010. Fire protection sprinkler system for extremely corrosive industrial duct environments. *Process Safety Progress* 29 (1), 70–78.

Wade, C., Spearpoint, M., Bittern, A. and Tsai, K.W.-H. 2007. Assessing the sprinkler activation predictive capability of the BRANZFIRE fire model. *Fire Technology* 43 (3), 175–193.

Widmann, J.F., Sheppard, D.T. and Lueptow, R.M. 2001. Non-intrusive measurements in fire sprinkler sprays. *Fire Technology* 37 (4), 297–315.

Wieczorek, C.J., Ditch, B. and Bill, R.G., Jr. 2011. Environmental Impact of Automatic Fire Sprinklers: Part 2. Experimental Study. *Fire Technology* 47 (3), 765–779.

Yang, D., Huo, R., Hu, L., Li, S. and Li, Y. 2008. A fire zone model including the cooling effect of sprinkler spray on smoke layer. *Fire Safety Science*, 919–930.

You, Y.-H., Li, Y.-Z., Huo, R., Hu, L.-H., Wang, H.-B., Li, S.-C. and Sun, X.-Q. 2008. Characteristics in cabin fire under sprinkler. *Ranshao Kexue Yu Jishu/Journal of Combustion Science and Technology* 14 (2), 137–142.

Yu, H.-Z., Pounder, D.B. and Fischer, M. 2004. Fire performance evaluation of a K-16.8 suppression-mode upright sprinkler. *Journal of Fire Protection Engineering* 14 (2), 101–124.

Zhou, X., D'Aniello, S.P. and Yu, H.-Z. 2012. Spray characterization measurements of a pendent fire sprinkler. *Fire Safety Journal* 54, 36–48.

6

Carbon Dioxide Gas Fire Extinguishing Systems

6.1 Introduction

Carbon dioxide (CO_2) is an effective fire suppression agent applicable to a wide range of fire hazards. Carbon dioxide works quickly, with no residual cleanup associated with a system discharge, which translates into minimal business interruption. Other benefits of carbon dioxide fire protection system include:

- Fast and effective: within seconds, CO_2 penetrates the entire hazard area to smother the combustion
- Electrically nonconductive, for a wide range of applications
- Versatile: CO_2 is effective on flammable and combustible materials and approved for class A, B, and C hazards

This book covers the minimum requirements for carbon dioxide systems, with which this chapter is concerned, and which are designed to provide a piped supply of carbon dioxide for the extinction of fire.

6.2 Characteristics and Uses of Carbon Dioxide

CO_2 is a colorless, odorless, electrically nonconductive gas that is a suitable medium for extinguishing fires. CO_2 extinguishes a fire by reducing the concentration of the oxygen in the air to the point where combustion stops.

CO_2 is 1.5 times heavier than air and therefore tends to settle in lower levels of the plant. This can be life-threatening due to oxygen displacement.

Materials used in generators and large motors are costly, and many are combustible. CO_2 is an effective way of limiting fire damage, since the extinguishing medium can be applied quickly and the concentration can be maintained sufficiently to extinguish deep-seated fires. In many cases, windings can be repaired after a fire and the unit returned to service fairly quickly.

CO_2 does not damage other materials in the generator, is nonconducting, and results in very little cleanup work, since CO_2 does not produce any debris. Toxic fumes are minimized while CO_2 is suppressing a fire. The cost of a CO_2 fire-extinguishing system is well justified by the savings realized by reduced damage to the machine and rapid resumption of power delivery.

Reclamation CO_2 systems are designed to establish CO_2 concentrations of between 30% and 50% by volume. These concentrations are significantly above that deemed injurious to humans. Table 6.1 illustrates the risks. With 30% to 50% concentrations of CO_2 in air housing after a discharge, personnel are certainly at risk should they become exposed.

It is common belief that the only risk to people from CO_2 is that of asphyxiation. However, CO_2 is not a truly inert gas, is toxic, and causes injuries and death by interfering with the functions of the central nervous system.

Discharge of CO_2 in concentrations used in reclamation units causes serious hazards to personnel, including suffocation and reduced visibility during and after the discharge period. There is also some risk of frostbite.

Carbon dioxide for use in fire extinguishing system should comply with the requirements of Table 6.1 when using the appropriate test method.

Carbon dioxide at atmospheric pressure is a colorless, odorless, and electrically nonconducting inert gas that is almost 1.5 times as dense as air. It is stored as a liquid under pressure, and 1 kg of liquid carbon dioxide expanded to atmospheric pressure will produce about 0.56 m^3 of free gas at a temperature of 30°C.

Carbon dioxide extinguishes fire by reducing the oxygen content of the atmosphere to a point where it will not support combustion. Reducing the oxygen content from the normal 21% in air to 15% will extinguish most surface fires, though for some materials a greater reduction as low as 5% is

TABLE 6.1

Carbon Dioxide Requirements

Property	Requirement
Purity, % (v/v) minimum	99.5%
Water content, % (mass/mass) maximum	0.015
Oil content, ppm by mass, maximum	5
Total sulfur compounds content, expressed as sulfur, ppm by mass, maximum	1

necessary. In some applications the cooling effect of carbon dioxide assists extinction.

Carbon dioxide should be used to fight fires of classes A and B. Class C fires can also be extinguished by carbon dioxide but in these cases the risk of explosion after extinction should be carefully considered.

Carbon dioxide is ineffective on fires involving materials such as metal hydrides, reactive metals such as sodium, potassium, magnesium, titanium, and zirconium, and chemicals containing oxygen available for combustion, such as cellulose nitrate.

Carbon dioxide is suitable for use on fires involving live electrical apparatus.

Carbon dioxide is present in the atmosphere at an average concentration of about 0.03% by volume. It is also a normal end product of human and animal metabolism. Carbon dioxide influences certain vital functions in a number of important ways, including control of respiration, dilation, and constriction of the vascular system, particularly the cerebrum and the pH of body fluids.

The concentration of carbon dioxide in the air governs the rate at which carbon dioxide is released from the lungs and thus affects the concentration of carbon dioxide in the blood and tissues. An increasing concentration of carbon dioxide in air can, therefore, become dangerous due to a reduction in the rate of release of carbon dioxide from the lungs and decreased oxygen intake.

Carbon dioxide obtained by converting dry ice to liquid will not usually comply with these requirements unless it has been processed to remove excess oil and waste.

6.3 Use and Limitations

Carbon dioxide fire extinguishing systems are useful in extinguishing fires in specific hazards or equipment, and in occupancies where an inert electrically nonconductive medium is essential or desirable, where cleanup of other media presents a problem, or where they are more economical to install than systems using other media.

All areas or parts of a hazard to which or from which a fire will spread should be simultaneously protected.

Some of the more important types of hazards and equipment that carbon dioxide systems will satisfactorily protect include

a. Flammable liquid materials. (Caution: Gas fire should normally be extinguished by shutting off the gas flow. Extinguishing the gas flame with carbon dioxide is desirable where necessary to permit immediate access to valves to shut off the gas supply. Extinguishment of gas fires indoors with carbon dioxide will create an explosion hazard and should not be used.)

b. Electrical hazards, such as transformers, oil switches, and circuit breakers, rotating equipment, and electronic equipment.

c. Engines utilizing gasoline and other flammable liquid fuels.

d. Ordinary combustibles such as paper, wood, and textiles.

e. Hazardous solids.

The discharge of liquid carbon dioxide is known to produce electrostatic charges which, under certain conditions, could create a spark. Carbon dioxide fire extinguishing systems protecting areas where explosive atmospheres could exist should utilize metal nozzles and be properly grounded. In addition, objects exposed to discharge from carbon dioxide nozzle should be grounded to dissipate possible electrostatic charges.

Carbon dioxide will not extinguish fires where the following materials are actively involved in the combustion process:

a. Chemicals containing their own oxygen supply, such as cellulose nitrate

b. Reactive metals such as sodium, potassium, magnesium, titanium, and zirconium

c. Metal hydrides such as $HNa-H_3Al$

6.4 System Components

All devices need to be designed for the service they will encounter and should not be readily rendered inoperative or susceptible to accidental operation. Devices should normally be designed to function properly from $-30°C$ to $55°C$ or be marked to indicate their temperature limitations.

Where the pressure of a permanent gas from pilot containers is used as a means of releasing the remaining containers, the supply and discharge rate should be designed for releasing all of the remaining containers. The pilot supply should be continuously monitored and a fault alarm given in the event of excessive pressure loss.

Where the pressure of a liquefied gas is used as a means of releasing the remaining containers, duplicate containers, each of which is capable of operating the system, should be used.

Various operating devices are necessary to control the flow of the extinguishing agent to operate the associated equipment. These include container valves, distribution valves, automatic and manual controls, delay devices, pressure trips and switches, and discharge nozzles.

All devices, especially those having external moving parts, should be located, installed, or suitably protected so that they are not subject to mechanical, chemical, or other damage that would render them inoperable.

6.5 Types of Systems

There are four types of systems recognized in this section:

1. Total flooding systems
2. Local application systems
3. Manual hose reel systems
4. Standpipe system and mobile supply

In the selection of a carbon dioxide extinguishing system, account should be taken of

a. The field of usefulness of the four systems
b. Operating requirements dictating either manual or automatic operation
c. The nature of the hazard
d. The location and degree of enclosure of the hazard
e. The degree of hazard to personnel arising from the CO_2 discharge

A carbon dioxide system should be used to protect one or more hazards or groups of hazards by means of directional valves. Where two or more hazards are simultaneously involved in fire by reason of their proximity, each hazard should be either

a. Protected with an individual system, with the combination arranged to operate simultaneously
b. Protected with a single system that should be sized and arranged to discharge on all potentially involved hazards simultaneously

6.6 Package Systems (Kits)

Package systems consist of system components designed to be installed according to pretested limitations as approved or listed by a testing laboratory.

Package systems should incorporate special nozzles, flow rates, methods of application, nozzle placement, and quantities of carbon dioxide that differ from those detailed elsewhere in this standard since they are designed for very specific hazards. All other requirements of the standard apply.

Package systems should be installed to protect hazards within the limitations that have been established by the testing laboratories where listed.

6.7 Total Flooding Systems

Total flooding systems are used for rooms, ovens, enclosed machines, and other enclosed spaces containing materials extinguishable by carbon dioxide. For effective total flooding, the space must be reasonably well enclosed. Openings must be arranged to close automatically and ventilation equipment to shut down automatically no later than the start of the discharge. Otherwise, additional carbon dioxide must be provided to compensate for the leakage.

Automatic closing devices for openings must be able to overcome the discharge pressure of the carbon dioxide. Conveyors, flammable liquid pumps, and mixers associated with an operation may be arranged to shut down automatically on actuation of the protection system. A typical arrangement of a total flooding carbon dioxide system is shown in Figure 6.1.

A total flooding system consists of a fixed supply of carbon dioxide permanently connected to fixed piping, with fixed nozzles arranged to discharge carbon dioxide into an enclosed space or enclosure about the hazard.

Figure 6.2 shows a simplified version of Figure 6.1.

6.7.1 Uses

This type of system should be used where there is a permanent enclosure about the hazard that is adequate to enable the required concentration to be built up, and to be maintained for the required period of time to ensure the complete and permanent extinguishment of the fire in the specific combustible material or materials involved.

1. Examples of hazards that are successfully protected by total flooding systems include rooms, vaults, enclosed machines, ducts, ovens, containers, and the contents thereof
2. Fires that can be extinguished or controlled by total flooding methods are
   a. Surface fires involving flammable liquids, gases, and solids
   b. Deep-seated fires involving solids subject to smoldering

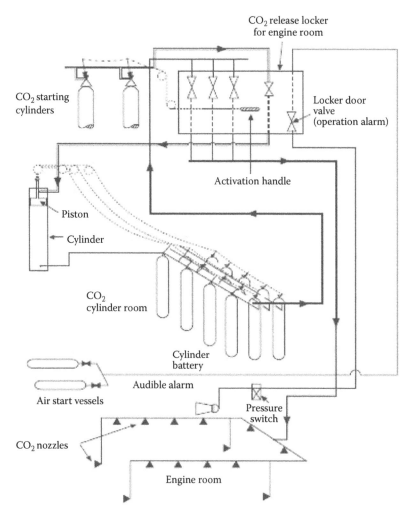

FIGURE 6.1
CO_2 total flooding system.

6.7.2 General Design

The quantity of carbon dioxide, which will vary according to the hazard and permitted openings, should be sufficient to reduce the oxygen content of the atmosphere within the enclosure to a point where combustion can no longer be sustained.

The distribution of the carbon dioxide should be so arranged that it is evenly and thoroughly mixed with the existing atmosphere. Special venting may be required to avoid excessive pressure buildup resulting from the volume of carbon dioxide discharged into the hazard area.

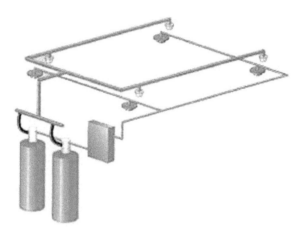

FIGURE 6.2
A simplified CO_2 total flooding system.

The system should be designed for either

a. Automatic and manual operation
b. Manual operation only

Note that this may be dependent on the requirements of the authorities concerned.

6.7.3 Enclosure

The protected volume should be enclosed by elements of construction having a fire resistance of not less than 30 minutes and classified as noncombustible. Where opening can be closed, these should be arranged to close before or at the start of CO_2 gas discharge.

Where carbon dioxide can flow freely between two or more interconnected volumes, the quantity of carbon dioxide should be the sum of quantities calculated for each volume using respective volume and material conversion factors. If one volume requires higher than normal concentration, the higher concentration should be used in all interconnected volumes.

The volume of the enclosure should be the gross volume. The only permitted reductions should be permanent, impermeable building elements within the enclosure.

A well-enclosed space is required to maintain the extinguishing concentration of carbon dioxide.

6.7.3.1 Area of Opening Required for Venting

The venting of flammable vapors and release of pressure caused by the discharge of quantities of carbon dioxide into closed spaces should be considered, and provision should be made for venting where necessary. The pressure venting consideration involves such variables as enclosure strength and injection rate.

Leakage around doors, windows, ducts, and dampers, though not apparent or easily determined, should provide sufficient venting relief for normal carbon dioxide systems without special provisions being made.

In many instances, particularly when hazardous materials are involved, relief openings are already provided for explosion venting. These and other available openings often provide adequate venting.

To prevent fire from spreading through openings to adjacent hazards or work areas that are possible reignition sources, such openings should be provided with automatic closures or local application nozzles. The gas required for such protection should be in addition to the normal requirement for total flooding. When neither method is practical, protection should be extended to include these adjacent hazards or work areas.

In the case of process and storage tanks where safe venting of flammable vapors and gases cannot be realized, the use of external local application systems are required.

Any openings that cannot be closed at the time of extinguishment should be compensated for by the addition of a quantity of carbon dioxide equal to the anticipated loss at the design concentration during a 1-minute period. This amount of carbon dioxide should be applied through the regular distribution system.

For ventilating systems that cannot be shut down, additional carbon dioxide should be added to the space through the regular distribution system in an amount computed by dividing the volume moved during the liquid discharge period by the flooding factor. This should be multiplied by the material conversion factor (determined in Figure 6.3) when the design concentration is greater than 34%.

For applications where the normal temperature of the enclosure is above 93°C, a 1% increase in the calculated total quantity of carbon dioxide should be provided.

For applications where the normal temperature of the enclosure is below −18°C, a 1% increase in the calculated total quantity of carbon dioxide should be provided for each degree below 18°C.

Under normal conditions, surface fires are usually extinguished during the discharge period. Except for unusual conditions, it will not be necessary to provide extra carbon dioxide to maintain the concentration.

A flooding factor of 0.48 m³/kg should be used in ducts and covered trenches. If the combustibles represent a deep-seated fire, it should be treated.

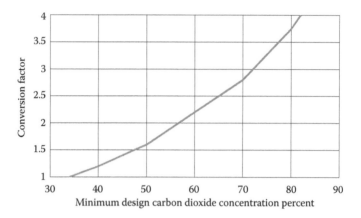

FIGURE 6.3
Material conversion factors.

6.8 Carbon Dioxide for Surface Fires

6.8.1 Flammable Materials

Proper consideration should be given to the determination of the design concentration of carbon dioxide required for the type of flammable material involved in the hazard. The design concentration is determined by adding a suitable factor 20% to the minimum effective concentration. In no case should a concentration less than 34% be used.

Table 6.2 gives the theoretical minimum carbon dioxide concentration and the suggested minimum design carbon dioxide concentration to prevent ignition of some common liquids and gases.

For materials not given in Table 6.2, the minimum theoretical carbon dioxide concentration should be obtained from some recognized source or determined by test. The theoretical minimum extinguishing concentrations in air for the materials in Table 6.2 were obtained from a compilation of U.S. Bureau of Mines Limits of Flammability of Gases and Vapors (Bulletins 503 and 627).

6.8.2 Volume Factor

The volume factor used to determine the basic quantity of carbon dioxide to protect an enclosure containing a material requiring a design concentration of 34% should be in accordance with Table 6.3.

In figuring the net cubic capacity to be protected, due allowance should be made for permanent nonremovable impermeable structures materially reducing the volume.

TABLE 6.2

Minimum Carbon Dioxide Concentration for Extinguishment

Material	Theoretical Minimum CO_2 % Concentration	Minimum Design CO_2 % Concentration
Acetylene	55	66
Acetone	27	34
Aviation gas grades 115/145	30	36
Benzol, benzene	31	37
Butadiene	34	41
Butane	28	34
Carbon disulfide	60	72
Carbon monoxide	53	64
Coal or natural gas	31	37
Cyclopropane	31	37
Diethyl ether	33	40
Dimethyl ether	33	40
Dowtherm	38	46
Ethane	33	40
Ethyl alcohol	36	43
Ethyl ether	38	46
Ethylene	41	49
Ethylene dichloride	21	34
Ethylene oxide	44	53
Gasoline	28	34
Hexane	29	35
Higher paraffin hydrocarbons (cn H2 m + 2 m – 5)	28	34
Hydrogen	62	75
Hydrogen sulfide	30	36
Isobutane	30	36
Iso-Butane	31	37
Isobutylene	26	34
Isobutyl formate	26	34
Jp-4	30	36
Kerosene	28	34
Methane	25	34
Methyl acetate	29	35
Methyl alcohol	33	40
Methyl butene i	30	36
Methyl ethyl ketone	33	40
Methyl formate	32	39
Pentane	29	35
Propane	30	36
Propylene	30	36
Quench, lube oils	28	34

TABLE 6.3

Flooding Factors

A	B		C
Volume of Space (m³ Incl.)	Volume Factor		Calculated Quantity (kg)
	m³/kg CO₂	kg CO₂/m³	Not Less Than
Up to 3.96	0.86	1.15	–
3.97–14.15	0.93	1.07	4.5
14.16–45.28	0.99	1.01	15.1
45.29–127.35	1.11	0.9	45.4
127.36–1415.0	1.25	0.8	113.5
OVER 1415.0	1.38	0.77	1135

As the average small space has proportionately more boundary area per enclosed volume than a larger space, greater proportionate leakages are anticipated and accounted for by the graded volume factors in Table 6.3. The least gas quantities for the smallest volumes are tabulated in order to clarify the intent of column B and thus avoid possible overlapping at borderline volumes.

In two or more interconnected volumes where free flow of carbon dioxide can take place, the carbon dioxide quantity should be the sum of the quantities calculated for each volume, using its respective volume factor from Table 6.3. If one volume requires greater than normal concentration, the higher concentration should be used in all interconnected volumes.

6.8.3 Material Conversion Factor

For materials (see Table 6.2) requiring a design concentration over 34%, the basic quantity of carbon dioxide calculated from the volume factor given in Table 6.3 should be increased by multiplying this quantity by the appropriate conversion factor given in Figure 6.3.

6.8.3.1 Special Conditions

Additional quantities of carbon dioxide should be provided to compensate for any special condition that can adversely affect the extinguishing efficiency.

6.8.4 Compensation for Abnormal Temperatures

Where there are abnormal temperatures, additional quantities of gas should be provided as follows:

a. Where the normal temperature of the enclosure is above 100°C, 2% carbon dioxide should be added for each additional 5°C over 100°C

b. Where the normal temperature of the enclosure is below –20°C, 2% carbon dioxide should be added for each 1°C below –20°C

6.9 Carbon Dioxide for Deep-Seated Fires

The quantity of carbon dioxide for deep-seated type fires is based on fairly tight enclosures. After the design concentration is reached, the concentration should be maintained for a substantial period of time, but not less than 20 minutes. Any possible leakage should be given special consideration since no allowance is included in the basic flooding factors.

6.9.1 Combustible Materials

For combustible materials capable of producing deep-seated fires, the required carbon dioxide concentrations cannot be determined with the same accuracy possible with surface-burning materials. The extinguishing concentration will vary with the mass of material present because of the thermal insulating effects. Flooding factors have therefore been determined on the basis of practical test conditions.

The design concentrations listed in Table 6.4 should be achieved for the hazards listed. Generally, the flooding factors have been found to provide proper design concentrations for the rooms and enclosures listed.

Flooding factors for other deep-seated fires should be justified to the satisfaction of the authorities before use.

Proper consideration should be given to the mass of material to be protected because the rate of cooling is reduced by the thermal insulating effects.

6.9.2 Special Conditions

Additional quantities of carbon dioxide should be provided to compensate for any special condition that will adversely affect the extinguishing efficiency.

TABLE 6.4

Flooding Factors for Specific Hazards

Design Concentration (%)	Flooding Factor		Specific Hazard
	m³/kg CO_2	kg CO_2/m³	
50	0.62	1.6	Dry electrical hazards in general (spaces 0–57 m³)
50	0.75	1.33 (91 kg CO_2 minimum)	Spaces greater than 57 m³
65	0.5	2	Record (bulk paper) storage, ducts, and covered trenches
75	0.38	2.66	For storage vaults, dust collectors

- Any openings that cannot be closed at the time of extinguishment should be compensated for by the addition of carbon dioxide equal in volume to the expected leakage volume during the extinguishing period.
- For deep-seated fires, such as will be involved with solids, unclosable openings should be restricted to those bordering or actually in the ceiling, if the size of openings exceeds the pressure relief venting requirements set forth in 2.6.2.1 of NFPA Volume 1, Section 12.

6.10 Rates of Application

For surface fires, the design concentration should be achieved within 1 minute. For deep-seated fires, the design concentration should be achieved within 7 minutes but the rate should be not less than that required to develop a concentration of 30% in 2 minutes.

The times specified above are considered adequate for the usual surface or deep-seated fire. Where the materials involved are likely to give a higher spread of fire, rates higher than the minimum should be used. Where a hazard contains materials that will produce both surface and deep-seated fires, the rate of application should be at least the minimum required for surface fires.

6.10.1 Extended Discharge

Where leakage is appreciable and the design concentration has to be obtained quickly and maintained for an extended period of time, carbon dioxide provided for leakage compensation should be applied at a reduced rate. This method is particularly suited to enclosed rotating electrical apparatus such as generators and alternators, but it can also be used on normal room flooding systems where suitable.

6.10.2 Rotating Electrical Machinery

For enclosed rotating electrical machinery, a minimum concentration of 30% should be maintained for the deceleration period of the machine. This minimum concentration should be held for the deceleration period or 20 minutes, whichever is longer.

Table 6.5 should be used as a guide to estimate the quantity of gas needed for the extended discharge to maintain the minimum concentration. The quantities are based on the internal volume of the machine and the

TABLE 6.5

Extended Discharge for Enclosed Recirculation: Rotating Electrical Equipment

Carbon Dioxide Required (kg)	Deceleration Time							
	5 min	10 min	15 min	20 min	30 min	40 min	50 min	60 min
	Volume Enclosed by the Machines (m³)							
45	34	28	23	17	14	11	9	6
68	51	43	34	28	21	17	14	11
91	68	55	45	37	28	24	18	14
113	93	69	57	47	37	30	23	17
136	130	88	68	57	47	37	28	20
159	173	116	85	71	57	47	34	26
181	218	153	108	89	71	57	45	34
204	262	193	139	113	88	74	60	45
227	306	229	173	142	110	93	79	62
250	348	269	210	173	139	119	102	88
272	394	309	244	204	170	147	127	110
295	436	348	279	235	200	176	156	136
319	479	385	314	266	230	204	181	159
340	524	425	350	297	259	232	207	184
363	566	464	385	329	289	261	232	207
386	609	503	421	360	320	289	258	229
408	651	541	456	391	350	317	285	255
431	697	581	491	422	379	346	312	278
454	739	620	527	453	411	374	337	303
476	782	666	564	484	442	402	364	326
499	824	697	596	515	470	430	389	351
522	867	736	632	547	501	459	416	374
544	912	773	667	578	532	487	442	399
567	954	813	702	609	562	515	457	422
590	1000	852	738	641	592	544	494	447
612	1042	889	773	673	623	572	521	472
635	1087	929	809	705	654	600	548	496
658	1130	968	844	736	685	629	575	520
680	1172	1008	879	767	715	657	600	544

deceleration time assuming average leakage. For dampered, nonrecirculating type machines, 35% should be added to the quantities given in Table 6.5.

6.10.3 Distribution Systems

6.10.3.1 Design

Piping for total flooding systems should be designed to deliver the required rate of application at each nozzle.

High-pressure storage temperatures can range from $-18°C$ to $55°C$ without requiring special methods of compensating for changing flow rates. Storage temperatures outside those limits require special design considerations to ensure proper flow rates.

6.10.3.2 Nozzle Selection and Distribution

Rooms with ceiling heights above 7.5 m should have discharge nozzles at two or more levels, depending on the height. Nozzles used in total flooding systems should be of the type most suitable for the intended purpose, and they should be properly located to achieve the best results. The lower ring of nozzles should be located approximately one-third of the height from the floor but no higher than 2.5 m.

The nozzles should be arranged in the protected space in a manner that will ensure adequate, prompt, and equal distribution of the carbon dioxide. Special consideration should be given to areas within the space that are of particular danger.

The type of nozzle selected and the disposition of the individual nozzles should be such that the discharge will not splash flammable liquids, dislodge ceiling tiles, or create dust clouds that can extend the fire, create an explosion, or otherwise adversely affect the contents of the enclosure. Nozzles vary in design and discharge characteristics and should be selected on the basis of their adequacy for the use intended.

6.11 Local Application Systems

6.11.1 Description

A local application system consists of a fixed supply of carbon dioxide permanently connected to a system of fixed piping with nozzles arranged to discharge directly into the fire.

6.11.2 Uses

Local application systems should be used for the extinguishment of surface fires in flammable liquids, gases, and shallow solids where the hazard is not enclosed or where the enclosure does not conform to the requirements for total flooding.

Examples of hazards that are successfully protected by local application systems include

a. Coating machines
b. Dip tanks

c. Quench tanks

d. Printing presses

e. Spray booths

f. Fume ducts

g. Process machinery

h. Oil-filled electric transformers and switchgear

Open cable or pipe trenches (covered perhaps with checker plate or similar) crossing, or adjacent to, a hazardous area should also be considered.

6.11.3 Carbon Dioxide Requirements

The quantity of carbon dioxide required for local application systems should be based on the total rate of discharge needed to blanket the area or volume protected and the time that the discharge must be maintained to assure complete extinguishment.

For systems with high-pressure storage, the computed quantity of carbon dioxide should be increased by 40% to determine nominal cylinder storage capacity since only the liquid portion of the discharge is effective. This increase in cylinder storage capacity is not required for the total flooding portion of combined local application-total flooding systems.

The quantity of carbon dioxide in storage should be increased by an amount sufficient to compensate for liquid vaporized in cooling the piping.

6.11.3.1 Rate of Discharge

Nozzle discharge rates should be determined by either the surface method or the volume method, depending on the type of risk.

The total rate of discharge for the system should be the sum of the individual rates of all the nozzles or discharge devices used on the system.

For low-pressure systems, if a part of the hazard is to be protected by total flooding, the discharge rate for the total flooding part should be sufficient to develop the required concentration in not more than the discharge time used for the local application part of the system.

For high-pressure systems, if a part of the hazard is to be protected by total flooding, the discharge rate for the total flooding part should be computed by dividing the quantity required for total flooding by the factor 1.4 and by the time of the local application discharge in minutes.

$$Q_F = \frac{W_F}{1.4T_L}$$

where

Q_F = Rate of flow for the total flooding portion in kg/min

W_F = Total quantity of carbon dioxide for the total flooding portion in kg

T_L = Liquid discharge time for the local application portion in minutes

6.11.3.2 Duration of Discharge

The minimum effective discharge time for computing quantity should be 30 seconds. The minimum time should be increased to compensate for any hazard condition that would require a longer cooling period to assure complete extinguishment.

Where there is a possibility that metal or other material become heated above the ignition temperature of the fuel, the discharge time should be increased to allow adequate cooling time.

Where the fuel has an autoignition point below its boiling point, such as paraffin wax and cooking oil, the effective discharge time should be increased to permit cooling of the fuel to prevent reignition. The minimum discharge time should be 3 minutes.

6.11.4 Rate by Area Method

The area method of system design is used where the fire hazard consists primarily of flat surfaces or low-level objects associated with horizontal surfaces.

System design should be based on listing or approval data for individual nozzles. Extrapolation of such data above or below the upper or lower limits should not be permitted.

6.11.4.1 Nozzle Discharge Rates

The design discharge rate through individual nozzles should be determined on the basis of location or projection distance in accordance with specific approvals or listings.

The discharge rate for overhead type nozzles should be determined solely on the basis of distance from the surface each nozzle protects.

The discharge rate for tank side nozzles should be determined solely on the basis of throw or projection required to cover the surface each nozzle protects.

6.11.4.2 Area per Nozzle

The maximum area protected by each nozzle should be determined on the basis of location or projection distance and the design discharge rate in accordance with specific approvals or listings.

The same factors used to determine the design discharge rate should be used to determine the maximum area to be protected by each nozzle.

The portion of the hazard protected by individual overhead type nozzles should be considered as a square area.

The portion of the hazard protected by individual tank side or linear nozzles should be either a rectangular or a square area in accordance with spacing and discharge limitations stated in specific approvals or listings.

When coated rollers or other similar irregular shapes are to be protected, the projected wetted area should be used to determine nozzle coverage.

Where coated surfaces are to be protected, the area per nozzle should be increased by 40% over the areas given in specific approvals or listings. Coated surfaces are defined as those designed for drainage that are constructed and maintained so that no pools of liquid will accumulate over a total area exceeding 10% of the protected surface. This subsection does not apply where there is a heavy buildup of residue.

When deep layer flammable liquid fires are to be protected, a minimum freeboard of 152 mm should be provided unless otherwise noted in approvals or listings of nozzles.

6.11.4.3 Location and Number of Nozzles

A sufficient number of nozzles should be used to adequately cover the entire hazard area on the basis of the unit areas protected by each nozzle.

Tank side or linear type nozzles should be located in accordance with spacing and discharge rate limitations stated in specific approvals or listings.

Overhead type nozzles should be installed perpendicular to the hazard and centered over the protected area by the nozzle. They should also be installed at angles between 45° and 90° from the plane of the hazard surface. The height used in determining the necessary flow rate and area coverage should be the distance from the aiming point on the protected surface to the face of the nozzle measured along the axis of the nozzle.

6.11.4.4 Nozzles Installed at an Angle

When installed at an angle, nozzles should be aimed at a point measured from the near side of the area protected by the nozzle, the location of which is calculated by multiplying the fractional aiming factor in Table 6.6 by the width of the area protected by the nozzle (see Figure 6.4).

6.11.4.5 Rate by Volume Method

The volume method of system design is used where the fire hazard consists of three-dimensional irregular objects that cannot be easily reduced to equivalent surface areas.

TABLE 6.6

Equivalent Orifice Sizes

Orifice Code No.	Equivalent Single Orifice Diameter (mm)	Equivalent Single Orifice Area (mm²)
1	0.79	0.49
1.5	1.19	1.11
2	1.59	1.98
2.5	1.98	3.09
3.	2.38	4.45
3.5	2.78	6.06
4	3.18	7.94
4.5	3.57	10
5	3.97	12.39
5.5	4.37	14.97
6	4.76	17.81
6.5	5.16	20.9
7	5.56	24.26
7.5	5.95	27.81
8	6.35	31.66
8.5	6.75	35.74
9	7.14	40.06
9.5	7.54	44.65
10	7.94	49.48
11	8.73	59.87
12	9.53	71.29
13	10.32	83.61
14	11.11	96.97
15	11.91	111.29
16	12.7	126.71
18	14.29	160.32
20	15.88	197.94
22	17.46	239.48
24	19.05	285.03
32	25.4	506.45
48	38.4	1138.71
64	50.8	2025.8

6.11.4.6 Assumed Enclosure

The total discharge rate of the system should be based on the volume of an assumed enclosure entirely surrounding the hazard.

    a. The assumed enclosure should be based on an actual closed floor unless special provisions are made to take care of bottom conditions

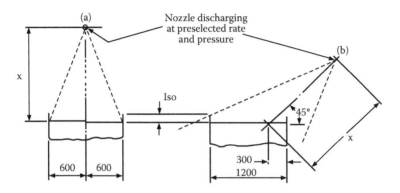

FIGURE 6.4
Aiming position for angled discharge nozzles. The diagram shows nozzles discharging at (a) 90° with the aiming point at the center of the protected surface, and at (b) 45° with the aiming point at 0.25 of the width of the protected surface, into a tray containing liquid fuel with a freeboard of 150 mm. X is the preselected height used to determine the flow rate required.

b. The assumed walls and ceiling of this enclosure should be at least 0.6 m from the main hazard unless actual walls are involved and should enclose all areas of possible leakage, splashing, or spillage

c. No deductions should be made for solid objects within this volume

d. A minimum dimension of 1.2 m shall be used in calculating the volume of the assumed enclosure

e. If the hazard is subjected to winds or forced drafts, the assumed volume should be increased to compensate for losses on the windward sides

6.11.4.7 System Discharge Rate

The total discharge rate for the basic system shall be equal to 16 kg/min/m³ of assumed volume.

6.11.5 Carbon Dioxide Extinguishers

Carbon dioxide extinguishers are filled with carbon dioxide, a nonflammable gas under extreme pressure. These extinguishers put out fires by displacing oxygen or taking away the oxygen element of the fire triangle. Because of its high pressure, when you use this extinguisher pieces of dry ice shoot from the horn, which also has a cooling effect on the fire.

You can recognize this type of extinguisher by its hard horn and absent pressure gauge.

CO_2 cylinders are red and range in size from 5 to 100 pounds or larger.

CO_2 extinguishers are designed for classes B and C (flammable liquid and electrical) fires only.

The following points should be noted:

- CO_2 is not recommended for class A fires because they may continue to smolder and reignite after the CO_2 dissipates
- Never use CO_2 extinguishers in a confined space while people are present without proper respiratory protection

6.11.5.1 Locations

Carbon dioxide extinguishers will frequently be found in industrial vehicles, mechanical rooms, offices, computer labs, and flammable liquid storage areas. Figure 6.5 shows a typical carbon dioxide extinguisher.

FIGURE 6.5
A carbon dioxide extinguisher.

6.12 Manual Hose Reel Systems

Manual hose reel systems consist of a hose reel or rack, hose, and discharge nozzle assembly connected by fixed piping to a supply of carbon dioxide.

A separate carbon dioxide supply can be provided for hand hose reel use or carbon dioxide can be piped from a central storage unit that should be supplying several hose reels or fixed manual or automatic systems.

6.12.1 Uses

Manual hose reel systems should be used to supplement fixed fire protection systems or to supplement first aid fire extinguishers for the protection of specific hazards for which carbon dioxide is a suitable extinguishing agent. These systems should not be used as a substitute for other fixed carbon dioxide fire extinguishing systems equipped with fixed nozzles, except where the hazard cannot adequately or economically be provided with fixed protection. The decision as to whether hose reels are applicable to the particular hazard should rest with the authorities concerned.

6.12.2 Location

Manual hose reel stations should be placed such that they are easily accessible and within reach of the most distant hazard that they are expected to protect. In general, they should not be located such that they are exposed to the hazard nor should they be located inside any hazard area protected by a total flooding system.

6.12.3 Spacing

If multiple hose stations are used, they should be spaced so that any area within the hazard may be covered by one or more hose lines.

6.12.4 Rate and Duration of Discharge

The rate and duration of discharge and consequently the amount of carbon dioxide should be determined by the type and potential size of the hazard. A manual hose reel system should have sufficient quantity of carbon dioxide to permit its effective (liquid phase) use for at least 1 minute.

6.12.5 Simultaneous Use of Hose Reels

Where simultaneous use of two or more hose reels is possible, a sufficient quantity of carbon dioxide should be available to supply the maximum number of nozzles that are likely to be used at any one time for at least 1 minute.

6.13 Standpipe Systems and Mobile Supply

A standpipe is a fixed total flooding, local application, or manual hose reel system without a permanently connected carbon dioxide supply. The carbon dioxide supply is mounted on a mobile vehicle that can be towed or driven to the scene of a fire and quickly coupled to the standpipe system protecting the involved hazard. Mobile supply is primarily fire brigade or fire department equipment requiring trained personnel for effective use.

6.13.1 Uses

A standpipe system and mobile supply may be used to supplement complete fixed fire protection systems or may be used alone for the protection of the specific hazards. A mobile supply may be used as a reserve to supplement a fixed supply, and may also be outfitted with manual hose reel for the protection of scattered hazards. These systems should be installed only with the approval of the authorities concerned.

6.14 Storage Containers

Storage containers and accessories should be located and arranged so that inspection, testing, recharging, and other maintenance is facilitated and interruption to protection is held to a minimum.

Storage containers should be located as near as possible to the hazard or hazards they protect, but should not be exposed to fire in a manner that is likely to impair system performance.

Storage containers should not be located so as to be subject to severe weather conditions or be subject to mechanical, chemical, or other damage. Where excessive climatic or mechanical exposures are expected, suitable guards or enclosures should be provided.

Containers should also be protected from direct sunlight through windows and from interference by unauthorized persons. The area selected should be dry and ventilated.

6.14.1 Exits

Adequate means of egress from a protected space should be provided. Doors at exits should open outward and be self-closing. All exit doors should open readily from the inside and any that have to be secured should be fitted with panic bolts or latches.

The means of egress from a protected space should be kept clear at all times.

6.14.2 Temperatures

The general ambient storage temperatures should not exceed the following limits unless the system is designed for proper operation with storage temperatures outside the appropriate range:

a. For total flooding systems, not more than 46°C or less than –18°C

b. For local application systems, not more than 46°C or less than 0°C

External heating or cooling may be used to keep the temperature within the range.

6.14.3 High-Pressure Storage Containers

The carbon dioxide supply should be stored in rechargeable containers designed to hold pressurized carbon dioxide in liquid form at atmospheric temperatures corresponding to a nominal pressure of 58.6 bars at 21°C.

Note that a change in ambient temperature from 10°C to 21°C will raise the pressure from 44 bar to 59 bar (see NFPA 12-24, Figure A-1.5.1).

Manifolded cylinders should be adequately mounted and suitably supported in a rack provided for the purpose, including facilities for convenient individual servicing or content weighings. Automatic means should be provided to prevent the loss of carbon dioxide from the manifold if the system is operated when any cylinder is removed for maintenance.

6.14.4 Low-Pressure Storage Containers

Low-pressure storage containers should be designed to maintain the carbon dioxide supply at a nominal pressure of 20.7 bars corresponding to a temperature of approximately –18°C. All containers in any one battery should be of the same size and contain the same mass of carbon dioxide.

6.14.5 Flexible Hoses

The use of flexible piping or hoses in a carbon dioxide system introduces a number of things to be considered that do not affect rigid piping. One of these is the nature of any changes of direction. The minimum radius of curvature for any flexible hose to be used in a carbon dioxide system should not be less than indicated by the manufacturer's data, usually shown in the listing information for a particular system. Other areas of concern are resistance to the effects of vibration, flexure, tension, torsion, temperature, flame, compression, and bending. It is also necessary for the hose to have the strength to contain the carbon dioxide during discharge, and be made of materials that will be resistant to atmospheric corrosion.

A dirt trap consisting of a tee with a capped nipple at least (51 mm) long should be installed at the end of each pipe run.

6.15 Discharge Nozzles

Discharge nozzles should be suitable for the use intended and be listed or approved for discharge characteristics. The discharge nozzle consists of the orifice and any associated horn, shield, or baffle.

Discharge nozzles should be permanently marked to identify the nozzle and to show the equivalent single orifice diameter regardless of shape and number of orifices. This equivalent diameter should refer to the orifice diameter of the standard single orifice type nozzle having the same flow rate as the nozzle in question. The marking should be readily discernible after installation. The standard orifice is an orifice having a rounded entry with a coefficient of discharge not less than 0.98 and flow characteristics as given in Tables 6.7 and 6.8.

TABLE 6.7

Discharge Rate per Square Millimeter of Equivalent Orifice Area for Low-Pressure Storage (20.7 Bar)

Orifice Pressure (Bar)	Discharge Rates (kg/min/mm^2)
20.7	2.97
20	2.041
19.3	1.671
18.6	1.443
17.9	1.284
17.2	1.165
16.5	1.073
15.9	0.992
15.2	0.918
14.5	0.851
13.8	0.792
13.1	0.737
12.4	0.688
11.7	0.642
11	0.600
10.3	0.559

TABLE 6.8

Discharge Rate per Square mm of Equivalent Orifice
Area for High-Pressure Storage (51.7 Bar)

Orifice Pressure (Bar)	Discharge Rates (kg/min/mm²)
51.7	3.258
50	2.706
48.3	2.403
46.5	2.174
44.8	1.995
43.1	1.84
41.4	1.706
39.6	1.59
37.9	1.488
36.2	1.397
34.5	1.309
32.8	1.224
31	1.14
29.3	1.063
27.6	0.985
25.9	0.908
24.1	0.83
22.4	0.76
20.7	0.69

For examples of equivalent orifice diameters, see Table 6.6. The orifice code numbers indicate the equivalent single orifice diameter in 0.8 mm increments. Orifice sizes other than those shown in Table 6.6 can be used and should be marked as decimal orifice equipment.

6.16 Additional Requirements for All Systems

6.16.1 Manifold Venting

In systems using stop valves, (accidental) release of the carbon dioxide from the storage containers should activate a device that gives a visual warning to indicate that carbon dioxide has been released and is trapped in the manifold. In addition to the pressure relief device, a manually operated vent valve should be fitted to the manifold so that the trapped carbon dioxide can be safely vented to the atmosphere. The vent valve should normally be kept in the locked shut position.

6.16.2 Audible and Visual Alarms

Clear visual indication should be provided at each entrance to a protected space to show whether the gas flooding system is on automatic or manual control.

An audible alarm, clearly distinguishable from the normal fire alarm, should be provided within a protected space and any associated hazard areas to sound in conjunction with the discharge of the gas. If a local application system is fitted with a time delay, the alarm should sound during the time delay period before the gas is released.

A continuing visual alarm should be given until the space has been ventilated and the atmosphere rendered safe. The following lamps and wording should be used to identify system conditions:

- Red lamp: CO_2 discharged
- Amber lamp: automatic and manual control
- Green lamp: manual control

Optional condition indicators should be provided as follows:

- Red (flashing): fire
- Amber: system disabled
- Green: supply healthy

System condition indicators as described above are normally required only for total flooding systems but are necessary in local application systems. Duplication of the lamps will mitigate the effects of bulb failure.

6.16.3 Automatic Detection

Automatic detection should be by any listed or approved method or device that is capable of detecting and indicating heat, flame, smoke, combustible vapors, or an abnormal condition in the hazard such as process trouble that is likely to produce fire.

6.16.4 Confined Spaces

Entry into confined spaces poses additional hazards because of restrictions on freedom of movement, ventilation, and on escape or rescue. Automatic operation of the system should be prevented before entry into floor or ceiling voids, ducts, process vessels, or similarly confined spaces that are protected by a gas flooding system.

Entry into confined spaces for any purpose should be controlled by a permit-to-work system. Provision should be made for ensuring that the atmosphere within the space is safe for entry and will remain so for the duration of entry.

In cases where effective ventilation cannot be ensured, the permit should specify the respiratory protective equipment to be used and any other special precautions to be observed to ensure safe working conditions.

6.16.5 Warning Signs

Appropriate signs should be prominently displayed at each manual control point and at each entrance to the area protected by the system.

6.16.6 Exits

Adequate means of egress from a protected space should be provided. Doors at exits should open outward and be self-closing. All exit doors should open readily from the inside and any that have to be secured be fitted with panic bolts or latches.

The means of egress from a protected space should be kept clear at all times.

6.16.7 Manual Hose Reels

A notice with the wording "Only for use by trained personnel" should be mounted on or adjacent to manual hose reels.

The use of manual hose reels for the application of fire-smothering gas present a hazard to personnel. This method of fire control should be used only by trained personnel who have been adequately instructed and trained in the use of the equipment and in the safety precautions to be adopted.

All persons other than those fighting the fire should be evacuated prior to the use of manual hose reels. Particular precautions are required where ventilation is restricted in order to guard against hazards that will arise from the fire or the extinguishing medium.

6.16.8 Area Ventilation after Discharge

A means of mechanically or naturally ventilating area after discharge of carbon dioxides should be provided.

Consideration should be given to adding an odor to the carbon dioxide to assist in the detection of hazardous atmospheres and in their effective ventilation.

The means provided for ventilation should not form part of the normal building ventilation system, and should incorporate extraction arrangements at a low level in the protected area.

Care should be taken to ensure that the post-fire atmosphere is not ventilated into other parts of the building. Provision should be made for the prompt discovery and rescue of persons rendered unconscious.

Before reentry to an area after discharge, the atmosphere therein should be tested by a responsible person as being safe for entry without breathing apparatus. This should also apply to adjoining areas into which the agent has dispersed. Carbon dioxide will tend to collect in low-level spaces such as pits and ducts.

6.16.9 Electrostatic Discharge

Carbon dioxide systems should not be designed, installed, or used for inerting explosive atmospheres. Carbon dioxide systems should not be test-discharged into areas containing explosive atmospheres.

The discharge of carbon dioxide is known to produce electrostatic charges which, under certain conditions, could create a spark.

6.16.10 Electrical Clearances

All system components should be located so as to maintain minimum clearances from live parts as shown in Table 6.9.

As used in this standard, "clearance" is the air distance between equipment, including piping and nozzles, and unenclosed or uninsulated live electrical components at other than ground potential.

The minimum clearances listed in Table 6.9 are for the purpose of electrical clearance under normal conditions; they are not intended for use as "safe" distances during fixed system operation.

The clearances given are for altitudes of 1000 m or less. At altitudes in excess of 1000 m the clearance should be increased at the rate of 1% for each 100-m increase in altitude above 1000 m.

TABLE 6.9

Clearance from Carbon Dioxide Equipment to Live Uninsulated Electrical Components

Nominal System Voltage (kV)	Maximum System Voltage (kV)	Design BIL (kV)	Minimum Clearances (mm)[a]
To 13.8	14.5	110	178
23	24.3	150	254
34.5	36.5	200	330
46	48.3	250	432
69	72.5	350	635
115	121	550	1067
138	145	650	1270
161	169	750	1473
230	242	900	1930
345	362	1050	2134
500	550	1500	3150
765	800	2050	4242

[a] For voltages up to 161 kV, the clearances are taken from NFPA 70, National Electrical Code. For voltage 230 kV and above, the clearances are taken from Table 124 of ANSI C-2, National Electrical Safety Code.

The clearances are based on minimum general practices related to design basic insulation level (BIL) values. To coordinate the required clearance with the electrical design, the design BIL of the equipment being protected should be used as a basis, although this is not material at nominal line voltages of 161 kV or less.

Up to electrical system voltages of 161 kV, the design BIL kV and corresponding minimum clearances, phase to ground, have been established through long usage. At voltages higher than 161 kV, uniformity in the relationship between design BIL kV and the various electrical system voltages has not been established in practice. For these higher system voltages it has become common practice to use BIL levels dependent on the degree of protection that is to be obtained. For example, in 230-kV systems, BILs of 1050, 900, 825, 750, and 650 kV have been utilized.

Required clearance to ground should also be affected by switching surge duty, a power system design factor that along with BIL must correlate with selected minimum clearances. Electrical design engineers should be able to furnish clearances dictated by switching surge duty. The selected clearance to ground should satisfy the greater of switching surge or BIL duty rather than be based on nominal voltage.

Note that the BIL values in Table 6.9 are expressed as kilovolts, the number being the crest value of the full wave impulse test that the electrical equipment is designed to withstand. For BIL values not listed in the table, clearances may be found by interpolation.

Possible design variations in the clearance required at higher voltages are evident in Table 6.9, where a range of BIL values is indicated opposite the various voltages in the high-voltage portion of the table. However, the clearance between uninsulated energized parts of the electrical system equipment and any portion of the carbon dioxide system should be not less than the minimum clearances provided elsewhere for electrical system insulations on any individual component.

When the design BIL is not available, and when nominal voltage is used for the design criteria, the highest minimum clearance listed for this group should be used.

6.17 Safety Requirements

The steps and safeguards necessary to prevent injury or death to personnel in areas whose atmospheres will be made hazardous by the discharge of carbon dioxide should include the following:

- Provision of adequate aisleways and routes of exit and keeping them clear at all times

- Provision of the necessary additional or emergency lighting, or both, and directional signs to ensure quick, safe evacuation
- Provision of alarms within such areas that will operate immediately upon activation of the system on detection of the fire, with the discharge of the carbon dioxide and the activation of automatic door closures delayed for sufficient time to evacuate the area before discharge begins
- Provision of only outward swinging, self-closing doors at exits from hazardous areas, and where such doors are latched, provision of panic hardware
- Provision of continuous alarms at entrances to such areas until the atmosphere has been restored to normal
- Provision of adding an odor to the carbon dioxide so that hazardous atmospheres in such areas can be recognized
- Provision of warning and instruction signs at entrances to and inside such areas
- Provision of such other steps and safeguards necessary to prevent injury or death as indicated by a careful study of each particular situation

6.17.1 Total Flooding Systems

6.17.1.1 Normally Occupied Areas

The automatic discharge of the system should be prevented by means of an automatic/manual or manual-only changeover device when persons are or will be present within the protected space or any adjacent area that could be rendered hazardous by discharge of the gas.

Provision should be made for the manual operation of the fire extinguishing system by means of a control situated outside the protected space or adjacent to the main exit from the space.

While the connection between the fire detection system and the gas release is interrupted, the operation of the fire detector should activate the fire alarm.

In order to guard against accidental release of the gas from the storage containers, the supply of carbon dioxide should be isolated by means of a monitored, normally closed valve in the feed line that will open on a signal from the detection system or manual release system.

The manual release push button or pull handle should be housed in a box and protected by a glass front or other quick access front that can be broken manually to gain access to the button or handle.

Entry into a protected space should only be made when the total flooding system has been placed under manual control. The system should be returned to fully automatic control only when all persons have left the space.

For greater protection the manual release could be key-operated with the operating key being retained in an adjacent frangible glass or other quick-access fronted box.

For areas not normally occupied but which will be entered, one of the following should be provided to prevent the automatic release of carbon dioxide when the area has been entered by personnel:

a. An automatic/manual or manual-only changeover device that renders the system capable of manual operation only

b. A manual stop valve sited in the supply line from the storage vessel(s)

During periods of entry, the automatic discharge of carbon dioxide, however brief, should be prevented. The system should be returned to automatic control as soon as all persons have left the space.

6.17.2 Local Application Systems

When unusual circumstances make it impossible for personnel to leave the space protected by a system within the period of the predischarge alarm (e.g., during difficult maintenance work), the automatic operation of the system should be prevented.

A local application system normally presents a lower risk to personnel than a total flooding system since the final developed concentration of extinguishant throughout the space will be lower. However, during the period of discharge it is necessary to produce an extinguishing concentration of gas around the protected area with a risk of high local concentrations. There is a further risk or higher concentrations of gas occurring in pits, wells, shaft bottoms, and similar low areas.

The system should normally be on automatic control if, after considering the geometry of the area in which a local application system is used, it can be established that there is not a foreseeable risk of a hazardous concentration of carbon dioxide being produced in any occupied part.

In assessing the degree of risk to personnel of automatically controlled systems, the need to approach close to the point of discharge or to work within the confines of the protected area should be considered. If it is necessary for personnel to work within an area that is likely to be quickly to be enveloped with CO_2 gas, consideration should be given to providing a predischarge alarm that gives sufficient warning to allow personnel to move away from the protected area before CO_2 is released.

6.17.3 Hazards to Personnel

The discharge of carbon dioxide in fire extinguishing concentration creates serious hazards to personnel, such as suffocation and reduced visibility during and after the discharge period. Consideration should be given to the possibility of carbon dioxide drifting and settling into adjacent places outside of the protected space. Consideration should also be given to where the carbon dioxide will migrate or collect in event of a discharge from a safety relief device of a storage container.

6.17.4 Warning Signs

Appropriate warning signs should be affixed outside of those spaces where concentrations of carbon dioxide gas can accumulate, not only in protected spaces but in adjacent areas where the carbon dioxide could migrate. Typical signs are shown below.

- A typical sign in a protected space:

  WARNING!

  CARBON DIOXIDE GAS

  WHEN ALARM OPERATES

  VACATE IMMEDIATELY

- A typical sign at an entrance to a protected space:

  WARNING!

  CARBON DIOXIDE GAS

  WHEN ALARM OPERATES DO NOT

  ENTER UNTIL VENTILATED

- A typical sign in a nearby space:

  CAUTION!

  CARBON DIOXIDE DISCHARGE

  INTO A NEARBY SPACE MAY COLLECT HERE

  WHEN ALARM OPERATES

  VACATE IMMEDIATELY

- Appropriate warning signs should be placed at every location where manual operation of the system should occur. Below is a typical sign at each manual actuation station:

  WARNING!

  ACTUATION OF THIS DEVICE WILL CAUSE

  CARBON DIOXIDE TO DISCHARGE

  BEFORE ACTUATING, BE SURE PERSONNEL

  ARE CLEAR OF THE AREA

In any use of carbon dioxide, consideration should be given to the possibility that personnel could be trapped in or enter into an atmosphere made hazardous by carbon dioxide discharge. Suitable safeguards should be provided to ensure prompt evacuation, to prevent entry into such atmospheres, and provide means for prompt rescue of any trapped personnel.

Personnel training should be provided. Predischarge alarms should be provided wherever necessary. Self-contained breathing apparatus should be provided for rescue purposes.

All persons that would at any time enter a space protected by carbon dioxide should be warned of the hazards involved.

The predischarge warning signal should provide a time delay of sufficient duration to allow for evacuation under worst-case conditions. Dry runs should be made to determine the minimum time that should be allowed for persons to remove themselves from the hazard area after allowing time to identify the warning signal.

Audible predischarge signals should be provided. Visual signals should be provided if the ambient noise level is high, or if persons with hearing impairment are involved.

All personnel must be acquainted with the fact that discharge of carbon dioxide gas from either high- or low-pressure systems directly at a person may endanger their safety by eye injury, ear injury, or even falls due to loss of balance on the impingement of the high-velocity discharging gas. Contact with carbon dioxide in the form of dry ice can cause frostbite.

To prevent accidental or deliberate discharge, a lockout should be provided when persons not familiar with the systems and their operation are present in a protected space. When protection is to be maintained during the lockout period, a person(s) should be assigned as a fire watch with suitable portable or semiportable firefighting equipment, or means to restore protection. The fire watch should have a communication link to a constantly monitored location. Authorities responsible for continuity of fire protection should be notified of lockout and subsequent restoration of the system.

Further Readings

BS 5306 Part 4. 2001. Fire Extinguishing Installations and Equipment on Premises; Specification for Carbon Dioxide Systems.

Dieken, D. 2008. Fighting fire with CO_2. *Power Engineering (Barrington, Illinois)* 112 (4), 80–82.

Hatakeyama, T., Aida, E., Yokomori, T., Ohmura, R. and Ueda, T. 2009. Fire extinction using carbon dioxide hydrate. *Industrial and Engineering Chemistry Research* 48 (8), 4083–4087.

Hirsch, D., Hshieh, F.-Y., Beeson, H. and Pedley, M. 2002. Carbon dioxide fire suppressant concentration needs for international space station environments. *Journal of Fire Sciences* 20 (5), 391–399.

Liu, C. 2012. Fire fighting of wind extinguisher with CO_2 gas assisted. *Applied Mechanics and Materials* 130–134, 1054–1057.

McGuire, C. 1982. Carbon dioxide fire protection systems. *Plant Engineering (Barrington, Illinois)* 36 (10), 151–155.

NFPA 12. 2000. Standard on Carbon Dioxide Extinguishing Systems Errata.

Son, Y., Zouein, G., Gokoglu, S. and Ronney, P.D. 2006. Comparison of carbon dioxide and helium as fire extinguishing agents for spacecraft, ASTM Special Technical Publication 1479 STP, pp. 253–258.

Takahashi, F., Linteris, G.T. and Katta, V.R. 2008. Extinguishment of methane diffusion flames by carbon dioxide in coflow air and oxygen-enriched microgravity environments. *Combustion and Flame* 155 (1–2), 37–53.

7

Dry Chemical Fire Extinguishing Systems

7.1 Introduction

Dry chemicals remain one of the most efficient extinguishing media available. On a weight basis, a dry chemical system frequently outperforms any gaseous system, including halon, although its applicability to manned areas is not as wide. This chapter specifies the minimum requirements for dry chemical powder fire-extinguishing systems that discharge powder from a container or centrally grouped containers through pipework or a hose by means of expellant gas to a nozzle or nozzles within the hazard area. It covers both stored pressure- and gas-contained systems, with maximum working pressures not exceeding 25 bar at maximum ambient temperature.

7.2 General Information and Requirements

Dry chemical fire suppression systems provide protection for many flammable and combustible liquid (class B) fire hazards. They are particularly suited for outdoor environments where concerns about freezing prevent the installation of water-based systems. Figure 7.1 shows a typical dry chemical fire suppression system.

Dry chemical is a powder composed of very small particles usually of sodium bicarbonate, potassium bicarbonate, urea-based potassium bicarbonate, or monoammonium phosphate with added particulate material supplemented by special treatment to provide resistance to packing, resistance to moisture absorption (caking), and the proper flow capabilities.

Fire-extinguishing mechanisms include

- Interruption of the chain reaction sequence
- Heat absorption effects

FIGURE 7.1
A typical dry chemical fire suppression system.

Multipurpose dry chemical is usually monoammonium-phosphate-based and is effective on fires in ordinary combustibles, such as wood or paper, as well as on fires in flammable liquids and so forth.

The type of dry chemical used in the system should not be changed unless proved to be changeable by a testing laboratory recommended by the manufacturer of the equipment and approved by the oil, gas, and petrochemical company's authorities. Systems are designed on the basis of the flow and extinguishing characteristics of a specific make and type of dry chemical. Types of dry chemical should not be mixed. Mixtures of certain dry chemicals will generate dangerous pressures and will form lumps.

Dry chemical piped fire suppression systems are designed to provide full-time fire protection in high-risk areas where large, potentially disastrous fires could occur. They provide a means of detection and suppression for complex hazards that are too difficult and inaccessible for manual means of firefighting or for areas where fires could be too large and hazardous for firefighters to enter and fight.

Characteristics:

1. Best applicable to fire-extinguishing systems for protection of dangerous and associate articles that involve serious hazard and the danger of quick spread of fire

2. Easy to clean after application of dry chemicals

3. Economical, as they are less contaminative

4. Excellent insulation permits application of dry chemicals to high-tension electric installations such as transformers

5. Handling of the control valve is simple and plain, ensuring easy cleaning of piping after use

Dry chemical agent definitions:

1. *Dry chemical powder fire extinguishing system.* When introduced directly to a fire, dry chemical powder extinguishes flames almost immediately. Dry chemical is a powder composed of very small particles usually of sodium bicarbonate, potassium bicarbonate, urea-based potassium bicarbonate, or monoammonium phosphate with added particulate material supplemented by special treatment to provide resistance to packing, resistance to moisture absorption (caking) and the proper flow capabilities. Multipurpose dry chemical is usually monoammonium phosphate-based and is effective on fires in ordinary combustibles, such as wood or paper, as well as on fires in flammable liquids, etc.

2. *Sodium bicarbonate based dry chemical.* This agent consists primarily of sodium bicarbonate ($NaHCO_3$) and is suitable for use on all types of flammable liquid and gas fires (class B) and also for fires involving energized electrical equipment (class C). Sodium bicarbonate based dry chemical is not generally recommended for the extinguishment of fires in ordinary combustibles (class A), although it may have a transitory effect in extinguishing surface flaming of such materials.

    [Principle]

$$2NaHCO_3 \rightarrow Na_2CO_3 + CO_2 + H_2O$$

$$Na_2CO_3 \rightarrow Na_2O + CO_2$$

3. *Potassium bicarbonate based dry chemical.* This agent consists primarily of potassium bicarbonate ($KHCO_3$) and is suitable for use on all types of flammable liquid and gas fires (class B) and also for fires involving energized electrical equipment (class C). Dry chemicals based on the salts of potassium are not generally recommended for the extinguishment of fires in ordinary combustible (class A), although they may have a transitory effect in extinguishing surface flaming of such materials.

$$2KHCO_3 \rightarrow K_2CO_3 + CO_2 + H_2O$$

$$K_2CO_3 \rightarrow K_2O + CO_2$$

7.2.1 Use and Limitations

7.2.1.1 Use

Types of hazards and equipment that can be protected using dry chemical extinguishing systems include the following:

a. Flammable or combustible liquids and combustible gases. Extinguishment of uncontrolled discharge of flammable liquids or combustible gases can result in a subsequent explosion hazard.

b. Combustible solids having burning characteristics similar to naphthalene and pitch, which melt when involved in fire.

c. Electrical hazards such as transformers or oil circuit breakers.

d. Ordinary combustibles such as wood, paper, or cloth using multipurpose dry chemical when it can reach all surfaces involved in combustion.

e. Restaurant and commercial hoods, ducts, and associated cooking appliance hazards such as deep-fat fryers.

f. Some plastics, depending upon the type of material and its configuration of hazard. For more specific information, consult the manufacturer of the equipment.

7.2.1.2 Limitations

Dry chemical extinguishing systems should not be considered a satisfactory protection for the following:

a. Chemicals containing their own oxygen supply, such as cellulose nitrate

b. Combustible metals such as sodium, potassium, magnesium, titanium, and zirconium

c. Deep-seated or burrowing fires in ordinary combustibles where the dry chemical cannot reach the point of combustion

Before dry chemical extinguishing equipment is considered for use to protect electronic equipment or delicate electrical relays, the effect of residual deposits of dry chemical on the performance of this equipment should be evaluated.

Multipurpose dry chemical should not be considered satisfactory for use on machinery such as delicate electrical equipment because upon exposure to temperatures in excess of 121°C or relative humidity in excess of 50%, deposits will be formed that will be corrosive, conductive, and difficult to remove.

Dry chemical, when discharged, will drift from the immediate discharge area and settle on surrounding surfaces. Prompt cleanup will minimize possible staining or corrosion of certain materials that may take place in the presence of moisture.

7.2.2 Pre-Engineered Systems

Pre-engineered systems are those having predetermined flow rates, nozzle pressures, and quantities of dry chemical. These systems have the specific pipe size, maximum and minimum pipe lengths, flexible hose specifications, number of fittings, and number and types of nozzles prescribed by a testing laboratory. The hazards protected by these systems are specifically limited as to type and size by a testing laboratory based on actual fire tests.

Limitations on hazards that can be protected by these systems are contained in the manufacturer's installation manual that is referenced as part of the listing.

7.3 Powder

Powder used for the initial supply should be suitable for the system and for the intended use.

7.3.1 Classification

Powders are classified according to their potential applications as follows:

- ABC powders: suitable for use on class A, class B, or class C fires
- BC powders: suitable for use on class B or class C fires and that should also be effective on surface fires of class A materials
- D powders: suitable for use on class D fires

The classes of fires are defined as follows:

- Class A: fires involving solid materials, usually of an organic nature, in which combustion normally takes place with the formation of glowing embers
- Class B: fires involving liquids or liquefiable solids
- Class C: fires involving gases or materials in contact with energized electrical power
- Class D: fires involving metals

Figure 7.2 shows a sample of ABC/BC dry chemical fire-extinguishing powder.

This part of the chapter is concerned only with class B fires involving most liquids including gasoline (but excluding liquids of a low fire point such as carbon disulfide) and surface-burning of class A materials that can be detected quickly enough to be extinguished by BC powders.

While powder systems are a suitable means of extinguishing class C fires, the risk of explosion should be carefully considered. Where appropriate and possible the gas flow should be isolated before or as soon as possible after extinction.

FIGURE 7.2
ABC/BC dry chemical fire-extinguishing powder. (Reprinted with permission from Nasco.)

Dry powder chemical may contribute to damage of electrical equipment and should not be used on fires involving computers or delicate instrumentation.

A supply of extinguishing powder that requires special engineering consideration and when approved for use on magnesium fire should be kept within easy reach of each operator performing a machining, grinding, or other operation on magnesium. The powder should be kept in suitable containers with easily removable covers and a hand scoop should be provided at each container for application of the powder. Approved portable extinguishers designed for use with these powders should replace the scoop and container.

The amount of extinguishing powder to be provided will depend on the amount of chips of turnings involved. Where conditions permit the development of a fire requiring a large quantity of powder, that quantity should be provided and long-handled shovels provided for its application. Heat-resistant gloves and face guards should be available for protection of the personnel applying the powder.

Containers of extinguishing powder should be plainly labeled.

Extinguishing powder should be applied by making a ring around the fire with the powder and then spreading the powder evenly over the surface of the fire to a depth sufficient to smother it. Care should be taken to avoid scattering the burning metal. If smoking continues in spots, more powder should be added as required. Where burning magnesium is on a combustible surface, a 25- to 50-mm layer of powder should be spread out nearby after the fire has been covered as described above, and the burning metal shoveled onto this layer with additional powder added as required.

Powder systems are not suitable as a means of inerting explosive atmospheres.

7.3.2 Chemical Composition

The composition of the powder should be suitable for the intended application.

Extinguishing powders are composed of very small particles of a solid extinguishing medium treated with selected flow additives to give resistance to packing, moisture absorption, and caking during storage and to give free-flowing qualities when discharged through pipework or hoses and nozzles.

Most BC powders are based on alkali metal salts, usually bicarbonate. For general use, sodium bicarbonate is less effective than potassium bicarbonate, but because sodium salts are less expensive they generally prove more cost effective or are preferred for use on deep-fat fryers where a surface layer of powder and saponified fat is produced that reduces the possibility of reignition. Carbonate powders, which are mixtures or compounds of urea and bicarbonate, are more effective than simple bicarbonates.

Monoammonium phosphate is the chemical most generally used for ABC powders; on BC fires it is no less effective than sodium bicarbonate, but it is not as effective as the potassium salts.

A wide range of chemicals has been used for class D fires, some for use on radioactive metals and some for use on nonradioactive metals.

The following should be noted:

- The preferred name for sodium bicarbonate is sodium hydrogen carbonate.
- The preferred name for potassium bicarbonate is potassium hydrogen carbonate.
- The preferred name for monoammonium phosphate is ammonium dihydrogen orthophosphate.
- Containers used for one powder should not be refilled with a different powder. It is most important that mixing or cross-contamination of different types of powders be avoided. Some mixtures can react, sometimes after a long delay, producing water and carbon dioxide with consequent caking of the powder, and in closed containers, a pressure rise. This rise in pressure could cause sealed containers to explode.

7.3.3 Particle Size

The particle size of the powder should be adequate to achieve extinction and suitable for discharge by the system. The finer the particles of extinguishing medium the more effective it is in extinguishing fire. The effect is not equally marked for all powders; carbonate powders show little increase in effectiveness with finer size, since their effectiveness depends on decrepitating within the flame. Finer powders are more difficult to discharge from the container and to project for any distance from the nozzle. They also tend to clog and pack more easily in pipework, and because the bulk density is less, less can be held in a given container or vessel.

7.4 Dry Chemical and Expellant Gas Supply

7.4.1 Quantity

The amount of dry chemical in the system should be at least sufficient for the largest single hazard protected or for the group of hazards that are to be protected simultaneously.

7.4.2 Quality

The dry chemical used in the system should be supplied by the manufacturer of the equipment. The characteristics of the system are dependent on the composition of the dry chemical and the type of expellant gas, as well as other factors, and therefore it is imperative to use the dry chemical provided by the manufacturer of the system and the type of expellant gas specified by the manufacturer of the system.

Where carbon dioxide or nitrogen is used as the expellant gas, it should be of good commercial grade and free of water and other contaminants that might cause container corrosion.

7.4.3 Reserve Supply

Where a dry chemical system protects multiple hazards by means of selector valves, sufficient dry chemical and expellant gas should be kept on hand for one complete recharge of the system. For single-hazard systems, a similar supply should be kept on hand if the importance of the hazard is such that it cannot be shut down until recharges can be procured.

7.4.4 Storage

Storage of charging supplies of dry chemical should be in a constantly dry area and the dry chemical should be contained in metal drums or other containers that will prevent the entrance of moisture even in small quantities. Prior to charging the dry chemical chamber, the dry chemical should be carefully checked to determine that it is in a flowing condition.

7.4.5 Container

The dry chemical container and expellant gas assemblies should be located near the hazard or hazards protected, but not where they will be exposed to a fire or explosion in these hazards.

The dry chemical container and expellant gas assemblies should be located so as not to be subjected to severe weather conditions, or to mechanical, chemical, or other damage. When excessive climatic or mechanical exposures are expected, suitable enclosures or guards should be provided.

TABLE 7.1

Propellants

Material	Maximum Water Content (% Mass)
Air	0.006
Argon	0.006
Carbon dioxide	0.015
Helium	0.006
Nitrogen	0.006

The dry chemical container and expellant gas assemblies utilizing nitrogen should be located where the ambient temperature is normally, between −40°C and 48.9°C. Assemblies utilizing carbon dioxide should be located where the ambient temperature is normally between 0°C and 48.9°C. If temperatures are outside these limits, the equipment should be listed for such temperatures or methods should be provided for maintaining the temperatures within the ambient ranges given.

The dry chemical container and expellant gas assemblies should be located where they will be easy to inspect, maintain, and service.

7.4.6 Propellant

The propellant should be one or more of the gases listed in Table 7.1.

7.5 General Design Principles

7.5.1 Steady Flow

Systems should maintain nonsurging two-phase flow of powder and propellant without separation.

System manufacturers and designers have reference data for flow of a particular powder through pipework and nozzles that take account of the general principles of two-phase flow.

Separation occurs more readily at low velocity (low flow rate at constant pipe diameter). Pipe diameters should be small enough to give powder flow rates of not less than 0.05 kg/mm² cross-sectional area (equivalent to 1.5 kg/s in 25-mm pipe). More or less constant velocity is achieved in balanced systems by reducing pipe diameters by a factor of $\sqrt{2}$ at each junction (see Figure 7.3a).

Separation will occur at points where the direction of flow changes and the following principles should be observed to minimize and compensate for this effect:

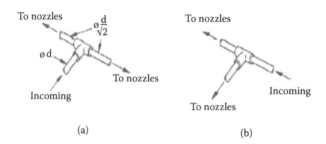

FIGURE 7.3
Direction of flow-through tees. (a) Correct and (b) incorrect.

   a. Changes of direction of pipe runs should be achieved only by the use of elbows (preferably 90°), not by the use of bends.
   b. Pipe runs should be divided only by the use of right angle tees in the configuration shown in Figure 7.3a or other equally effective means. The configuration shown in Figure 7.3b will lead to unequal division of the flow and irregular flow and should not be used.
   c. Separation should occur at elbows but recombination of the phases will take place within a length of pipework equal to 20 pipe diameters; if division of flow in the plane of the elbow occurs within this distance, flow irregularities should occur.
   d. Nozzles should be mounted in elbows on the pipe.

7.5.2 Flow Distribution

Powder systems should provide the intended design flow at each nozzle.
   Even distribution of flow is more easily achieved in balanced systems. Unbalanced systems are more difficult to design, are more likely to have variable and unpredictable performance, and are more likely to have fluctuating rates of discharge from the nozzles.
   Correspondingly greater care should be taken in the design of unbalanced systems; wherever possible a test discharge should be carried out to verify correct function. The simplest form of a balanced system is a symmetrical array of pipework nozzles, all of the same size and at the same elevation, as illustrated in Figure 7.4.
   The pipework from one outlet of any tee is balanced by an identical array at the other outlet with the same length of pipework and the same number and arrangement of fittings.
   In practice the symmetry need not be exact. Effective balance can be achieved provided that the equivalent lengths of the various corresponding arrays are within ±5%, which allows for some variation of the actual length of pipework but for little difference in the number of fittings.

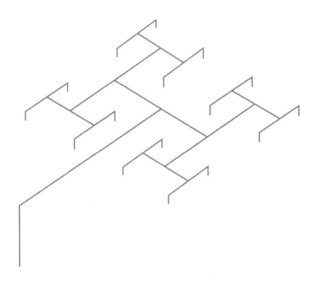

FIGURE 7.4
Symmetrical balanced system (all nozzles discharge at the same rate, R).

Where nozzles of differing discharge rates are used, the system will be nonsymmetrical. Such a system can be balanced by selecting the larger nozzles to discharge at 2, 4, 8, and so forth times the rate R of the smallest. The orifice diameters will be less than 2, 4, 8, and so forth times that of the smallest nozzle to provide additional flow resistance (pressure drop) to balance the resistance in the corresponding longer nonidentical pipe array.

7.5.3 Calculation and Design

The process of computing with the use of equations, graphs, or tables, the system characteristics such as flow rate, pipe size, area, or volume protected by each nozzle, nozzle pressure, and pressure drop, is not required for listed pre-engineered systems since these systems must be installed in accordance with their pretested limitations described in the manufacturer's installation manual.

7.5.4 Hazards to Personnel

Powders should not present a toxic hazard. In extinguishing concentrations, class B powders generally available are of minimal chemical toxicity, but some types of powder, in particular class D powders, need special precautions in use. Powders are not pleasant to inhale and some will cause severe irritation, albeit temporary, particularly if breathed for a considerable period. A powder cloud in a confined space severely reduces visibility and persons within the space may suffer loss of orientation and consequent collision with obstructions within the space.

7.5.5 Contamination

Precautions should be taken to minimize the effects on other materials of contamination by powder.

The powder discharged from a system will cover all exposed surface in the vicinity. If this is cleaned up within a few hours there are normally no problems. Powders exposed to air absorb moisture, and prolonged contact with this damp powder may cause corrosion of some metals.

Particular attention should be paid to the cleaning of open machinery and electrical equipment.

7.6 Types of Systems

For the purposes of this part of Chapter 7, powder systems should be classified as one of the following types:

1. Total flooding systems
2. Local application systems
3. Manual hose reel systems
4. Monitor systems

Systems are also distinguished as either stored pressure systems or gas container systems. Systems are sometimes elsewhere referred to as engineered (designed especially for a particular hazard), or pre-engineered (of a size and design that allows them to be installed for any hazard within certain limits of size and type).

7.6.1 Systems Protecting One or More Hazards

If more than one hazard is involved simultaneously in fire by reason of their proximity, the hazards should be protected by individual systems installed to operate simultaneously or by a single system designed to protect all hazards that may be simultaneously involved. Any hazard that will allow fire propagation from one area to another should constitute a single fire hazard.

7.6.2 Hose Reel

Where a manual hose reel system is used on a hazard that is also protected by a fixed system, separate dry chemical supplies should be provided. An exception is

- If a single dry chemical supply is used for both a hose reel system and a fixed nozzle system, the hazards protected by the two systems should be separated so that the hose reel cannot be simultaneously used on the hazard protected by the fixed monitor system.

7.7 Total Flooding Systems

A total flooding system means a supply of dry chemical permanently connected to fixed piping with fixed nozzles arranged to discharge dry chemical into an enclosed space or enclosure about the hazard.

This type of system should be used only where there is a permanent enclosure about the hazard that is adequate to enable the required concentration to be built up.

The leakage of dry chemical from the protected space should be minimized since the effectiveness of the flooding system depends on obtaining an extinguishing concentration of dry chemical.

In total flooding systems, the rate of application should be such that the design concentration in all parts of the enclosure can be obtained within 30 seconds.

7.7.1 Use of Total Flooding Systems

The design methods of this section should be used only for fixed nozzle systems where there is a permanent enclosure about the hazard.

Fires that can be extinguished by total flooding are surface fires involving flammable liquids and solids.

7.7.2 Design Conditions

The quantity of powder discharged and the rate of application should be sufficient to build up and maintain the specified concentration throughout the enclosure with an adequate margin of safety to compensate for any uncloseable openings and for any ventilation system that is not shut down or closed off on operation of the system.

Loss of powder from the enclosure generally reduces effectiveness and should in most cases be minimized by closing openings and shutting off ventilation systems; however, where extraction ductwork forms part of the hazard, it may be preferable to leave the ventilation system running to facilitate extinction in the ductwork.

7.7.2.1 Minimum Quantity

The minimum quantity needed, M (in kilograms), should be assessed on the basis that

$$M = M1 + M2 + M3 + M4$$

where

M1   is the basic quantity (in kilograms) directly related to enclosure volume

M2   is an additional quantity (in kilograms) to compensate for openings, each less than 5% of the bounding area, where the aggregate area of all such openings exceeds 1% of the bounding area

M3   is an additional quantity (in kilograms) to compensate for openings each of area not less than 5% of the bounding area

M4   is an additional quantity (in kilograms) to compensate for any ventilation system that is not shut or closed down, determined as an addition to the volume enclosure equal to the volume of the air entering or removed from the enclosure during the discharge

The quantities M1 and M2 should be evenly distributed throughout the enclosure; the quantity M3 should be applied across the whole area of each relevant opening in proportion to its area; and the quantity M4 should be applied at the points of air entry into the enclosure.

Special venting may be necessary to avoid excessive pressure buildup resulting from the amount of propellant discharged.

The powder manufacturer should be consulted for the appropriate design criteria. For good-quality sodium-bicarbonate-based powder, the following may be used:

M1 (in kg) = 0.65 × enclosure volume (in m³)

M2 (in kg) = 2.5 × area of openings (in m²) (each less than 5%)

M3 (in kg) = 5.0 × area of opening (in m²) (each not less than 5%)

M4 (in kg) = 0.65 × ventilation rate (in m³/s) × discharge time of system (in s)

Example 7.1

An enclosure 5 m × 10 m × 3 m is to be protected by a sodium bicarbonate powder total flooding system. There is no ventilation system and the aggregate area of uncloseable openings is 1.5 m.

Enclosure volume 5 × 10 × 3 = 150 m³

Bounding area 2 (5 × 10) + 2 (3 × 10) + 2 (3 × 5) = 190 m²

The area of uncloseable openings is less than 1% of the bounding area, so no compensatory powder (M2 or M3) is needed. The minimum quantity needed of M is

M = M1 = 0.65 × 150 = 97.5 kg

Example 7.2

The enclosure described in Example 7.1 is fitted with a ventilation system discharging 15 m³/min that is not shut down when the system operates. The additional quantity M4 needed, if the discharge time is to be 25 s, is

$$M4 = 0.65 \times 15 \times 25/60 = 4.1 \text{ kg}$$

This is to be applied at the point of air entry.

Example 7.3

The enclosure described in Example 7.1 has two uncloseable openings:

  a. 2 m × 5 m = 10 m²
  b. 2 m × 1 m = 2 m²

Since opening a exceeds 5% of the bounding area, additional powder M3 to be applied at the opening is

$$M3 = 5 \times 10 = 50 \text{ kg}$$

Since opening b exceeds 1% of the bounding area, but is less than 5%, additional powder M2 to be applied with M1 is

$$M2 = 2.5 \times 2 = 5.0 \text{ kg}$$

7.7.2.2 Minimum Rate

The minimum rate of discharge R (in kg/s) should be not less than that given by the equation

$$R = M/30$$

where
  M is the quantity discharged (in kg)

Where no additional quantities (M2, M3, or M4) are to be provided, the minimum rate corresponds to a rate of 0.022 times the enclosure volume (in m³) for a good-quality sodium bicarbonate powder.

To give safe design conditions, factors of safety are applied as follows:

  Minimum design quantity = twice the experimental minimum quantity
  Minimum design rate = twice the experimental critical rate

The design conditions are usually presented in the form

Minimum quantity = constant × volume of enclosure
Minimum rate = constant × volume of enclosure

Additional allowance is made for loss of powder through openings and by ventilation.

7.7.3 Enclosure

The volume used in calculation should be the gross volume of the enclosure, less only the volume of any permanent, impermeable, noncombustible building elements within the enclosure.

The protected volume should be enclosed by rigid elements of construction and classified as noncombustible. The area of these elements should be not less than 55% of the bounding area of the enclosure. A fire-resistance test should be in accordance with BS 476 pt 21.

Where openings can be closed, these should be arranged to close before or at the start of discharge.

The area of any drop curtains used in the enclosure should not exceed 30% of the bounding area.

The aggregate area of uncloseable openings, whether in the sides, bottom, or top, should not exceed 15% of the bounding area.

7.7.4 Nozzle Selection and Distribution

The nozzle should be positioned to provide the extinguishing concentration of powder throughout the entire hazard during discharge, and to cover any uncloseable openings of area not less than 5% of the bounding area.

The type of nozzle selected and the disposition of the individual nozzles should be such that the discharge will not splash flammable liquids, dislodge fittings such as ceiling tiles, or create dust clouds that might extend the fire or create an explosion.

7.7.5 Leakage and Ventilation

- The leakage of dry chemical from the protected space should be minimized since the effectiveness of the flooding system depends on obtaining an extinguishing concentration of dry chemical
- Where possible, openings such as doorways and windows should be arranged to close before, or simultaneously with, the start of the dry chemical discharge

- Where forced-air ventilating systems are involved, they should either be shut down and/or closed before or simultaneously with the start of the dry chemical discharge

7.8 Local Application Systems

Local application systems should be used for the extinguishment of fires in flammable or combustible liquids, gases, and shallow solids such as paint deposits, where the hazard is not enclosed or where the enclosure does not conform to the requirements for total flooding. Application of dry chemical should be from nozzles mounted on the tank side or overhead.

1. *Area method:* Applicable to superficial fire, and the amount of extinguishing agent depends on the hazardous area
2. *Volume method:* Applicable to cubical fire, and the amount of extinguishing agent depends on the volume of the object in danger

The hazard should include all areas that are or may become coated by combustible or flammable liquids or shallow solid coatings, such as areas subject to spillage, leakage, dripping, splashing, or condensation, and all associated materials or equipment such as freshly coated stock, drain boards, hoods, ducts, and so forth, that might extend fire outside or lead fire into the protected area.

The design methods of this section should be used for fixed nozzle systems where the hazard is not enclosed, or where the enclosure is large.

Fire that can be extinguished or controlled by local extinguishing systems are surface fires involving flammable liquids and solids.

Examples of hazards that can be protected by local application systems are

a. Dip tanks
b. Quench tanks
c. Spray booths
d. Process machinery
e. Deep-fat fryers
f. Vent stacks
g. Pressure relief vents
h. Vehicle fueling areas

7.8.1 Design Conditions

The quantity, discharge rate, and time of application of powder should provide an extinguishing concentration around the hazard for a time sufficient to extinguish the fire with an adequate margin of safety.

Local application systems should be of the overhead type with nozzles above the hazard, or of the tank side type where the nozzles are positioned to discharge across the surface of the hazard, or a combination of the two arrangements. Different conditions apply for indoor and outdoor use.

Extinction is dependent on the achievement of a sufficient concentration of powder across the area of the fire for a minimum time.

The fixed minimum discharge time for a given area gives a maximum design rate that increases in proportion to the design quantity and is approximately half the experimental limiting maximum value.

The limiting conditions for extinction depend on the method of application (overhead or tank side nozzles), whether the fire is indoors or outdoors, as well as the type of powder and the area of fire. Considerable experimental work may be necessary to establish the safe design conditions for a particular powder. Powder systems are most effective in still air conditions. Wind affects overhead nozzles more than tank side nozzles.

The design methods of this specification apply for more or less still air conditions for overhead nozzles and for wind speeds up to 10 m/s for tank side nozzles.

The use of screens, or the application of powder to an area larger than the hazard, should be considered where wind speeds may be higher.

The powder manufacturer should be consulted for appropriate design criteria.

7.9 Monitor and Hose Reel Systems

The design methods of this section should be used for systems in which the nozzles are movable and each can be directed manually by one person.

Monitor and hose reel systems should be used to supplement fixed fire protection systems or portable fire extinguishers for the protection of specific hazards for which extinguishing powder is suitable.

Wherever possible, powder hose reels should be sited adjacent to the powder container.

Where powder monitors or hose reels are installed in addition to fixed systems, the powder supply for the monitors and/or hose reel should be separate from that for the fixed systems.

7.9.1 Location and Spacing of Monitors and Hose Reels

Monitors and hose reels should be located so that their use is not impeded by a fire in the protected area.

All parts of the protected area should be covered by one or more monitor(s) or hose reel(s).

7.9.2 Rate of Discharge

The rate of discharge of a hose reel should be not less than 1.5 kg/s.

Typical monitors have discharge rates up to 10 kg/s. Discharge rates up to 3 kg/s are used for typical manual hose reel nozzles. At high discharge rates or high pressures, reaction forces make it difficult to control the hose and nozzle. Hose reels discharging at rates above 3 kg/s should only be used where essential because of the size and nature of the hazard and where fixed nozzle and monitor systems cannot be used.

7.9.3 Minimum Quantity

The powder content of a monitor or hose reel system should be sufficient for 30-second continuous operation at the maximum discharge rate of the maximum number of monitors or hose reels that may be used simultaneously.

7.9.4 Hose Reel Design

Where the powder container pressure may exceed 20 bar at any temperature within the operating range, the system should incorporate a device that limits the inlet pressure to the hose to not more than 20 bar.

7.9.5 Monitor Design

Monitors should be suitable for operation by one person directly or from a remote location.

7.9.6 Jetty Deck

One or two fixed installed dry chemical powder package units for each berth may be installed on the jetty deck. The dry powder monitor of the units should be located to cover the ship's manifold and the loading/unloading facilities. The capacity of each unit should be 1500 kg to supply a monitor having a discharge rate of 20–50 kg/s and a throw of 30–50 m, and two hose reels with nozzles having a discharge rate of 2.5 kg/s and a throw of 15 m.

7.10 Alarms and Indicators

An alarm or indicator should be provided to show that the system has operated, that personnel response may be needed, and that the system is in need of recharge.

The extinguishing system should be connected to the alarm system, if provided, in accordance with the requirements of the appropriate signaling system standard so that actuation of the dry chemical system auto/manual will sound the fire alarm as well as provide the function of the extinguishing system.

Two sources of electrical power should be provided. These should consist of a primary (main) supply and a secondary (standby) supply. The primary (main) power supply should have a high degree of reliability, adequate capacity for the intended service, and should consist of one of the following:

a. Light and power service

b. Engine-driven generator or equivalent

7.10.1 Secondary (Standby) Power Supply Capacity and Sources

The secondary (standby) supply should be provided to supply the energy to the system under the maximum normal load for 24 hours and then be capable of receiving one fire alarm signal persisting for 5 minutes in the event of a total power failure or low-voltage condition (less than 85% of the nameplate voltages) of the primary (main) power supply.

The secondary (standby) power supply should automatically transfer to operate the system within 30 seconds of the loss of the primary (main) power supply. The secondary (standby) power supply should consist of one of the following:

a. A storage battery with 24-hour capacity

b. An engine-driven generator

c. Multiple automatic-starting engine-driven generators capable of supplying the energy required with the largest generator out of service

Secondary (standby) power should not be required to operate the evacuation alarm-indicating appliances or other supplemental functions not essential to the receipt of signals at the main control unit.

These systems should be electrically supervised so the occurrence of a single open or single ground fault condition of its installation wiring that prevents the normal operation of the system or failure of the primary electric power supply will be indicated by a distinctive trouble signal.

Alarms indicating failure of supervised devices or equipment should give prompt and positive indication of any failure and should be distinctive from alarms indicating operation or hazardous conditions.

Total flooding and local application systems should give an audible and visible alarm on operation, and where the premises are provided with a main fire alarm system should operate that alarm system.

7.11 Safety Precautions

Suitable provision should be made to safeguard persons in areas where the atmosphere may be made hazardous by the discharge, either planned or accidental, of the fire extinguishing system.

7.11.1 System Blowdown

The system should have a facility to allow residual powder to be blown out of the pipework after system discharge.

Suitable safeguards should be provided to ensure prompt evacuation of contaminated locations, and also to provide means for prompt rescue of any trapped personnel. Safety items to be considered should include but not be limited to personnel training, warning signs, discharge alarms, predischarge alarms, and respiratory protection.

7.11.2 Discharge Prevention during Maintenance

The system should have a device to prevent discharge during system inspection and servicing, which can then be carried out in safety, and also during times when the protected area is undergoing alterations or extensive maintenance.

When dry chemical pressure containers are not attached to piping or hand hose lines, the discharge outlet should be provided with a protective diffusing safety cap to protect personnel from recoil and high flow discharge in case of accidental actuation. Such protective caps should also be used on empty pressure containers to protect threads. These caps should be provided by the manufacturer of the equipment.

7.11.3 Visual Indicators of System Status

Visual indicators should be present at each entrance to

a. An enclosure protected by a total flooding system
b. An area in which personnel are normally present, protected by or adjacent to a local application system

There should be a system status lamp unit having the following indications:

a. Red lamp: system discharged
b. Green lamp: manual control
c. Amber lamp: automatic and manual control

7.11.4 Venting Indication

Systems with closed pipework that is not normally pressurized should be fitted with a device that will indicate the accidental release of propellant or powder into the closed pipework.

7.11.5 Electrical Earthing

All exposed metalwork in systems that are housed near or in buildings or premises with electrical installations should be efficiently earthed to prevent the metalwork becoming electrically charged.

7.11.6 Electrical Hazards

Where exposed electrical conductors are present, clearances should be provided, where practicable, between the electrical conductors and all parts of the system that will be approached during maintenance. Where these clearance distances cannot be achieved, warning notices should be provided and a safe system of maintenance work should be adopted.

7.11.7 Areas Not Normally Occupied but Are to Be Entered

Systems should be provided with a device that is used to prevent automatic discharge of the system while retaining the manual operation, detection, and alarm facilities.

The discharge of large amounts of dry chemical will create hazards to personnel such as reduced visibility and temporary breathing difficulty.

Further Readings

British Standards Institution (BSI), BS 5306 pt 7, Specification for Powder Systems 2009.

Elmore, K.S. 1998. Using dry chem on class B fires. *Fire Engineering* 151 (2), 79–82.

Ewing, C.T., Faith, F.R., Hughes, J.T. and Carhart, H.W. 1989. Flame extinguishment properties of dry chemicals: Extinction concentrations for small diffusion pan fires. *Fire Technology* 25 (2), 134–149.

Ewing, C.T., Faith, F.R., Romans, J.B., Siegmann, C.W., Ouellette, R.J., Hughes, J.T. and Cathart, H.W. 1995. Extinguishing class B fires with dry chemicals: Scaling studies. *Fire Technology* 31 (1), 17–43.

Fu, X., Cai, C., Shen, Z., Ma, S. and Xing, Y. 2009. Superfine spherical hollow ammonium dihydrogen phosphate fire-extinguishing particles prepared. *Drying Technology* 27 (1), 76–83.

Kuang, K., Chow, W.K., Ni, X., Yang, D., Zeng, W. and Liao, G. 2011. Fire suppressing performance of superfine potassium bicarbonate powder. *Fire and Materials* 35 (6), 353–366.

Leng, N., Wang, S. and Han, P. 2012. Development of new fire extinguishing agent for grassland. *Advanced Materials Research* 550–553, 62–70.

National Fire Codes (NFC) (NFPA), Volume 1, Section 17, Dry Chemical Extinguishing Systems, Volume 7 code 480 Dry Powder for Magnesium 2010.

Ni, X., Chow, W.K., Li, Q. and Tao, C. 2011. Experimental study of new gas-solid composite particles in extinguishing cooking oil fires. *Journal of Fire Sciences* 29 (2), 152–176.

Ni, X., Chow, W. and Liao, G. 2008. Discussions on applying dry powders to suppress tall building fires. *Journal of Applied Fire Science* 18 (2), 155–191.

Xing, J., Du, Z.-M., Chen, D.-S. and Li, R.-X. 2011. Preparation of superfine ammonium phosphate dry chemical fire extinguishing agent by using vibratory milling method and surface modification. *Journal of North University of China (Natural Science Edition)* 32 (5), 613–618.

Ye, M.-Q., Han, A.-J., Ma, Z.-Y. and Li, F.-S. 2005. Application of superfine particle and its composite technology to cold aerosol fire extinguishing agents. *Nanjing Li Gong Daxue Xuebao/Journal of Nanjing University of Science and Technology* 29 (2), 236–239.

8

Foam Generating and Proportioning Systems

8.1 Introduction

Fire-fighting foam is a stable mass of small bubbles of lower density than most flammable liquids and water. Foam is a blanketing and cooling agent that is produced by mixing air into a foam solution that contains water and foam concentrate.

This chapter specifies the requirements for foam-producing and liquid concentrates employed for fire extinguishment, and gives designs of fixed and semifixed systems for applying low, medium, and high expansion foam to fires in buildings, industrial plants, and storage facilities. It also covers

- Classes of foam concentrates
- Foam spray systems
- Total flooding systems
- Local application systems
- Wetting agents

8.2 Foams

Foam extinguishes flammable or combustible liquid fires in four ways (see Figure 8.1):

1. Excludes air from the flammable vapors
2. Eliminates vapor release from fuel surface
3. Separates the flames from the fuel surface
4. Cools the fuel surface and surrounding metal surfaces

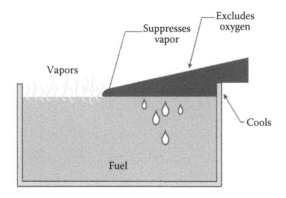

FIGURE 8.1
A schematic for extinguishing flammable or combustible liquid fire using foam. (Reproduced with permission from VDM Imaginary.)

Foam concentrates are categorized into three expansion ranges, as follows:

1. *Low-expansion:* Expansion ratio up to 20:1. Foams designed for flammable liquids. Low expansion foam has proven to be an effective means of controlling, extinguishing, and securing most flammable liquid (class B) fires. Foam has also been used successfully on class A fires where the cooling and penetrating effect of the foam solution is important.

2. *Mid-expansion:* Expansion ratio from 20:1 to 200:1. Medium expansion foams may be used to suppress the vaporization of hazardous chemicals. Foams with expansions between 30:1 and 55:1 have been found to produce the optimal foam blanket for vapor mitigation of highly water-reactive chemicals and low-boiling organics.

3. *High-expansion:* Expansion ratio above 200:1. High expansion foams are designed for confined space firefighting. High expansion foam concentrate is a synthetic, detergent-type foaming agent used in confined spaces such as basements, mines, and onboard ship when used in combination with a high expansion foam generator.

Foam extinguishment is applicable to oils, fats, and highly flammable liquids.

Foam for fire protection is an aggregate of air-filled bubbles formed from aqueous solution and is lower in density than the lightest flammable liquids. It is principally used to form a coherent floating blanket on flammable and combustible liquids lighter than water and prevents or extinguishes fire by excluding air and cooling the fuel. Reignition is excluded by suppression of flammable vapors. It adheres to surfaces and provides protection from adjacent fires. Foam is supplied by fixed pipe systems or portable foam-generating units.

All fixed foam installations should be designed to reduce time lag in charging service lines to a minimum, and provision made for draining all services after use.

Subsurface foam injection facilities should be provided for fixed roof tanks.

8.2.1 Classification of Foam Concentrates

Foam concentrates are liquids, usually aqueous solutions, which are mixed with water to produce the foam solution used to make foam. Foam concentrates are generally classified by composition, and for the purposes of this book are as described in this section.

8.2.1.1 Protein Foam

Protein foam concentrates are aqueous solutions of hydrolyzed protein and are generally used at 3% and 6% concentrations. Regular protein foams (RPs) are intended for use on hydrocarbon fuels only. They produce a homogeneous, stable foam blanket that has excellent heat resistance, burnback, and drainage characteristics.

RP foams have slow knockdown characteristics; however, they provide superior postfire security at very economical cost. RP foams may be used with freshwater or seawater. They *must* be properly aspirated and should not be used with nonaspirating structural fog nozzles.

Protein foams were the first types of mechanical foam to be marketed extensively and have been used since World War II. These foams are produced by the hydrolysis of granulized keratin protein (protein hydrolysate) such as hoof and horn meal, chicken feathers, and so forth. In addition, stabilizing additives and inhibitors are included to prevent corrosion, resist bacterial decomposition, and to control viscosity.

8.2.1.2 Fluoroprotein Foam

Fluoroprotein (FP) foam concentrates are protein foam concentrates with added fluorinated surface active agents. The foam is generally more fluid than protein foam, gives faster control and extinction of the fire, and has greater ability to reseal if the foam blanket is disturbed. FP is resistant to contamination by hydrocarbon liquids and is generally used at 3% or 6% concentration. FPs have fluorochemical surfactants that greatly enhance performance with fast knockdown, improved resistance to fuel pickup, and dry chemical compatibility. They are intended for use on hydrocarbon fuels and select oxygenated fuel additives. As with protein, they have excellent heat resistance burnback and postfire security. FPs may be used with freshwater or seawater. They *must* be properly aspirated and should not be used with nonaspirating structural fog nozzles.

FP foams are made by the addition of special fluorochemical surfactants to protein foam. This enhances the properties of protein foam by increasing foam fluidity and improves the properties of RP foam by providing faster knockdown and excellent fuel tolerance.

8.2.1.3 Film-Forming FP Foam

Film-forming FP (FFFP) foam concentrates are foam concentrates with added fluorinated surface active agents. The foam is more fluid than both protein and standard FP foams. The foam is resistant to contamination by hydrocarbon liquids. The solution is film-forming on some liquid hydrocarbon fuel surfaces and is generally used at 3% or 6% concentration. FFFPs are a combination of fluorochemical surfactants with PF. They are designed to combine the fuel tolerance and burnback resistance of an FP foam with an increased knockdown power. FFFP foams release an aqueous film on the surface of the hydrocarbon fuel.

8.2.1.4 Synthetic Detergent Foam (Mid- and High-Expansion)

Synthetic foam concentrates are solutions of hydrocarbon surface active agents. Fluorinated surface active agents if present are present in amounts that do not lead to film-forming on hydrocarbon liquids. Synthetic foam concentrates are generally used at concentrations between 1% and 6%. They are not generally used in low-expansion foam systems and are not considered in this standard.

Effective on class A fires, high expansion is very useful for confined space firefighting and as a wetting agent. High expansion can be used on small-scale class B hydrocarbon fires. Synthetic foams are a mixture of synthetic foaming agents and stabilizers. Mid-expansion of synthetic detergent based foam is used for suppressing hazardous vapors. Specific foams are required depending on the chemicals involved. High-expansion foams can be used on fixed installations to provide total flooding of warehouses or other enclosed rooms containing class A materials such as wood, paper, plastic, and rubber. Care must be taken with regard to any electrical power source in the area. Fire extinguishment in these cases is rather different from low-expansion foam. High-expansion fire extinguishment really amounts to smothering the fire area and cooling the fuel.

8.2.1.5 Aqueous Film-Forming Foam

Aqueous film-forming foam (AFFF) concentrates are generally based on mixtures of hydrocarbon and fluorinated hydrocarbon surface active agents. Foam solutions made from fluorochemical concentrates are film-forming on some liquid hydrocarbon fuel surfaces and are generally used at 1%, 3%, or 6% concentration. The AFFF family of foams are designed to provide the fastest possible knockdown on hydrocarbon fuels. Their fluidity allows them to quickly flow around obstacles, wreckage, and debris. Different percentages may be selected depending on the user's proportioning hardware. Standard AFFFs are premixable, dry powder compatible, and can be used with either freshwater or seawater. AFFFs may be used through nonaspirating devices; however, for optimum performance aspirating nozzles should be used.

AFFFs are a combination of fluorochemical surfactants and synthetic foaming agents. AFFFs extinguish fires by forming a aqueous film. This film is a thin layer of foam solution that spreads rapidly across the surface of a hydrocarbon fuel, causing dramatic fire knockdown.

The aqueous film is produced by the action of the fluorochemical surfactant reducing the surface tension of the foam solution to a point where the solution can actually be supported on the surface of the hydrocarbon.

8.2.1.6 Alcohol-Resistant Foam

Alcohol-resistant (AR) foam concentrates are formulated for use on foam-destructive liquids; the foams produced are more resistant than ordinary foams to breakdown by the liquid. They may be of any of the classes and may be used on fires of hydrocarbon liquids with a fire performance generally corresponding to that of the parent type. Film-forming foams do not form films on water-miscible liquids. AR foam concentrates are generally used at 6% concentration on water-miscible fuels and at 3% or 6% concentration on hydrocarbon fuels.

AR-AFFF foams are produced from a combination of synthetic detergents, fluorochemicals, and polysaccharide polymer. Polar solvents (or water-miscible) fuels such as alcohols are destructive to nonalcohol-resistant type foams. AR-AFFF foams act as a conventional AFFF on hydrocarbon fuels, forming an aqueous film on the surface of the hydrocarbon fuel. When used on polar solvents (or water-miscible fuels), the polysaccharide polymer forms a tough membrane (see Figure 8.2) that separates the foam from the fuel and prevents the destruction of the foam blanket.

While some concentrates are designed for use on hydrocarbon fuels at 3% and polar solvents at 6%, today's newer formulations are designed to be

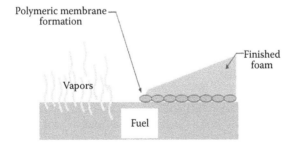

FIGURE 8.2
Formation of polymeric membrane.

used at 3% on both fuel groups. These newer formulations provide more cost-effective protection of alcohol type fuels, using half the amount of concentrate of a 3%/6% agent. The use of a 3 × 3 AR-AFFF also simplifies setting the proportioning percentage at an incident, since it is always 3%. Overall, AR-AFFFs are the most versatile type of foam available today, offering good burnback resistance, knockdown, and high fuel tolerance on both hydrocarbon and polar-solvent (or water-miscible) fires.

8.2.2 Foam Characteristics

To be effective, a good foam must contain the right blend of physical characteristics:

Knockdown speed and flow: This is time required for a foam blanket to spread across a fuel surface or around obstacles and wreckage in order to achieve complete extinguishment

Heat resistance: The foam must be able to resist the destructive effects of heat radiated from any remaining fire from the liquid's flammable vapor and any hot metal wreckage or other objects in the area

Fuel resistance: An effective foam minimizes fuel pickup so that the foam does not become saturated and burn

Vapor suppression: The vapor-tight blanket produced must be capable of suppressing the flammable vapors and so minimize the risk of reignition

Alcohol resistance: Due to alcohol's affinity to water and because a foam blanket is more than 90% water, foam blankets that are not alcohol-resistant will be destroyed

Table 8.1 summarizes the properties and comparisons of fire-fighting foam types.

TABLE 8.1

Properties and Comparisons of Fire-Fighting Foam Types

Property	Protein	FP	AFFF	FFFP	AR-AFFF
Knockdown	Fair	Good	Excellent	Good	Excellent
Heat resistance	Excellent	Excellent	Fair	Good	Good
Fuel resistance (hydrocarbons)	Fair	Excellent	Moderate	Good	Good
Vapor suppression	Excellent	Excellent	Good	Good	Good
Alcohol resistance	None	None	None	None	Excellent

8.2.2.1 Foam Concentrations

Foam concentrates are designed to be mixed with water at specific ratios; 6% concentrates are mixed with water at a ratio of 94 parts water to 6 parts foam concentrate. For example, if you were going to "premix" a batch of foam concentrate with water to make 100 gallons of foam solution, you would mix 6 liters of foam concentrate with 94 liters of water. When using a 3% foam you would mix 3 liters of foam concentrate with 97 liters of water. Once proportioned (mixed) with water, the resulting foam solutions of a 3% foam or a 6% foam are virtually the same with regard to performance characteristics.

A 3% concentrate is more concentrated than a 6%, therefore requiring less product to produce the same end result. The trend of the industry is to reduce the proportioning percentages of foam concentrates as low as possible.

Lower proportioning rates allow the user to minimize the amount of space required to store the concentrate. By switching from a 6% foam to a 3% foam you can either double your fire-fighting capacity by carrying the same number of gallons or cut your foam supply in half without compromising suppression capacity.

Lower proportioning rates can also reduce the cost of foam system components and concentrate transportation. AR foam concentrates that have two percentages on the pail label are designed to be used at two different ratios. For example, a 3%/6% foam concentrate is designed to be used on hydrocarbon fuels at 3% and polar-solvent fuels at 6%.

This is due to the amount of active ingredient that provides the foam blanket with alcohol resistance. Newer formulations of AR-AFFFs have improved alcohol resistance so that they can be used at 3% on either hydrocarbons or polar solvents. Wetting agents and class A foam concentrates are less complicated mixtures of ingredients that can be proportioned at rates lower than 1%, typically 0.1%–1.0%. A premix at 0.5% is one-half liter of concentrate to 99.5 liters of water.

8.2.3 Basic Guidelines for Foam

8.2.3.1 Storage

If manufacturer recommendations are followed, protein or synthetic foam concentrates should be ready for active service after many years of storage.

8.2.3.2 Water Temperature and Contaminants

Foams in general are more stable when generated with lower-temperature water. Although all foam liquids will work with water in excess of 100°F, preferred water temperatures are 35°F–80°F. Either freshwater or seawater may be used. Water containing known foam contaminants such as detergents, oil residues, or certain corrosion inhibitors may adversely affect foam quality.

8.2.3.3 Combustible Products in Air

It is desirable to take clean air into the foam nozzle at all times, although the effect of contaminated air on foam quality is minor with low-expansion foams.

8.2.3.4 Water Pressures

Nozzle pressures should be held between 50 and 200 psi. If a proportioner is used, proportioner pressure should not exceed 200 psi. Foam quality deteriorates at higher pressures. Range falls off at lower pressures.

8.2.3.5 Unignited Spills

Where flammable liquids have spilled, fires can be prevented by prompt coverage of the spill with a foam blanket. Additional foam may be necessary from time to time to maintain the blanket for extended periods until the spill has been cleaned up.

8.2.3.6 Electrical Fires

Foam should be considered nearly the same as water when used on electrical fires, and is therefore not generally recommended. If it is used, a spray rather than a straight stream is safer; however, because foam is cohesive, even a dispersed (spray) foam stream is more conductive than a water fog.

Note that electrical systems should be de-energized via manual or automatic shutdowns before applying water or foams.

8.2.3.7 Vaporized Liquids

Foam is not recommended for use on materials that may be stored as liquids but are normally vapor at ambient conditions, such as propane, butadiene, and vinyl chloride. Fire-fighting foam is not recommended for use on materials that react with water such as magnesium, titanium, potassium, lithium, calcium, zirconium, sodium, and zinc.

8.2.4 Expansion

Finished foam is a combination of foam concentrate, water, and air. When these components are brought together in proper proportions and thoroughly mixed, foam is produced. Figure 8.3 shows how foam is made through typical proportioning equipment.

Foams are classified by their expansion ratio (expansion), the ratio of the volume of the made foam to the volume of the solution from which it is made, as follows:

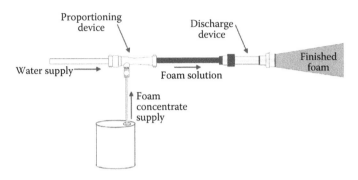

FIGURE 8.3
Foam formation through typical proportioning equipment.

a. Low-expansion foams, with expansions between 1 and 20, are intended primarily for application to the surface of flammable liquid fires.

b. Medium-expansion foams, with expansions between 21 and 200, are intended for surface application or for application to fires that require a certain depth of foam to obtain coverage (e.g., up to depths of 4 m).

c. High-expansion foams, with expansions between 201 and 1000, are intended for filling enclosures within which a number of fires are burning at different levels up to 10 m. High-expansion foam concentrate is a careful blend of premium-grade surface-active agents and synthetic, detergent type foaming agents. This concentrate is proportioned at about 1½% with water then expanded with air to form high-expansion foam. When used in combination with a high-expansion foam generator, it makes a superior foam with an average expansion of 500 to 1, but can be used with generators producing foams from 200:1 to 1000:1. The foam liquid produces a foam with a smooth texture and uniform bubble that has excellent fluidity to flow around and over obstructions.

Its tough adhesiveness provides a foam of superior stability. This foam liquid has a high expansion capability and good stability when used with either freshwater or seawater. In addition to being a superior foaming agent, it also has wetting ability to increase the penetrating effect of water on deep-seated class A fires. Its all-purpose use eliminates the need for stocking wetting agents.

8.2.5 Foam Solution

Protein foams are made from concentrates based on proteinaceous products alone and with the addition of fluorinated additives to give FP foams.

Another development is a mixture of hydrocarbon and fluorinated surfactant materials with stabilizers to form AFFF type concentrates. These are used for the extinction of fires in hydrocarbons and other water-immiscible flammable liquids.

Modifications of FP foams and AFFFs have been developed that are suitable for use on fires in water-miscible flammable liquids (e.g., alcohols and ketones). These are the AR foams or general-purpose foams. Medium-expansion foams can be made from low-expansion foam concentrates, but it is more usual to make them from a surfactant concentrate based on ammonium lauryl ether sulfate. High-expansion foams are also made from this concentrate.

The use of low-expansion foams is restricted largely to the extinction of flammable liquid fires, but there is no reason why they should not also be applied to solid fuels, provided adequate coverage can be obtained to exclude air as much as possible. In practice, low-expansion foams may either be applied to the surface of a burning liquid (surface application) or beneath the surface so that the foam stream floats to the surface and spreads to form a protective layer or blanket on it (subsurface application). The first method is more common and can be used against spill fires, fires in bunded areas, fuel tanks, and so forth using the appropriate equipment.

Medium-expansion foams are generally applied to the surface of flammable liquid fires, either by handheld foam-making branches or from fixed foam makers. This medium can also be used effectively on solid fuel fires, or mixed fires of solid and liquid combustibles. Typical examples would be on a fire in a gas-turbine-driven generating set, a fuel fire in the engine room of a ship, heat-treatment baths, or places where fuel spillages can occur (e.g., in pit areas, garages or overhaul shops).

High-expansion foams are similar in action to medium-expansion foams, but they require generators in which air is supplied by a fan in order to achieve the rate of flow necessary for their production. They work by blanketing or smothering a fire, but the degree of cooling available is much less than for medium-expansion foams due to their lower water content. They can, however, produce much greater foam depths of at least 10 m and can therefore smother a fire in goods stored on high racks. For this, the depth of the foam needs to increase rapidly in order to match or overtake the upward rate of development of the fire.

8.2.6 Low-Expansion Foam: Types and Requirements

8.2.6.1 System Description

A system consists of an adequate water supply, a supply of foam concentrate, suitable proportioning equipment, a proper piping system, foam makers, and discharge devices designed to adequately distribute the foam over the hazard. Some systems may include detection devices.

These systems are of the open outlet type in which foam discharges from all outlets at the same time, covering the entire hazard within the confines of the system.

Self-contained systems are those in which all components and ingredients, including water, are contained within the system. Such systems usually have a water supply or premix solution supply tank pressurized by air or inert gas. The release of this pressure places the system into operation.

There are four basic types of systems:

1. Fixed
2. Semifixed
3. Mobile
4. Portable

8.2.6.2 Fixed System

Fixed systems are complete installations piped from a central foam station and discharging through fixed delivery outlets to the hazard area to be protected. Any required pumps are permanently installed.

8.2.6.3 Semifixed Systems

Semifixed systems can be either

a. The type in which the hazard is equipped with fixed discharge outlets connected to piping that terminates at a safe distance. The fixed piping installation may or may not include a foam maker. The necessary foam-producing materials are transported to the scene after the fire starts and are connected to the piping.
b. The type in which foam solutions are piped through the area from a central foam station, the solution being delivered through hose lines to portable foam makers such as monitors, foam towers, hose lines, and so forth.

8.2.6.4 Mobile Systems

Mobiles systems include any foam-producing unit that is mounted on wheels and is self-propelled or towed by a vehicle. These units should be connected to a suitable water supply or utilize a premixed foam solution.

8.2.6.5 Portable Systems

Portable systems are those in which the foam-producing equipment and materials, hose, and so forth, are transported by hand.

8.2.7 Uses

The requirements of this section apply to low-expansion foam and foam systems suitable for extinguishing fires on a generally horizontal flammable liquid surface.

Extinction is achieved by the formation of a blanket of foam over the surface of the burning liquid. This provides a barrier between the fuel and air, reducing the rate of emission of flammable vapors to the combustion zone and cooling the liquid.

Low-expansion foam is not generally suitable for the extinction of running fuel fires (e.g., fuel running from a leaking container or from damaged pipework or pipe joints), however, low-expansion foam can control any pool fire beneath the running fire that is then extinguished by other means.

Low-expansion foam, except for the AR type, is generally not suitable for use on foam-destructive liquids that cause rapid breakdown of the foam.

Low-expansion foam is not suitable for use on fires involving gases or liquefiable gases with boiling points below 0°C or cryogenic liquids. The advice of the manufacturer should, therefore, be sought for this application.

8.2.8 Application Method

Low-expansion foams should be applied

a. Gently to the surface of the burning liquid (as pourer or semisubsurface systems)
b. Forcefully to the surface of the burning liquid (as in monitor and branch pipe systems)
c. Below the surface so that they float to the surface under their own buoyance (as in subsurface systems)

8.2.9 Potential Hazards

Foam systems should include provision to minimize the danger when foam is applied to liquids above 100°C, energized electrical equipment, or reactive materials. Since all foams are aqueous solutions, where liquid fuel temperatures exceed 100°C they are ineffective, and particularly where the fuel depth is considerable (e.g., tanks), can be dangerous in use. The foam and drainage of the water from the foam can cool the flammable liquid but boiling of this water may cause frothing or slopover of the burning liquid.

Boilover, which may occur even where foam is not applied, is a more severe and hazardous event. Large-scale expulsion of the burning contents of a tank is caused by the sudden and rapid boiling of water in the base of the tank or suspended in the fuel. It is caused by the eventual contact of the upper layer of liquid fuel in the tank, heated to above 100°C by the fire, with the water layer.

Particular care should be taken when applying foam to high-viscosity liquids, such as burning asphalt or heavy oil, above 100°C. Because foams are made from aqueous solutions they are dangerous to use on materials that react violently with water, such as sodium or potassium, and should not be used where they are present. A similar danger is presented by some other metals such as zirconium or magnesium only when they are burning.

Low-expansion foam is a conductor and should not be used on energized electrical equipment in situations where this would be a danger to personnel.

8.2.10 Compatibility with Other Extinguishing Media

The foam produced by the system should be compatible with any media provided for application at or about the same time as foam.

Certain wetting agents and some extinguishing powders are incompatible with foams, causing a rapid breakdown of the latter. Only media that are substantially compatible with a particular foam should be used in conjunction with it.

Use of water jets or sprays should adversely affect a foam blanket. They should not be used in conjunction with foam unless account is taken of any such effects.

8.2.11 Compatibility of Foam Concentrates

Foam concentrate (or solution) added or put into a system should be suitable for use and compatible with any concentrate (or solution) already present in the system. Foam concentrates or foam solutions, even of the same class, are not necessarily compatible, and it is essential that compatibility be checked before mixing two concentrates or premixed solutions.

8.2.12 Foam Destructiveness

For the purposes of this standard, when taking into account foam destructiveness, flammable liquids are considered as falling into two groups:

1. Hydrocarbons and those nonhydrocarbon liquids that are not more foam-destructive than hydrocarbons
2. Foam-destructive liquids that are generally water soluble and are much more foam-destructive than hydrocarbons

Special types of concentrates are used for foam-destructive liquids. Higher rates of application are specified for foam-destructive liquids than for hydrocarbons and it is usually essential to use gentle application methods.

The degree of foam destructiveness varies, however, and isopropyl alcohol, butyl alcohol, isobutyl methyl ketone, methyl methacrylate monomer, and mixtures of water-miscible liquids in general may require higher application rates.

Protection of products such as amines and anhydrides, which are particularly foam-destructive, requires special consideration.

8.3 System Design

The system should be designed to suit the particular hazard. The following should be considered:

a. Full details of the flammable liquid, its storage, handling, and location
b. The most suitable class of foam concentrate and concentration
c. The most suitable solution application rate
d. The most suitable equipment for making and delivering foam
e. Required system operation time
f. Quantity of foam concentrate required for extinction
g. The most suitable proportioning method(s)
h. Pipework sizes and pressure losses
i. Water supply quantity, quality, and pressure
j. Method of system operation and any fire or gas detection equipment required
k. Any special considerations, such as the use of electrical equipment in areas where flammable vapors may be present
l. Reserve foam concentrate supply
m. Drainage and bunds
n. Environmental conditions

8.3.1 Foam Supply System

8.3.1.1 Fixed Roof Tanks

Fires on this type of tank should be extinguished with the aid of mobile foam monitors fed by a fire truck supplying a quantity of foam solution to 0.1 dm³/s per m² of liquid surface to be extinguished. The foam concentrate storage capacity and quantity of monitors and fire trucks should be determined in relation to the diameter of the tank.

Fixed foam equipment on tankage is not required, other than base injection equipment on fixed roof tanks and the dry rising main to the gager's platform on some floating roof tanks for fighting rim fires. For subsurface foam systems.

Experience has indicated that fixed installed foam chambers at the top course of the tank wall are not reliable and should no longer be applied because of the following:

- Corrosion, which can cause rupture of the gas seal plate, allowing gas to pass to the foam solution inlet connection at the road side and thus creating a hazardous situation
- An explosion preceding a fire in a fixed roof tank will most likely rupture the foam chamber supply piping, making the equipment nonoperational at the critical moment

8.3.1.2 Floating Roof Tanks

Fire protection for floating roof tanks is based on rim fires only, because this type of tank has an excellent fire record and if maintained properly the possibility of a large fire is remote. Overfilling the tank should be avoided under all circumstances.

A high alarm and an independent high-high alarm are installed to warn the plant operating personnel before overfilling levels are reached. When required, the tank filling valve(s) can be made to close automatically by the level alarm signal. Where access is provided and situations permit, rim fires should be fought from the wind girder, otherwise they will be fought from the gager's platform or from ground level.

For tanks less than 48 m in diameter a 100-mm single dry rising main should be provided from outside the bund to the gager's platform, terminating in a fixed foam pourer/generator and in two 65-mm feed hose couplings complete with isolating valves and air vents for quick expulsion of entrapped air. The riser is provided to enable foam solution to be pumped up to the gager's platform and allow foam to be produced either via the fixed foam pourer or at a branch pipe on the end of a hose.

For tanks 48 m in diameter and above, the 100-mm dry riser should terminate in a 100-mm ring main to follow the line of the wind girder, with single, valved hose connections at approximately 46-m intervals. Foam dams will be provided as required for such tanks if fitted with pantograph seals (see Figure 8.4).

A rim plate should be welded to the floating roof to form a foam dam (see Figure 8.5). Based on a rim plate of about 0.3 m high at a distance of 1 m from the tank wall, the content of the foam dam in dm^3 is approximately $1000D$ where D = diameter in meters of the largest floating roof tank.

Assuming an expansion factor of 7, the required foam solution in dm^3 is $1000D/7$.

A detector alarm system within the tank rim seal with the alarm indication terminating in a manned control room should be installed.

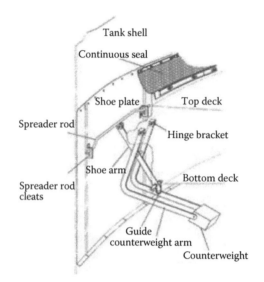

FIGURE 8.4
Typical pantograph seal.

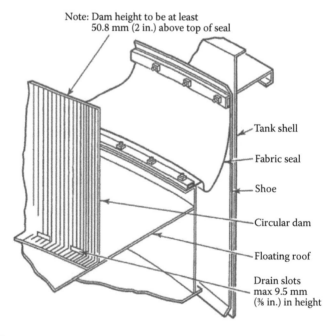

FIGURE 8.5
Typical foam dam for floating roof tank protection.

8.4 Subsurface Foam Systems

The requirements of this section are applicable to systems used for the protection of fixed roof storage tanks containing low-viscosity hydrocarbon liquids, in which foam is injected (through a product line or through a special foam line) at the base of the tank and rises to the surface through the liquid in the tank.

Experience with fuel storage tank firefighting has shown that the main problems are operational (i.e., difficulty in delivering the foam relatively gently to the fuel surface at an application rate sufficient to effect extinguishment). A properly engineered and installed subsurface foam system offers the potential advantages of less chance for foam-generation equipment disruption as a result of an initial tank explosion or the presence of fire surrounding the tank and the conduct of operations a safe distance from the tank. Thus, opportunity for establishing and maintaining an adequate foam application rate is enhanced.

The typical arrangement shown in Figure 8.3 should be used where it can be ensured that the isolating stop valve at the base of the tank is normally open.

Where a subsurface system has been installed for tank protection, arrangements may be made to allow the system to discharge into the surrounding bund in order to supplement other methods of bund protection. Subsurface systems are not suitable for foam-destructive fuels or for some high-viscosity fuels. Subsurface systems are not used for the primary protection of floating roof tanks because the roof will prevent complete foam distribution.

Only FP, FFFP, and AFFF foams that will tolerate severe mixing with fuel are suitable for subsurface application.

Protein foam is not suitable. Subsurface systems are sometimes called base injection systems.

Consideration should be given to the following when selecting one of the variations of this type of system:

a. The total foam output reaches the tank.

b. With large tanks, inlets are suitably distributed to allow even foam spread over surface of the fuel.

c. The system is essentially simple, and being at ground level, is less likely to be damaged by fire or explosion than overhead systems.

d. The rising foam stream induces vertical circulation of cold fuel from the base of the tank to the burning surface dissipating hot fuel layers at the burning surface and assisting extinction.

e. Essential equipment and operating personnel can be located at a safe distance from the fire.

f. The system is easy to check and maintain.

g. The high back-pressure foam generator and foam solution supply should be fixed or portable for connection to foam inlet pipes or product line connections outside the bunded area.

h. A high back-pressure foam generator is used to produce foam at a pressure sufficient to overcome the high-pressure head of fuel as well as all frictional losses in the foam pipework. Frictional losses with foam differ from those with foam solution.

i. Where the foam is injected through the product line, it is essential that automatic closure fire valves are not fitted.

8.4.1 Discharge Rate

The discharge rate (in L/min) should be not less than four times the area of the tank (in m²). With the same fuels, where there has been a long pre-burn prior to the application of foam, a hot zone will exist near the burning surface at temperatures in excess of 100°C. In order to avoid frothing and slop-over, continuous application of foam should be avoided in the initial stages. Intermittent application of the foam can induce circulation of the fuel in the tank, thereby bringing the cooler layers of fuel to the surface. The foam injected intermittently will disperse without sufficient steam formation to produce frothing. The rates of foam discharge from each outlet should be approximately equal.

8.4.2 Duration of Discharge

The minimum duration of discharge of systems discharging at the minimum rate should be as given in Table 8.4 of this chapter. The minimum duration of discharge of systems discharging at higher than the minimum rate should be reduced in proportion but should be not less than 70% of the time given in Table 8.2.

TABLE 8.2

Minimum Discharge Times for Semisubsurface and Fixed Foam Pourer Systems (Except Open-Top Floating Roof Tanks) Discharging at the Minimum Rate

Hazard	Minimum Discharge Time (minutes)
Spillage	10 (all classes of foam concentrate)
Tanks Containing Liquid Hydrocarbons	
Flash point not above 40°C	55 (protein)
	45 (AFFF, FP, and FFFP)
Flash point above 40°C	30 (all classes of foam concentrate)
Tanks containing foam-destructive liquids	55 (AR)
Bunds	60

8.5 Semisubsurface Systems

The requirements of this section are applicable to systems used to apply foam to the surface of fixed roof storage tanks via a flexible hose rising from the base of the tank. The system is not normally considered appropriate for floating roof tanks with or without a fixed roof because the floating roof prevents foam distribution. The hose is initially contained in a sealed housing and is connected to an external foam generator capable of working against the maximum product head. On operation the end of the hose is released to float to the liquid surface.

Consideration should be given to the following when selecting this type of system:

a. The total foam output reaches the surface of the burning liquid.

b. With large tanks, the semisubsurface units can be arranged to produce an even distribution over the fuel surface.

c. Any type of concentrate suitable for gentle surface application to the particular fuel should be used.

d. Foam-generating equipment and operating personnel may be located at a distance from the fire.

e. The system should be used for the protection of foam-destructive liquids provided the flexible hose is not affected by them.

f. Certain high-viscosity fuels are not suitable for protection by this system.

g. Circulation of the cold fuel that could assist extinction is not induced.

h. The system is difficult to check, test, and maintain.

i. The foam generator has to produce foam at a pressure sufficient to overcome the high back-pressure of the head of the fuel as well as all frictional losses in the foam pipework. Frictional losses with foam differ from those with foam solution.

8.5.1 Discharge Rate

The discharge rate should be not less than the appropriate rate given in Table 8.3 multiplied by the liquid surface area in the tank.

8.5.2 Duration of Discharge

The minimum duration of discharge should be as given in Table 8.4.

8.5.3 Number and Position of Units

The number of units should be not less than that given in Table 8.5.

TABLE 8.3

Minimum Application Rates for Pourer Systems (Fixed Roof Tanks and Bunds) and Semisubsurface Systems

Foam Concentrate Class	Flammable Liquid	Minimum Application Rate (L/m²/min)
Any	Hydrocarbon flash point not above 40°C	4
AR	Foam-destructive liquids	6.5

TABLE 8.4

Minimum Discharge Times for Subsurface Systems Discharging at the Minimum Rate

Risk	Minimum Discharge Time (min)
Tanks Containing Liquid Hydrocarbons	
Flash point not above 40°C	45
Flash point above 40°C	30

TABLE 8.5

Minimum Number of Semisubsurface Units for Tanks

Tank Diameter (m)	Number of Semisubsurface Units
Up to 24	1
Over 24 up to and including 36	2
Over 36 up to and including 42	3
Over 42 up to and including 48	4
Over 48 up to and including 54	5
Over 54 up to and including 60	6
Over 60	6, plus one for each 450 m² of the liquid surface of the tank above 2820 m²

Where more than one unit is required, these should be spaced equally around the tank shell away from tank level indicator devices and swing-arm product pipes.

For backing up subsurface and semisubsurface foam injection systems, mobile/portable foam monitors should be available with a capacity of 240 m³/h water/foam solution, at a working pressure of 10 bar(ga).

8.6 Pressure Storage of C4 and Lighter Hydrocarbons

Fire protection of pressure storage spheres and vessels containing liquefied hydrocarbons is also based on safe spacing distances. Leakage from large spheres and vessels should be drained away from underneath the equipment via a sloping floor into a containment channel.

Spheres and vessels should be protected against engulfed fires and radiation from the fires of adjacent equipment by automatic water spray systems. The spray system should ensure an even distribution of water over the entire surface of the sphere or vessel regardless of wind forces or wind direction, but should be divided into separate sections in order to limit water consumption when applying radiation protection.

The minimum required water rate is 8.5 $dm^3/min/m^2$ of equipment surface for engulfed fires. Protection against heat radiation from adjacent equipment will require much less water and should be calculated on the basis of applicable distances.

8.7 LNG and NGL Atmospheric Refrigerated Storage Tanks

LNG and NGL storage tank fires are unlikely due to the inherent safety of the tank design. If they do occur they cannot be extinguished due to the intensity of the fire and consequently a complete burnout of the tank has to be accepted. Attempts to extinguish such a fire may produce a cloud of combustible vapor that can be a greater hazard than the tank fire itself. The basic approach to these fires is that of containment and control until ultimate extinction, prevention of escalation, and in maintaining the integrity of adjacent equipment particularly if it contains flammable material. In order to achieve this purpose, it is necessary to prevent tank roofs and appurtenances, tank walls, pipe bridges, and manifolds from exceeding their maximum design temperatures when an adjacent tank is on fire.

This should be achieved by

- Adequate tank spacing
- Tank design
- Adequate water spray and exposure protection facilities

Fire protection for LNG and NGL refrigerated storage tanks is highly specialized; therefore systems should be designed by competent technical persons.

8.8 Foam Monitors and Handlines

This section relates to systems in which foam is applied through fixed or portable monitor or hose streams. They are suitable when used alone for extinguishment of spill fires, diked area fires, and fires in vertical fixed roof atmospheric storage tanks. They are suitable as auxiliary protection in conjunction with fixed systems. Portable hose streams are suitable for extinguishment of rim fires in open-top floating roof tanks.

8.8.1 Use of Foam Handlines and Monitors

The limitations of foam handline and monitors are detailed below.

Monitor nozzles should not be considered as the primary means of protection for fixed roof tanks over 18 m in diameter. Foam handlines should not be considered as the primary means of protection for fixed roof tanks over 9 m in diameter or over 6 m high.

(NOTE: When the entire liquid surface was involved, fires in tanks up to 39 m in diameter have been extinguished with large-capacity foam monitors. Depending on the fixed roof tank outage and fire intensity, the updraft due to chimney effect should prevent sufficient foam from reaching the burning liquid surface for formation of a blanket.)

Foam should be applied continuously and evenly. Preferably, it should be directed against the inner tank shell so that it flows gently onto the burning liquid surface without undue submergence. This can be difficult to accomplish as adverse winds, depending on velocity and direction, will reduce the effectiveness of the foam stream. Monitors operated at grade usually are not recommended for floating roof rim fire extinguishment because of the difficulty of directing foam into the annular space. Fires in fixed roof tanks having ruptured roofs with only limited access for foam are not easily extinguished by monitor application from ground level. Fixed foam monitors are installed for protection of drum storage areas or diked areas).

8.8.2 Foam Application Rates

The minimum delivery rate for primary protection based on the assumption that all the foam reaches the area being protected should be as called for below. In determining total solution flow requirements, consideration should be given to potential foam losses from wind and other factors.

8.8.3 Tanks Containing Liquid Hydrocarbons

The foam solution delivery rate should be at least 6.5 L/min/m² of liquid surface area of the tank to be protected.

Note the following:

1. Included in this section are gasohols and unleaded gasolines containing no more than 10% alcohol by volume.

2. Flammable liquids having a boiling point of less than 37.8°C should require higher rates of application. Suitable rates of application may be determined by test. Flammable liquids with a wide range of boiling points should develop a heat layer after prolonged burning and then should require application rates of 8.1 L/min/m² or more.

3. Care should be taken in applying portable foam streams to high-viscosity materials heated above 93.3°C. Judgment should be used in applying foam to tanks containing hot oils, burning asphalts, or burning liquids that are above the boiling point of water. Although the comparatively low water content of foams may beneficially cool such fuels at a slow rate, it will also cause violent frothing and slop-over of the contents of the tank.

8.8.4 Tanks Containing Other Flammable and Combustible Liquids Requiring Special Foams

Water-soluble and certain flammable and combustible liquids and polar solvents that are destructive to regular foams require the use of AR foams. In general, alcohol type foams may be effectively applied through foam monitor or foam hose streams to spill fires of these liquids when the liquid depth does not exceed 25 mm.

For liquids in greater depths, monitor and foam hose streams should be limited for use with special alcohol type foams listed for the purpose. In all cases, the manufacturer of the foam concentrate and the foam-making equipment should be consulted as to limitations and for recommendations based on listings or specific fire tests. The data in Table 8.6 show the minimum application rates.

TABLE 8.6

Minimum Liquid Rates for Various Liquids

Type of Liquid	Solution Rate (L/min)/m²
Methyl and ethyl alcohol	6.5
Acrylonitrile	6.5
Ethyl acetate	6.5
Methyl ethyl ketone	6.5
Acetone	9.8
Butyl alcohol	9.8
Isopropyl ether	9.8

Products such as isopropyl alcohol, methyl isobutyl ketone, methyl methacrylate monomer, and mixtures of polar solvents in general should require higher application rates. Protection of products such as amines and anhydrides, which are particularly foam-destructive, require special consideration.

When using AR foam concentrate, consideration should be given to solution transit time. Solution transit time (the elapsed time between injection of the foam concentrate into the water and the induction of air) should be limited depending on the characteristics of the foam concentrate, the water temperature, and the nature of the hazard protected. The maximum solution transit time of each specific installation should be within the limits established by the manufacturer.

If application results in foam submergence, the performance of AR foams usually deteriorates significantly, particularly where there is a substantial depth of fuel. The degree of deterioration of performance will depend on the degree of water solubility of the fuel (i.e., the more soluble, the greater the deterioration).

8.8.5 Durability of Discharge

The equipment should be capable of operation to provide primary protection at the delivery rates for the following minimum periods of time:

a. Tanks containing liquid hydrocarbons:
   1. Flash point between 37.8°C and 93.3°C, 50 minutes
   2. Flash point below 37.8°C or liquids heated above their flash points, 65 minutes
   3. Crude petroleum, 65 minutes
b. Tanks containing other flammable and combustible liquids requiring special foams:

   Alcohol type foams require special application procedures. The operation time should be 65 minutes at specified application rates unless the manufacturer has established, by fire test, that a shorter time may be permitted.

Where the system's primary purpose is for spill fire protection, the minimum discharge time should be 10 minutes for fixed equipment and 15 minutes for portable equipment.

8.8.6 Protection of Hydrocarbon Spill Fires

The minimum foam solution delivery rate for the protection of the potential spill area should be 6.5 L/min/m^2 when protein or FP concentrate is used.

When AFFF or FFFP concentrate is used, the minimum rate should be 4.1 L/min/m².

8.9 Foam Concentrate and Solution

The nominal concentration of use should be not less than that recommended by the manufacturer. The actual concentration, for a fixed system operating at the design application rate should be

  a. For a nominal percentage concentration equal to or greater than 5% within plus or minus one percentage point of the nominal concentration, C ± 1

  b. For a nominal percentage concentration less than 5%, but not less than 3%, within plus one percentage point of, and no less than, the nominal concentration, C −0 + 1

  c. For a nominal percentage concentration less than 3%, within plus one quarter of percentage point of, and no less than, the nominal concentration, C −0 + 25

Premixed foam solution used in the system should have a concentration within the range of 0.9 to 1.1 times the value specified by the manufacturer.

Only AFFF, FFFP, or FP foam concentrate should be used in subsurface systems.

The nominal concentration of use for mixtures of foam concentrates should be not less than the higher or highest value recommended by the manufacturer or manufacturers.

Protein (P) foam is not suitable for subsurface systems but should be used in top application or semisubsurface systems.

AR foams are formulated for use against fires of foam-destructive liquids but are also suitable for use on hydrocarbon liquids. The fire performance of AR foams against hydrocarbon fuels generally corresponds to the performance of the parent concentrate. The high viscosity of some concentrates needs to be considered when specifying the proportioning system.

The solutions of some AR concentrates are required to be foamed within a specific time of the solution being mixed; it is essential that the solution transit time (the time for foam solution to flow from the point at which concentrate enters the water stream to the point at which air enters the stream, usually expressed in seconds) is less than this limiting time.

When applied forcefully to deep layers of foam-destructive liquids, all types of AR foams may show a significant loss of performance compared

TABLE 8.7

Minimum Values for Expansion and 25% Drainage (Aspirated Foam)

Application	Expansion	25% Drainage Time (min)
Surface or semisubsurface	5	2.0 (Protein, FP) or 1.5 (AFFF, FFFP)
Subsurface	2 (but not more than 4)	1.5

with results using gentle application. For flammable liquids that are only partially soluble in water, the loss of performance can only be slight but in some cases equipment designed to give very gentle application is necessary. In all cases tests should be conducted or advice sought from the suppliers regarding these liquids.

In portable, transportable, and semifixed systems the conditions of induction are not controlled by the system design. The actual concentration of use in such systems should be within the above limits when the equipment is used under the conditions specified by the manufacturer.

8.10 Foam Quality

The expansion and drainage time values of foam produced by an aspirating system should be not less than the values given in Table 8.7.

Values for AR foam should be not less than the values for the parent class (P, FP, FFFP, or AFFF).

The expansion and drainage time of nonaspirated foam is difficult to measure. Values for these are not given in this book.

8.11 Dike Area Protection

In general, portable monitors, or foam hose streams, or both, have been adequate in fighting diked area and other spill fires. In order to obtain maximum flexibility due to the uncertainty of location and the extent of a possible spill in process areas and tank farms, portable or trailer-mounted monitors are more practical than fixed foam systems in covering the area involved.

8.11.1 Drainage of Bunds (Dike)

Drains and interceptors in bunded areas should be of adequate capacity to carry the anticipated drainage of water used in firefighting.

8.12 Water Supplies and Pumps

8.12.1 Quantity, Pressure, and Flow Rate

The water supply should provide the total quantity, flow rate, and supply pressure specified for the foam system and for any other fire protection systems that should be used simultaneously with it, for the specified discharge times.

The supply is reduced by drought or by freezing, or where process water is used to maintain normal working conditions (e.g., for cooling reactors).

Where the primary source is not capable of meeting the system design requirements at all times, storage facilities should be used to meet the shortfall. Consideration should be given to duplication of the water supply pipework or the use of a ring main system so that the effects of interruptions in the main supply are minimized.

8.12.2 Quality

The water supply to foam systems may be hard or soft freshwater or saltwater, but should be of suitable quality so that adverse effects on foam formation or foam stability does not occur. No corrosion inhibitors, emulsion-breaking chemicals, or any other additives should be present without prior consultation with the foam concentrate supplier.

8.12.3 Quantity

The water supply should be adequate in quantity to supply all the devices that are used simultaneously for the specified time. This includes not only the volume required for the foam apparatus but also water to be used in other fire-fighting operations in addition to the normal plant requirements. Premixed solution-type systems need not be provided with a continuous water supply.

8.12.4 Pressure

The pressure available at the inlet to the foam system (foam generator, air foam maker, etc.) under required flow conditions should be at least the minimum pressure for which the system has been designed.

8.12.5 Temperature

Optimum foam production is obtained using water at temperatures between 4°C and 37.8°C. Higher or lower water temperatures will reduce foam efficiency.

8.12.6 Storage

Water supply or premixed solution should be protected against freezing in climates where freezing temperatures can be expected.

8.13 Storage

Foam concentrate or premixed solution should be stored in an accessible location not exposed to the hazard it protects against. The material of construction of any building should be noncombustible. Foam concentrate in shipping containers and in storage tanks should be stored in accordance with the manufacturer's recommendations. Exposure to extreme heat, cold, contamination, or mixing with other materials should be avoided.

Storage containers should be sited where they will be readily accessible for inspection, testing, recharging, or maintenance with the minimum of interruption of protection.

During the design stage it should be considered whether it is worthwhile to erect additional foam storage facilities near the tank farm when processing units and tank farm are located at such a distance that a rapid backup supply of foam concentrate would be hampered.

Means should be provided to ensure that the concentrate or premixed solution is kept within its design operating temperature range. Storage vessels should be clearly marked with the class of concentrate and its grade (concentration in the foam solution).

Storage tanks should have sufficient ullage to accommodate thermal expansion of the concentrate or solution.

Only suitable concentrates should be stored as premixed solutions. Not all foam concentrates are suitable for storage as a premixed solution and the manufacturer's advice should be sought and followed. High storage temperatures may accelerate any deterioration due to aging of the solution.

For smaller hazards a pressure tank is usually used to provide a quick-acting automatic system. Nitrogen, CO_2, or water is used to expel the contents.

8.13.1 Quantities of Foam Concentrate

The quantity of foam concentrate or foam solution available for immediate use in the system should be not less than

$$V = \frac{A \times R \times C \times T}{100}$$

TABLE 8.8

Minimum Number of Supplementary Branch Pipes for Tanks, and Minimum Discharge Time

Diameter of Largest Tank (m)	Minimum Number of Foam Branch Pipes	Minimum Discharge Time (min)
Up to 10	1	10
Over 10 up to 20	1	20
Over 20 up to 30	2	20
Over 30 up to 40	2	30
Over 40	3	30

or

$$V_1 = A \times R \times T$$

where

V_1 is the minimum quantity of foam solution (in L)
V is the minimum quantity of foam concentrate (in L)
A is the area of application (in m²)
R is the rate of application of foam solution (in L/m² per minute)
C is the nominal concentration (in %)
T is the duration of application (in min)

plus sufficient foam concentrate to permit operation of all extra branch pipes simultaneously with the primary means of fire protection and for the minimum discharge duration given in Table 8.8. Values for the area of application should not be less than

- For fixed roof tanks: the area of the bund
- For floating roof tanks: the area of the rim seal
- For spills: the area of the spill
- For bunds: the bund area except for branch pipe and monitor systems when the area of application is half the bund area

The area of a bund should be taken as the gross area less the area of any nonelevated tank or tanks within the bund.

8.13.2 System Discharge Rate

The discharge rates of portable branch pipe and monitor systems should be not less than the appropriate minimum application rate given in Table 8.9 multiplied by the area of the tank or spill, or by half the area of the bund, as appropriate.

TABLE 8.9

Minimum Application Rates for Monitor and Branch Pipe Systems

| Foam Concentrate Rate Class | Minimum Application Rates of Foam Solution (L/m²/min) | | | Hazard |
	Spill Fires	Tank Fires	Bund Fires (suitable when used alone for extinguishment of spill fires)	
AFFF	4	6.5	4	Hydrocarbon, flash point above 40°C
FFFP	4	6.5	4	
FP	5	6.5	5	
Protein	6.5	8	6.5	Hydrocarbon, flash point above 40°C
AR	By test	Not suitable	By test	Foam-destructive liquids, flash point above 40°C

Higher minimum rates may be needed if there is exceptional loss of foam because of wind or fire updraft.

Minimum rates for liquids with a flash point not above 40°C and for other liquids not listed in the table should be determined by specific test or from the concentrate manufacturer's data.

For bund protection systems, in addition to the above requirement, the system discharge rate and the actual area of application is not less than that given in Table 8.9.

The hazard requiring the greatest quantity of foam concentrate should be used to determine the amount to be held at immediate readiness.

Allowance should be made for the quantity of foam concentrate needed to fill the feed lines installed between the source and the most remote monitor or branch pipe.

A reserve supply of foam concentrate should be available to enable the system or systems to be put back into service within 24 hours of operation. This supply may be stored in separate tanks, in drums or cans on the premises, or be available from an outside source.

Adequate loading and transportation facilities should be assured at all times.

Other equipment necessary to recommission the system, such as bottles of nitrogen or carbon dioxide for premix systems, should also be readily available.

8.14 Foam Concentrate Pumps

Pumps for foam concentrate should be self-priming or flooded-suction pumps driven by a suitable prime mover that is constantly available.

TABLE 8.10

Minimum Number of Hydrants for
Supplementary Protection of Storage Tanks

Tank Diameter (M)	Minimum Number of Hydrants
Up to 20	1
Over 20	2

Pumps should have adequate capacity to meet the maximum system requirements. To ensure positive injection, the discharge pressure rating at design discharge capacity should be sufficiently in excess of the maximum water pressure under any condition at the point of injection of the concentrate.

8.15 Hydrants

In addition to a primary fixed piping system and any supplementary protection, foam hydrants should be provided for use with portable or mobile equipment or water hydrants with suitable foam-producing equipment in the event that a fixed discharge outlet on the primary protection system is damaged.

The number of hydrants that may have more than one outlet should be as shown in Table 8.10.

Each hydrant should be located between 15 and 75 m from the shells of the tanks being protected by the associated primary system.

The flow from hydrants should be sufficient for all the portable equipment to be used.

8.16 Foam Concentrate Facilities

One or more horizontal carbon steel vessel(s) for foam concentrate storage should be installed at a safe and readily accessible location, preferably near the fire station. The vessel(s) should be equipped with a pressure-vacuum valve set at both approximately 5 mbar vacuum- and pressure-installed on the manhole cover. On the same cover a 1½-inch inlet should be provided fitted with a threaded cap with a dipstick to measure the liquid level of the foam concentrate.

Via this inlet a sealing liquid should be supplied in order to avoid oxidation of the foam. For this purpose the foam level should also be maintained just under the manhole cover.

The elevation of the 2½-inch outlet hose connection should be approximately 3 m above ground level to enable filling of mobile equipment by gravity.

An electric-motor-driven pump should be installed to fill the vessel from drums. The pump capacity should be approximately 3 dm³/s at a discharge pressure of 3 bar gauge. An unloading rack with a foam collector underneath should be installed.

The pump should also be connected to the vessel for circulating the foam to avoid sediment formation.

8.17 Automatically Operated Systems

Automatic systems should incorporate a manually operated lockoff device that will prevent discharge of the system but will not prevent sounding off the alarm signal.

Operation of the lockoff device should be indicated at the plant or fire control center.

The lockoff device is for use when maintenance personnel are working on the system.

8.18 Detection and Alarm Equipment

Automatic detection and control equipment should give a positive warning of any fault or abnormality (e.g., loss of power or pressure that may render the detection and control system inoperative).

Automatic detection equipment should provide a local alarm at the control point of each automatic system as well as at the plant or central control point.

Automatic systems should include a facility for coincidental shutdown of any heat source or potential means of ignition or reigniting in the vicinity of the hazard. Detection and alarm equipment may be electrical, pneumatic, hydraulic, or mechanical (e.g., link line type).

8.19 Foam Spray Systems

The requirements of this section are applicable to systems discharging a spray of aspirated foam or nonaspirated foam solution to provide primary protection for flammable liquid spills.

The spray nozzles should be arranged to discharge downward in overhead systems and horizontally or upward as in ground-level pop-up sprayers.

Spray systems are particularly suitable both outdoors and indoors where flammable liquids may be spilled in large quantities. Typical examples include loading racks, horizontal tanks, pump rooms, dip tanks, and bunds.

Generally these systems are not suitable for use on water-miscible liquids exceeding 25 mm in depth.

Any type of foam concentrate should be used in aspirating systems, but for nonaspirating systems only AFFF or FFFP should be used. Nonaspirating systems should be regarded as water spray systems discharging foam solution.

Consideration should be given to the following when selecting one of the many variations of this type of system:

a. Hot surfaces in contact with the fuel can be effectively cooled by a spray discharge. Structures should also be protected from heat radiation by a spray discharge.

b. The system is particularly suitable for automatic operation. Automatic operation is for indoor or unmanned hazards.

c. Even distribution of the foam over the fuel surface is achieved but discharge is carried by the wind beyond the area of the fuel spill except where ground-level pop-up nozzles, which deliver foam at the seat of the fire, are used.

d. Foam sprayers have small passages susceptible to blockage.

e. Obstructions, such as vehicles or equipment temporarily positioned, may be present when the system is operated and may interfere with the discharge.

f. Pipework for overhead nozzles obstructs normal activities or imposes an undue load on the roof structure.

g. Overhead application needs supplementary low-level application to provide coverage below large obstructions, such as aircraft in hangars.

h. For hazards where a large spill area is likely to be involved, the foam spray system should be subdivided into zones, each protecting a specific floor area and individually actuated by a suitable fire detection system.

i. Nonaspirated nozzles can be used to apply a spray of water instead of foam solution, which can provide effective fire control of some flammable liquids.

The following issues should be noted:

1. Spray foam applied externally to tanks or vessels has the added advantages of cooling and insulating the tanks or vessels while the spill fire is being extinguished. Overhead pipework for overhead

TABLE 8.11

Minimum Application Rates for Sprayer Systems (Hydrocarbon Liquids Only)

Foam Concentrate Class	Height of Discharge Point above Lowest Point of Hazard		Minimum Application Rate
Protein	Up to and including 10	6.5	Not suitable
	Above 10	8	
FFFP	Up to and including 10	6.5	4
AFFF	Above 10	8	6.5
FP	Up to and including 10	6.5	Not suitable
	Above 10	8	

    applicators must neither obstruct normal operations nor impose an undue load on the roof structure. While foam is not considered an effective agent for extinguishing three-dimensional running flammable liquid fires, it can control the pool fire underneath the running fire, thus permitting control by other means.

2. These systems should also be used to protect small outdoor opentop tanks having a liquid surface area not exceeding 18.6 m².

8.19.1 Discharge Rate

Systems should deliver a foam solution at not less than the appropriate minimum application rate given in Table 8.11 multiplied by the area of the spill.

    Application rates for foam-destructive liquids should be determined by specific test or taken from the foam concentrate manufacturer's data.

8.19.2 Duration of Discharge

The minimum duration of discharge of systems discharging at the minimum rate should be as given in Table 8.12. The minimum duration of discharge of systems discharging at higher than the minimum rate may be

TABLE 8.12

Minimum Discharge Times for Sprayer Systems (Hydrocarbon Liquids Only) Discharging at the Minimum Rate

Hazard	Area of System or Zone (m²)	Minimum Discharge Time for All Classes of Foam Concentrate (min)
Indoor contained liquid hydrocarbon spills	50 or less	5
	More than 50	10
Indoor open-top process tanks containing liquid hydrocarbons	50 or less	5
	More than 50	10
Outdoor applications	Any area	10

reduced in proportion but should be not less than 70% of the time given in Table 8.12. The minimum duration of discharge for foam-destructive liquids should be determined by specific test or taken from the foam concentrate manufacturer's data.

8.19.3 Number and Location of Discharge Outlets

There should be not less than one discharge outlet per 10 m² of protected area.

Generally, sprayers should be spaced to provide even distribution over the whole area. For some hazards it may be advantageous to cluster sprayers in areas where fire is likely to originate.

8.20 Medium- and High-Expansion Foam Systems

High-expansion foam is an agent for control and extinguishment of class A and class B fires and is particularly suited as a flooding agent for use in confined spaces. The development of the use of high-expansion foams for fire-fighting purposes started with the work of the Safety in Mines Research Institutes regarding the difficult problem of fires in coal mines. It was found that by expanding an aqueous surface-active agent solution to a semistable foam of about 1000 times the volume of the original solution, it was possible to force the foam down relatively long corridors, thus providing a means for transporting water to a fire inaccessible to ordinary hose streams.

This work has led to the development of specialized high-expansion foam-generating equipment for fighting fires in mines, for application in municipal industrial firefighting, and for the protection of special hazard occupancies. Medium-expansion foam was developed to cover the need for a foam more wind-resistant than high-expansion foam for outdoor applications.

Medium- and high-expansion foams are aggregations of bubbles mechanically generated by the passage of air or other gases through net, screen, or other porous medium that is wetted by an aqueous solution of surface-active foaming agents. Under proper conditions, fire-fighting foams of expansions from 20:1 to 1000:1 can be generated.

Such foams provide a unique agent for transporting water to inaccessible places; for total flooding of confined spaces; and for volumetric displacement of vapor, heat, and smoke. Tests have shown that under certain circumstances, high-expansion foam when used in conjunction with water sprinklers will provide more positive control and extinguishment than either extinguishment system by itself. High-piled storage of rolled paper stock is an example. Optimum efficiency in any one type of hazard is dependent to some extent on the rate of application and also the foam expansion and stability.

Medium- and high-expansion foams, which are generally made from the same type of concentrate, differ mainly in their expansion characteristics.

Medium-expansion foam is used on solid fuel and liquid fuel fires where some degree of in-depth coverage is necessary (e.g., for the total flooding of small enclosed or partially enclosed volumes such as engine test cells or transformer rooms). It can provide quick and effective coverage of flammable liquid spill fires or some toxic liquid spills where rapid vapor suppression is essential. It is effective both indoors and outdoors.

High-expansion foam should also be used on solid and liquid fuel fires but the in-depth coverage it can give is greater than that for medium-expansion foam. It is therefore most suitable for filling volumes in which fires exist at various levels. For example, experiments have shown that high-expansion foam can be used effectively against high-rack storage fires provided that the foam application is started early and the depth of foam is rapidly increased. It can also be used for the extinction of fires in enclosures where it is dangerous to send personnel (e.g., in basement and underground passages). It can be used to control fires involving LNG and LPG and to provide vapor dispersion control for LNG and ammonia spills.

High-expansion foam is particularly suited for indoor fires in confined spaces. Its use outdoors may be limited because of the effects of wind and lack of confinement. Medium- and high-expansion foam have several effects on fires:

a. When generated in sufficient volume, they can prevent free movement of air, which is necessary for continued combustion.

b. When forced into the heat of a fire, the water in the foam is converted to steam, reducing the oxygen concentration by dilution of the air.

c. The conversion of the water to steam absorbs heat from the burning fuel. Any hot object exposed to the foam will continue the process of breaking the foam, converting the water to steam, and of being cooled.

d. Because of their relatively low surface tension, solution from the foams that is not converted to steam will tend to penetrate class A materials. However, deep-seated fires will require overhaul.

e. When accumulated in depth, medium- and high-expansion foam can provide an insulating barrier for protection of exposed materials or structures not involved in a fire and can thus prevent fire spread.

f. For LNG fires, high-expansion foam will not normally extinguish a fire but it reduces the fire intensity by blocking radiation feedback to the fuel.

g. Class A fires are controlled when the foam completely covers the fire and burning material. If the foam is sufficiently wet and is maintained long enough, the fire can be extinguished.

h. Class B fires involving high flash point liquids can be extinguished when the surface is cooled below the flash point. Class B fires involving low flash point liquids can be extinguished when a foam blanket of sufficient depth is established over the liquid surface.

8.20.1 Mechanisms of Extinguishment

Medium- and high-expansion foam extinguishes fire by reducing the concentration of oxygen at the seat of the fire, by cooling, by halting convection and radiation, by excluding additional air, and by retarding should be specifically evaluated to verify the applicability of medium- or high-expansion foam as a fire control agent.

Some important types of hazards that medium- and high-expansion foam systems may satisfactorily protect include

a. Ordinary combustibles
b. Flammable and combustible liquids
c. Combinations of a and b
d. LNG (high-expansion foam only)

Ability to control or extinguish a fire in a given hazard may depend on such factors as expansion, drainage, and fluidity. These factors will vary with the concentrate, equipment, water supply, and air supply.

Susceptibility of the protected hazard to water damage should be evaluated.

Medium- and high-expansion foam systems should not be used on fires in the following hazards unless competent evaluation, including tests, indicates acceptability:

a. Chemicals, such as cellulose nitrate, that release sufficient oxygen or other oxidizing agents to sustain combustion
b. Energized unenclosed electrical equipment
c. Water-reactive metals such as sodium (Na) and potassium (K)
d. Hazardous water-reactive materials, such as triethylaluminum and phosphorous pentoxide
e. Liquefied flammable gas

8.20.2 Expansion

Foams are arbitrarily subdivided into three ranges of expansion:

1. Low-expansion foam (LX): expansion up to 20
2. Medium-expansion foam (MX): expansion 21 to 200
3. High-expansion foam (HX): expansion 201 to 1000

8.20.2.1 Medium-Expansion Foam

Medium-expansion foam should have an expansion between 21 and 200.

8.20.2.2 High-Expansion Foam

High-expansion foam should have an expansion between 201 and 1000.

8.20.3 Application Method

8.20.3.1 Medium-Expansion Foam

Medium-expansion foams should be applied

a. Gently to the surface of a flammable liquid or solid combustible fire
b. By means of a medium-expansion foam branch pipe or monitor

The first method is suitable for fixed systems where the location, size, and shape of the hazard is known and the system can be designed to meet this requirement. The second method is more appropriate where the size and location of the hazard vary with circumstance, and needs to be dealt with by a more flexible approach.

8.20.3.2 High-Expansion Foam

High-expansion foams should be applied

a. By filling the volume in which the fire occurs
b. By guiding a wall of foam in the direction of a localized fire in order to submerge and suppress it

The foam may be introduced directly or through flexible ducting. High-expansion foam, by its nature, can only be applied gently to fires. Method a is generally preferable as the water content of the foam needs to be retained as far as possible to ensure heat resistance at the fire. Horizontal movement at floor level promotes water drainage and degrades the foam quality. To make high-expansion foam effective in large compartments and up to heights of 10 m, flexible barriers should be used to retain the foam in the required area and to permit its fast buildup to the required height. Wherever possible, foam should be applied at a high level (i.e., above the level of foam in the fire space).

8.20.4 System Design

The system should be designed to suit the particular hazard, and the following should be considered when preparing the design:

a. Full details of the solid combustibles and/or flammable liquids, their methods of storage and packaging, handling, and location

b. The most suitable class of foam concentrate, concentration, and solution application rate

c. The most suitable method of application of the foam, and the most suitable equipment to provide this method, including the method of proportioning

d. The quantity of foam concentrate required for extinction, including backup supplies where extended application is necessary for concealed or prolonged fires

e. The required system operation time, taking into account item d

f. The quantity of foam concentrate to be held in reserve

g. Water supply quantity, quality, and pressure

h. Pipework sizes and pressure losses

i. Method of system operation and any fire or gas detection equipment required; need for a manual override where personnel are present

j. Any special considerations (e.g., the need to use flameproof electrical equipment where flammable vapors are present)

k. Drainage and bunds

l. Environmental conditions

8.20.4.1 Compatibility with Other Extinguishing Media

The foam produced by the system should be compatible with any media provided for application at or about the same time as foam. Certain wetting agents and some extinguishing powders are incompatible with foams, causing rapid breakdown of the latter. Only media that are substantially compatible with a particular foam should be used in conjunction with it.

Water jets or sprays will adversely affect a foam blanket, but the simultaneous application of water from sprinklers can be beneficial provided that allowance is made for the increased breakdown of foam.

8.20.4.2 Compatibility of Foam Concentrates

Foam concentrate (or solution) added or put into a system should be suitable for use and compatible with any concentrate (or solution) already present in the system. Foam concentrates and foam solutions, even of the same class, are not necessarily compatible, and it is essential that compatibility be checked before mixing two concentrates or premixed solutions.

8.20.4.3 Foam Destructiveness

For the purposes of this standard, when taking into account foam destructiveness, flammable liquids are considered as falling into two groups:

1. Hydrocarbons and those nonhydrocarbon liquids that are not more foam-destructive than hydrocarbons
2. Foam-destructive liquids, which are generally water-soluble and much more foam-destructive than hydrocarbons

Special types of concentrate are used for foam-destructive liquids. Higher rates of application are specified for foam-destructive liquids than for hydrocarbons and it is usually essential to use gentle application methods.

The degree of foam destructiveness varies, however, and isopropyl alcohol, butyl alcohol and isobutyl methyl ketone, methyl methacrylate monomer, and mixtures of water-miscible liquids in general require higher application rates. Protection of products such as amines and anhydrides that are particularly foam-destructive require special consideration.

8.20.5 Water, Foam Concentrate, and Air Supply

8.20.5.1 Water Quantity

Water should be available in sufficient quantity and pressure to supply the maximum number of medium- and high-expansion foam generators likely to operate simultaneously in addition to the demands of other fire protection equipment.

8.20.5.2 Water Pumps

The pump should supply water to the inlet of the foam system within the range of flow and pressure for which the system is designed. Pumps providing a water supply to foam equipment should be correctly sized so that at maximum demand they operate below their overload characteristic.

They should be capable of operating satisfactorily following long periods of inactivity.

Where an alternative water supply is available, a single pump should be used, otherwise multiple pump arrangements are preferred to improve reliability.

Diesel engines are preferred to electric motors for driving pumps. The use of one diesel-driven and one electrically driven pump of appropriate size is an acceptable arrangement.

The electric power supply to a pump should be a separately switched circuit; where only electric pumps are used an alternative independent supply of electric power should be provided. Water supply should be protected against freezing.

8.20.5.3 Foam Concentrate Quantity

The amount of foam concentrate in the system should be at least sufficient for the largest single hazard protected or a group of hazards that are to be protected simultaneously.

8.20.5.4 Air Supply

Air from outside the hazard area should be used for foam generation unless data is provided to show that air from inside the hazard can be successfully employed. The data should be specific for the products of combustion to be encountered and should provide factors for increasing foam discharge rates.

Vents should be located to avoid recirculation of combustion products into the air inlets of the foam generators.

8.20.5.5 Foam-Generating Apparatus Location

Foam-generating apparatus should be located and arranged so that inspection, testing, recharging, and other maintenance is facilitated and interruption of protection is held to a minimum.

8.20.5.6 Protection against Exposure

Foam-generating equipment should be located as close as possible to the hazard(s) it protects, but not where it will be unduly exposed to a fire or explosion. Foam generators installed inside the hazard area should be constructed to resist or protect against fire exposure.

Such protection may be in the form of insulation, fire-retardant paint, water spray, or sprinklers. In certain applications additional generators should be substituted for fire exposure protection with the approval of the authorities concerned.

8.20.5.7 Ducts

Foam distribution and air inlet ducts should be designed, located, installed, and suitably protected so that they are not subject to undue mechanical, chemical, or other damage.

Duct closures such as selector valves, gates, or doors should be of the quick-opening type, allowing free passage of the foam. When located where they may be subjected to fire or heat exposure either inside or outside the area to be protected, special care should be taken to ensure positive operation.

Ducts should be designed and installed so that undue turbulence is avoided and the actual foam discharge rate should be determined by test or other method acceptable to the relevant authorities.

8.20.6 Foam Requirements for Medium-Expansion Foam Systems

8.20.6.1 Application Rate

a. Determination of Application Rate (Medium Expansion) and Foam Discharge Rate (High Expansion)

Apparatus:

Pressure gage, installed adjacent to the discharge point in the hydraulically most remote location with respect to the main foam solution supply line to the system

Calculation:

Medium-Expansion Foam

Calculate the overall foam solution flow rate (Q) (in L/min) where only one type of nozzle is used from the equation

$$Q = \sum_{}^{n} N \times K \times P^{0.5}$$

where
Q is the foam solution flow rate (in L/min)
K is the nozzle discharge coefficient
N is the number of nozzles fitted
P is the steady state nozzle pressure (in bar)
n is the number of types of nozzle

Calculate the application rate R (in L/m² per minute) from the equation

$$R = Q/A$$

where
A is the area covered by the system (in mm²)

Note that the discharge coefficients are determined by separate tests of the nozzles concerned measuring flow rates over the pressure range involved.

High-Expansion Foam

Calculate the foam solution flow rate in accordance with B.3.1. Calculate the foam discharge rate (in m³) from

$$F = Q \times E$$

where
E is the expansion, calculate the foam expansion from the equation

$$E = \frac{166.2}{W_2 - W_1}$$

where
  W_1 is the mass of the empty pan (in kg)
  W_2 is the mass of the full pan (in kg)

    The volume of foam is the volume of the pan, 166.25 L, and $W_2 - W_1$ is the volume of water (equal to the mass of the foam) contained in it.

b. Flammable Liquids

    The application rate should be not less than

(1) The rate, agreed with the user, shown to be effective by tests

(2) If test data is not available:

    4 L/m²/min for hydrocarbon liquids or 6.5 L/m²/min for foam-destructive liquids

c. Combustible Solids

    The application rate should be not less than the rate agreed with the user.

8.20.6.2 Duration of Discharge

The minimum duration of systems discharging at higher than the minimum rate may be reduced in proportion but should be not less than 70% of the time given in Table 8.13.

8.20.7 Foam Requirements for High-Expansion Foam Systems

8.20.7.1 Vent Design

The vent(s) should be positioned at the most remote point(s) from the foam inlets(s), and should be to the open air. The vent(s) should be of open design, or if normally closed should open automatically on actuation of the system.

    Correct positioning of the vent(s) is necessary to ensure that the submergence depth is achieved throughout the protected area. Venting is to the outside air to allow the safe dispersal of smoke and combustion products.

TABLE 8.13

Minimum Discharge Times for Medium-Expansion Foam Systems Discharging at the Minimum Rate

Hazard	Minimum Discharge Time (min)
Indoor and outdoor spill up to 100 m²	10
Other indoor hazards and outdoor protection	15

The area of the vent(s) should be sufficient to limit the venting velocity to not more than 300 m/min.

This will be achieved if the vent area (in m²) is not less than $F/300$, where F is the foam discharge rate in m³/min.

Venting is not usually necessary where air from within the enclosure is used to make the foam.

8.20.7.2 Submergence Depth

The system should produce, throughout the protected area, a depth of foam sufficient to cover and extinguish the highest hazard.

In unsprinklered enclosures of combustible construction the submergence depth should be sufficient to fill the enclosure.

For combustible solids, in enclosures that are sprinklered or are of non-combustible construction, the submergence depth should be sufficient to cover the highest hazard with 1 m or 0.1 times the height of the highest hazard, in meters, whichever is greater, of foam.

For flammable liquids the submergence depth should be determined by test, and should be considerably more than for combustible solids.

8.20.7.3 Submergence Time

The system should produce throughout the protected area a depth of foam not less than the submergence depth in not more than the appropriate maximum time given in Table 8.14.

In calculating the foam application rate, the volume of vessels, machinery, or other permanently located equipment may be deducted from the total volume to be protected. Volumes occupied by stored materials are not deducted from the volume of the area to be protected, since the quantity may vary with time.

TABLE 8.14

Maximum Submergence Times for High-Expansion Foam Systems

| | Maximum Submergence Times | |
Hazards	High-Expansion Foam Only (min)	High-Expansion Foam with Supporting Water Sprinklers (min)
Flammable liquids with flash points not above 40°C	2	3
Flammable liquids with flash points above 40°C	3	4
Low-density combustible solids, (e.g., foam rubber, foam plastics, rolled tissue, or crepe paper)	3	4
High-density combustible solids, (e.g., rolled paper, rubber types)	5	7

Provided appropriate attention is given to distribution the requirements for submergence time will be met if the discharge rate of the system is not less than

$$F = C_N \times C_L \times \frac{\left[F_S + (D \times A) - V_{eq} \right]}{T}$$

where

D is the submergence depth (in m)

F is the foam discharge rate (in m^3/min)

T is the submergence time (in min)

A is the floor area of the protected space (in m^2)

V_{eq} is the volume of any permanently installed equipment, vessels, or machinery, excluding the volume of any removable stored materials or equipment (in m^3)

$C_N = 1.20$, an empirical factor based on the average reduction in foam quantity due to solution drainage, fire, wetting of dry surfaces, and so forth

$C_L = 1.1$, an empirical factor compensating for the loss of foam due to leakage around doors and windows where these are closed but not sealed

F_S is the rate of foam breakdown by sprinkler discharge (in m^3/min). This factor should be determined either by test, or in the absence of specific test data, by the following formula:

$$F_S = 0.075 \times Q$$

where

Q is the estimated total discharge from the maximum number of sprinklers expected to operate (in L/min).

8.20.7.4 Quantity of Foam Concentrate

The quantity of foam concentrate (in liters) available for immediate use in the system should be not less than

a. For fire involving combustible solids:

$$250 \times \frac{FC}{E}$$

b. For fires involving flammable liquids:

$$150 \times \frac{FC}{E}$$

where
  F is the foam discharge rate (in m³/min)
  C is the concentration (in %)
  E is the expansion

The quantities specified allow system running times (whether continuously or intermittently) to 25 minutes for combustible solids and 15 minutes for flammable liquids. For flammable liquids it is usual for the system to run continuously, but for systems protecting combustible solids once submergence is achieved it is usual to run the system intermittently, in effect discharging foam at a rate equivalent to the breakdown rate, to maintain the submergence depth for the maximum time possible.

8.20.7.5 Types of Systems

The types of systems recognized in this standard include

- Total flooding systems
- Local application systems
- Portable foam-generating devices

8.20.8 Total Flooding Systems

A total flooding system consists of fixed foam-generating apparatus complete with a piped supply of foam concentrate and water arranged to discharge into an enclosed space or enclosure around the hazard.

8.20.8.1 Uses

This type of system should be used where there is a permanent enclosure around the hazard that is adequate to enable the required amount of fire extinguishing medium to be built up and to be maintained for the required period of time to ensure the control or extinguishment of the fire in the specific combustible material(s) involved.

Examples of hazards that are successfully protected by total flooding systems include rooms, vaults, storage areas, warehousing facilities, and buildings containing class A and class B combustibles either singly or in combination.

Fires that can be controlled or extinguished by total flooding methods are divided into three categories:

1. Surface fires involving flammable or combustible liquids and solids
2. Deep-seated fires involving solids subject to smoldering
3. Three-dimensioned fires in some flammable liquids

8.20.8.2 General Requirements

Total flooding systems should be designed, installed, tested, and maintained in accordance with the applicable requirements in Chapter 1, Clause 1-6 of NFPA IIA and with the additional requirements set forth in this section. Only listed or approved equipment and devices should be used in these systems.

8.20.8.3 Enclosure Specifications

Since the efficiency of the medium- or high-expansion foam system depends on the development and maintenance of a suitable quantity of foam within the particular enclosure to be protected, leakage of foam from the enclosure should be avoided.

Openings below design filling depth, such as doorways, windows, and so forth, should be arranged to close automatically before, or simultaneously with, the start of the foam discharge, with due consideration for evacuation of personnel. These should be designed to maintain a closure during a fire and be capable of withstanding pressures of foam and sprinkler water discharge. If any unclosable openings exist, the system should be designed to compensate for the probable loss of foam and should be tested to assure proper performance.

The venting so required should consist of suitable openings, either normally open or normally closed and arranged to open automatically when the system operates. When design criteria demand exhaust fans, they should be approved for high-temperature operation and installed with due consideration for protection of switches, wiring, and other electrical devices to ensure equal reliability of exhaust fan performance as for the foam generators. Where forced-air ventilating systems interfere with the proper buildup of foam, they should be shut down or closed off automatically.

8.20.8.4 Quantity

Sufficient high-expansion foam concentrate and water should be provided to permit continuous operation of the entire system for 25 minutes or to generate four times the submergence volume, whichever is less, but in no case less than enough for 15 minutes of full operation. The quantity for medium-expansion foam should be determined by suitable tests developed by an independent testing laboratory.

8.21 Local Application Systems

A local application system consists of fixed foam-generating apparatus complete with a piped supply of foam concentrate and water and arranged to discharge foam directly onto the fire or spill.

8.21.1 Uses

Local application systems are used for the extinguishment or control of fires in flammable or combustible liquids, LNG, and ordinary class A combustibles where the hazard is not totally enclosed. These systems are best adapted to the protection of essentially flat surfaces such as confined spills, open tanks, drainboards, curbed areas, pits, trenches, and so forth. For multiple-level or three-dimensional fire hazards where total building flooding is not practical, the individual hazard should be provided with suitable containment facilities acceptable to the authorities.

8.21.2 General Requirements

Local application systems should be designed, installed, tested, and maintained in accordance with the applicable requirements in total flooding and with the additional requirements set forth in this section. Only listed or approved equipment, devices, and agents should be used in these systems.

8.21.3 Hazard Specifications

8.21.3.1 Location of Hazard

Local application medium- and high-expansion foam systems should be used to protect hazards located indoors, partly sheltered, or completely out-of-doors. Provisions should be made to compensate for winds and other effects of weather.

8.21.3.2 Foam Requirements for Flammable and Combustible Liquids and Solids

Sufficient foam should be discharged at a rate to cover the hazard to a depth of at least 0.6 m within 2 minutes.

8.21.3.3 Quantity

Sufficient foam concentrate and water should be provided to permit continuous operation of the entire system for at least 12 minutes.

8.21.4 Foam Applications for LNG

High-expansion foam has been shown to be effective in controlling LNG spill test fires and in reducing downwind vapor concentration from unignited LNG spill test fires in confined areas up to 111 m^2.

8.21.4.1 System Design Considerations

Determination of the high-expansion foam system design depends on an analysis specific to the individual site. Since time to initiate actuation is a critical factor in LNG fire control, the analysis must consider effects of heat exposure on adjacent plant equipment. In many cases automatic alarms and actuation will be required for fixed systems.

8.21.4.2 Application Rate

As established by test, the application rate should be such that a positive and progressive reduction in radiation is attained within the time limitations established in the analysis. The application rates determined by the test should be increased by the necessary factor to account for the initial vaporization rate and the configuration of the hazard. After steady state control conditions have been reached, the application rates for maintenance of fire control established in the test should be used to maintain control.

8.21.4.3 Quantity

The initial quantity of foam concentrate should permit a continuous application at the initial design rate sufficient for fire control to reach steady state conditions. Additional foam concentrate supplies should be on hand to provide control maintenance for the calculated fire duration.

8.21.4.4 Foam System Arrangement

The foam system should have foam outlets so arranged to supply foam to cover the design fire area within the specified time.

8.22 Safety Hazards

Foam solutions are generally not considered toxic to humans but contact will cause skin or eye irritation. Read warning labels on foam concentrate containers. Effects and antidotal procedures will vary for each foam agent.

Fragments generated by the rupture of dry foam layers can cause sneezing and coughing. These effects are transient and will stop when exposure to the source of the fragments stops.

Prevent prolonged exposure to foam. Do not enter foam if full submergence would occur. Both vision and hearing are obscured. Some breathing apparatus can be adversely affected by foam submergence. If wading of moving through foam is necessary, be careful of tripping hazards and always wear a lifeline.

If possible, do not perform any action that will mechanically degrade that portion of the foam blanket covering the spill; that is walking or dragging hoses through foam. Such actions can create a localized vapor hazard. Do not depend on films to rapidly reform and prevent vapor release.

Where possible, the relative location of foam discharge points to building exits should be arranged to facilitate evacuation of personnel. Additional exits and other measures may be necessary to ensure safe evacuation of personnel.

Because foams are made from aqueous solutions, they will be dangerous to use on materials that react violently with water, such as sodium or potassium, and should not be used when these are present. A similar danger is presented by other metals, such as zirconium or magnesium, but only when they are burning.

Medium- and high-expansion foams are electrically conductive and should not be used on energized electrical equipment where this would be a danger to personnel.

8.23 Wetting Agents

Experience, as well as tests, has indicated that the addition of a proper wetting agent to water will, when properly applied, increase the extinguishing efficiency of that water with respect to quantity used as well as time saved. The value of such a factor has become of considerable importance, especially in rural areas where the amount of water available for firefighting is often inadequate. This is due to the fact that the addition of a proper wetting agent to the charge in a booster tank will increase the extinguishing efficiency of the water.

Certain types of fires, such as those in baled cotton, stacked hay, some rubber compounds, and some flammable liquids, which do not ordinarily respond to treatment with water, should be extinguished when a proper wetting agent is used. This property is attributed to an increase in the penetrating, spreading, and emulsifying powers of water due to such factors as lowering the surface tension. This decreased surface tension can be described as a disruption of the forces holding the surface film of water together, thereby permitting it to flow and spread uniformly over solid surfaces.

As a result, the treated water acquires the ability to penetrate into small openings and recesses that water would flow over by the simple bridging action of the surface film. Note that such solutions exhibit not only penetrating and spreading qualities, but increased absorptive speed and superior adhesion to solid surfaces.

When mixed with water and air, wetting agents having foaming characteristics as referred to in this standard produce a foam that retains the wetting and penetrating characteristics of the wetting agent and provides an efficient smothering action for the extinguishment of both class A and class B combustibles or provides a fluid insulation for protection against fire exposure.

The foam produced in this manner has the additional advantage of breakdown at approximately 79.4°C and returns to its original liquid state retaining the penetrating and wetting qualities. The breakdown of this foam when applied on class A combustibles automatically provides an efficient and adequate application rate for efficient extinguishment.

There are numerous chemicals that fulfill the primary function of a wetting agent, which is to lower the surface tension of water.

However, very few of these chemicals are suited to fire control work because application to this purpose is complicated by such considerations as toxicity, corrosive action on equipment, and stability in naturally occurring waters.

In view of this fact, therefore, these standards set forth certain basic requirements and limitations for the use of a wetting agent as an aid for fire extinguishment. The requirements are intended to ensure that the addition of a wetting agent to any natural water should not affect that water adversely with respect to fire-fighting properties, nor render it harmful to personnel, property, or equipment. It is further intended to establish standards for the evaluation of wetting agents as fire extinguishing mediums.

8.23.1 Uses

In general, this standard is intended to signify that a wetting agent that successfully meets the requirements herein set forth should not be limited in use or application except as herein specified.

The addition of proper wetting agents to water will increase its penetrating and emulsifying abilities, and should provide foaming characteristics as to extend the efficiency of water for the protection against fire exposure and the extinguishment of class A and class B fires in ordinary combustibles and combustible liquids that are insoluble in water and ordinarily stored at atmospheric temperatures and pressures.

In general, wetting agents can be effectively applied and used with all types of standard fire protection equipment where water is normally used. The degree of efficiency obtained will depend on utilizing the most efficient application methods, techniques, and devices for the hazard involved.

When water containing listed wetting agents is applied to a fire, some of the wetting agent should be expected to remain after extinguishment. This residual wetting agent is effective in reducing the surface tension of water that will subsequently be applied.

The authorities should be consulted in all cases where the use of wet water is considered for application through fixed equipment, such as water spray, sprinkler, or foam systems. The volume of extinguishing medium required will vary with each type of system and hazard. If applied as a liquid solution, the standard applicable to water systems will apply.

Effective exposure protection can be accomplished by the application of wet water foam directly to the exposed structure or equipment to reduce the heat transferred from the exposure fire.

This protection is afforded whether applied from portable or fixed equipment. Due to the cellular structure and reflective characteristics of wet water foam, the water requirements can be appreciably reduced.

The addition of wetting agents to water will increase the efficiency due to the spreading characteristics of the wetting agent, thus affording greater protection than water alone.

8.23.2 Limitations

The addition of wetting agents to water, which changes its physical characteristics, creates certain limitations for use that should be recognized.

- *Class A fires:* Wet water has the same limitations as water with respect to extinguishing fires involving chemicals that react with water to create new hazards.

- *Class B fires:* The effective use of wet water for the extinguishment of fires involving class B flammable or combustible liquids is limited to those materials not soluble in water, such as petroleum products. In water-soluble materials of the alcohol type, some control should be realized, but extinguishment is questionable.

- *Class C fires:* Wet water solutions can conduct electricity and have limitations similar to water in fighting fires involving energized electrical equipment so far as safety to fire-fighting personnel is concerned. Application as a straight stream should not be considered. Spray or fog application can be employed with usual caution.

Wet water should not be used on Class D fires.

8.23.3 Use of Wetting Agents with Wetting Agents Other Than Water

Admixing of wetting agents with other wetting agents or with mechanical or chemical foam liquids should be avoided. The mixing of these agents has adverse results and thus renders them ineffective for fire extinguishment.

The use of wetting agents in concentrations greater than those specified by the manufacturer and/or by the testing laboratory should be avoided. High concentrations cause adverse effects.

8.23.4 Basic Requirements

Wetting agents for firefighting should be listed by a testing laboratory and should be approved by the authorities.

Special equipment, such as proportioners, should be listed by a testing laboratory and should be approved by authorities.

8.23.5 Fire Department Supply Requirements

The wetting agent should be premixed in a booster tank in such concentration as specified by the manufacturer. Where such premixing is considered undesirable, an amount of wetting agent determined to be sufficient for the water contained in the portable tanks on the apparatus should be carried in a container that can readily be emptied into such tanks.

Where portable tanks are not a part of the apparatus, or where it is desired to carry the wetting agent separately for use either with water from portable tanks or with water from other sources of supply, the amount considered necessary should be carried in a suitable tank connected to appropriate proportioning equipment on the apparatus. Where such equipment is used also to take suction from a hydrant supplied by potable water, extra care should be exercised to prevent contamination of such potable water supplies with the wetting agent.

8.23.5.1 Additional Supplies

Additional supplies of the wetting agent will be needed to ensure continuity of operation and this should be carried on the apparatus. Further supply should be stocked in suitable storage facilities to recharge the apparatus.

8.23.5.2 Fixed Systems

Existing standards covering all fixed systems should be followed where the addition of a wetting agent to the system is contemplated. Such installations should be approved by the authorities with consideration being given primarily to limitations and to

a. The possibility of increased water damage due to the high absorption ability of wet water
b. The possibility of increased floor loads due to the retention of large volumes of wet water

8.24 Reduction of Soil Pollution by Reducing Foam Exploitation

8.24.1 Comparing Extinguishing Technologies

Pollution of above and underground water can be decreased by reducing the amount of extinguishing materials used and increasing the efficiency of their exploitation. Load of the sewage system can also be reduced by the foam amount used. The total amount of foam volume getting into the tank and into the dike area due to the aiming loss should be considered as pollution.

The rate of soil pollution depends on the foam exploitation of the certain technology as well as the method of foam introduction.

In order to be able to compare the different extinguishing technologies, we introduce the concept and the standard of the foam effectiveness factor. Defining its rate can be carried out by using easily adaptable evaluating methods. Having examined the relationship between the foam effectiveness factor and the foam intensity we realized that we can define the intensity of the optimal working point of certain extinguishing technology in concrete figures.

Based on theoretical assumptions, the relationship between the foam exploitation factor and the foam solution intensity is depicted in Figure 8.6.

The value of η in the event that the case of less-than-critical foam solution intensity is 0; the fire could not be put out. In the event that this low level of cumulative foam volume flow, the penetrational speed decreases to 0 in view of the advanced time spent on putting out the fire and spreading of the foam face stops. The fire continuously consumes the released foam amount; t (extinguishing) = ∞.

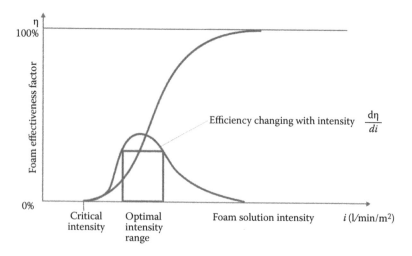

FIGURE 8.6
Estimation of foam efficiency factor.

The value of η is 100% if the driven foam amount is used to create the foam blanket without any waste with its full volume. This situation takes place in the case of cold foam spreading experiments, providing the adaptation of high-quality stable foam.

If we are aware of the relation between the foam exploitation factor and solution intensity of different extinguishing strategies, before starting extinguishing the fire we can calculate the full foam volume to be produced in a certain case, as well as the necessary foam concentrate amount taking the intensity into account resulting from the capacity of our technical background.

As a conclusion, we can state that fixed extinguishing equipment is favorable regarding preparation time.

The benefits of extinguishing with high-capacity foam intensity should be exploited to reduce extinguishing time. The appropriate extinguishing strategy should be chosen by taking the best foam utilization factor into account and it should be operated in an optimal working point.

The technologies are characterized by their foam introduction manner. The most appropriate among all the possible foam introductions is the curtain-like manner that is able to cool the inner part of the shell and stop the wall effect.

Based on these assumptions we developed a new extinguishing strategy of storage tank fires whose adaptability was tested in an experimental way.

8.25 The Dynamic Extinguishing Strategy

In the case of extinguishing, the first phase is to develop the necessary foam thickness for extinguishment. It requires as few as 10 seconds in the range of usual system. The traditional, static approach prescribed the same intensity value independently from the combustion surface.

Modifying this by instructing that the intensity value should be adjusted to the combustion surface, we can realize that for a bigger surface the foam should be exposed over a longer distance to the effects of the fire, as the attacked foam surface gets bigger. The impacts of the inevitable foam destruction can be outweighed by increasing intensity and decreasing extinguishing time. With the help of an increased intensity value, the foam rolls faster on the surface after being exposed to the drying thermal demolishing impacts of fire for a short time of some 10 seconds.

The second phase of the extinguishing procedure is the increasing of the foam blanket thickness. Its importance lies in preventing reignition. We should continue the accession of the foam to the surface of the liquid until it grows to an adequate thickness. We should be able to maintain the resistance

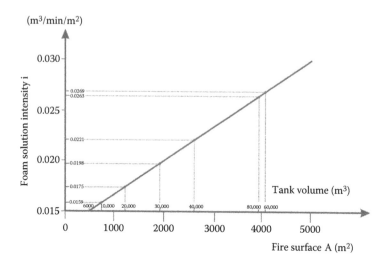

FIGURE 8.7
Relation between foam solution intensity and fire surface.

of the foam blanket even if a strong sidewind tries to make the blanket open. The bigger intensity value besides the fixed-foam introduction time results in a greater safety factor and the required bigger foam blanket thickness will be set by the end of the prescribed total foam introduction time.

Based on our measurements and the conclusions from calculating the foam spreading velocity to different tank sizes, the values indicated in Figure 8.7 are proposed for indicating the relation between the foam solution intensity and the fire surface.

8.26 Sample Calculation of Foam Compound Requirement for a Depot/Terminal

8.26.1 Foam Compound Calculation for the Single Largest Floating Roof or Coned Roof Tank in a Dike, Whichever Is Higher

The foam compound calculation for the single largest floating roof tank in a dike is as follows:

    Tank data:

       Total storage capacity in one dike area = 120,000 m³

       Number of tanks = 2

       Capacity of each tank = 60,000 m³

Diameter of each tank = 79 m

Height of each tank = 14.4 m

Foam compound requirement for tank:

Foam solution application rate = 12 lpm/m² of rim seal area of tank

Foam dam height = 800 mm

Foam solution required = 3.142 × 79 m × 0.8 m × 12 lpm/m²

Foam compound required (3%) = 0.03 × 2383 lpm = 71.49 lpm

Foam compound required for 65 minutes = 65 min × 71.49 lpm = 4647 liters

8.26.2 Foam Compound Calculation for the Single Largest Coned Roof Tank in a Dike

Tank data:

Total storage capacity in one dyke area = 50,000 m³

Number of tanks = 4

Capacity of each tank = 12,500 m³

Diameter of each tank = 37.5 m

Height of each tank = 12 m

Foam compound requirement for tank:

Foam solution application rate = 5 lpm/m² of liquid surface area of tank

Foam solution required = 3.142 × (18.75)² m² × 5 lpm/m²

Foam compound required (3%) = 0.03 × 5523 lpm = 165.69 lpm

Foam compound required for 65 minutes = 65 min × 165.69 lpm = 10,770 liters.

8.26.3 Foam Compound Calculation for Installation Having Aggregate Product Storage Capacity More than 25,000 Kl

Foam compound requirement for two portable foam monitors of 2400 lpm capacity:

Foam solution required = 2 × 2400 lpm

Foam compound required (3%) = 0.03 × 4800 lpm = 144 lpm

Foam compound required for 65 minutes = 65 min × 144 lpm = 9360 liters

8.26.4 Foam Compound Calculation for Two Hose Streams of Foam, Each with a Capacity of 1140 lpm

Foam compound requirement for two foam hose streams of 1140 lpm capacity:

Foam solution required = 2×1140 lpm
Foam compound required (3%) = 0.03×2280 lpm = 68.4 lpm
Foam compound required for 65 minutes = 65 min $\times 68.4$ lpm = 4446 liters

The aggregate quantity of foam concentrate should be the largest of the three above calculated values.

8.27 Brief Summary of Fire-Fighting Foam

8.27.1 Fire-Fighting Foam

Fire-fighting foam is a homogeneous mass of tiny air- or gas-filled bubbles of low specific gravity, which when applied in correct manner and in sufficient quantity forms a compact fluid and stable blanket capable of floating on the surface of flammable liquids and preventing atmospheric air from reaching the liquid.

8.27.2 Types of Foam Compound

Two types of foams are used for fighting liquid fires:

1. *Chemical foam:* When two or more chemicals are added, the foam generates due to chemical reaction. The most common ingredients used for chemical foam are sodium bicarbonate and aluminum sulfate with stabilizer. The chemical foam is generally used in fire extinguishers.
2. *Mechanical foam:* Produced by mechanically mixing a gas or air to a solution of foam compound (concentrate) in water. Various types of foam concentrates are used for generating foam, depending on the requirement and suitability. Each concentrate has its own advantages and limitations. A brief description of foam concentrates is given below.

8.27.3 Mechanical Foam Compound

Mechanical foam compound may be classified in to three categories based on its expansion ratio:

1. *Low-expansion foam:* Foam expansion ratio may be up to 50 to 1, but usually between 5:1 and 15:1 as typically produced by self-aspirating foam branch pipes. Low-expansion foam contains more water and has better resistance to fire. It is suitable for hydrocarbon liquid fires and is widely used in oil refinery, oil platforms, and petrochemical and other chemical industries.

2. *Medium-expansion foam:* Foam expansion ratio varies from 51:1 to 500:1 as typically produced by self-aspirating foam branch pipes with nets. This foam has limited use in controlling hydro-carbon liquid fire because of its limitations with regard to poor cooling, poor resistance to hot surface/radiant heat, and so forth.

3. *High-expansion foam:* Foam expansion ratio varies from 501:1 to 1500:1, usually between 750:1 and 1000:1 as typically produced by foam generators with air fans. This foam has also very limited use in controlling hydrocarbon liquid fire because of its limitations with regard to poor cooling, poor resistance to hot surface/radiant heat, and so forth. It is used for protection of hydrocarbon gases stored under cryogenic conditions and for warehouse protection.

8.27.4 Types of Low-Expansion Foam

8.27.4.1 Protein-Based Foam

In protein-based foam, the foam concentrate is prepared from hydrolyzed protein either from animal or vegetable sources. A suitable stabilizer and preservatives are also added.

The concentrate forms a thick foam blanket and is suitable for hydrocarbon liquid fires but not on water-miscible liquids. The effectiveness of foam is not very good on deep pools or low flash point fuels that have had lengthy preburn time unless applied very gently to the surface.

The concentrate is available for an induction rate of 3% to 6%, and the shelf life of the concentrate is 2 years.

8.27.4.2 FP Foam

FP foam is similar to protein-based foam with fluorochemicals, which makes it more effective than protein-based foam.

The concentrate forms a thick foam blanket and is suitable for hydrocarbon liquid fires but not on water-miscible liquids. The foam is very effective on deep pools of low flash point fuels that have had lengthy preburn time.

The concentrate is available for an induction rate of 3% to 6% and the shelf life is similar to that of protein-based foam.

8.27.4.3 AFFF

AFFF foam concentrate mainly consists of fluorocarbon surfactants, a foaming agent, and a stabilizer. This can be used with freshwater as well as with seawater. It produces very fluid foam that flows freely on a liquid surface. The aqueous film produced suppresses liquid vapor quickly. The foam has a quick fire knockdown property and is suitable for liquid hydrocarbon fires. As the foam has poor drainage rate, the effectiveness is limited on deep pool fires of low flash point fuels that have had lengthy preburn time.

The concentrate is available for an induction rate of 3% to 6% and the shelf life is more than 10 years. This can also be used with nonaspirating type nozzles.

8.27.4.4 Multipurpose AFFF

Multipurpose AFFF concentrate is a synthetic foaming liquid designed especially for fire protection of water-soluble solvents and water-insoluble hydrocarbon liquids. This can be used with freshwater as well as with seawater.

When applied it forms foam with a cohesive polymeric layer on a liquid surface, which suppresses the vapor and extinguishes the fire. The foam is also suitable for deep pool fires because of superior drainage rate and more resistive to hot fuels/radiant heat.

The 3% induction rate is suitable for liquid hydrocarbon fires and the 5% induction rate for water-miscible solvents. The shelf life of the concentrate is not less than 10 years. This can also be used with nonaspirating type nozzles.

8.27.4.5 FFFP

FFFP combines the rapid fire knockdown quality of conventional film-forming AFFF with the high level of postfire security and burnback resistance of FP foam. The concentrate can be used with freshwater as well as with seawater.

The foam is suitable for hydrocarbon liquid fires including deep pool fires of low flash point fuels that have had lengthy preburn time.

The concentrate is available for an induction rate of 3% to 6% and the shelf life is not less than 5 years. This can also be used with nonaspirating type nozzles.

8.27.5 Types of Medium- and High-Expansion Foam

Synthetic foam concentrate is used with suitable devices to produce medium- and high-expansion foams. This can be used on hydrocarbon fuels with a low boiling point. The foam is very light in weight and gives poor cooling effect in comparison to low-expansion foams. The foam is susceptible to easy breakdown by hot fuel layers and radiant heat.

The induction rate in water may vary from 1.5% to 3%. Many of the low-expansion foam concentrates can also be used with suitable devices to produce medium-/high-expansion foam.

Further Readings

Boulton, A.J., Moss, G.L. and Smithyman, D. 2003. Short-term effects of aerially-applied fire-suppressant foams on water chemistry and macroinvertebrates in streams after natural wild-fire on Kangaroo Island, South Australia. *Hydrobiologia* 498, 177–189.

British Standard Institution (BSI). 2009. 5306 6.1 Specification for Low Expansion Foam; 5306 6.2 Specification for Medium and High Expansion Foam System; 5306 Section 6.1 Specification for Low Expansion Foams.

Fleming, J.W. and Sheinson, R.S. 2012. Development of a cup burner apparatus for fire suppression evaluation of high-expansion foams. *Fire Technology* 48 (3), 615–623.

Harman, R. 2003. Focus on foam, *Fire Prevention* 369, 21–23.

Jayakody, C., Myers, D., Sorathia, U. and Nelson, G.L. 2000. Fire-retardant characteristics of water-blown molded flexible polyurethane foam materials. *Journal of Fire Sciences* 18 (6), 430–455.

Kim, A. 2003. Firefighting foams. *Canadian Consulting Engineer* 44 (3), 30–32.

Klein, R. 2004. Focus on foam. *Fire Prevention and Fire Engineers Journals* 64 (247), 24–25, 40–42.

Lattimer, B.Y. and Trelles, J. 2007. Foam spread over a liquid pool. *Fire Safety Journal* 42 (4), 249–264.

Lorenzetti, A., Modesti, M., Gallo, E., Schartel, B., Besco, S. and Roso, M. 2012. Synthesis of phosphinated polyurethane foams with improved fire behaviour. *Polymer Degradation and Stability* 97 (11), 2364–2369.

Luda, M.P., Nada, P., Costa, L., Bracco, P. and Levchik, S.V. 2004. Relevant factors in scorch generation in fire retarded flexible polyurethane foams: I. Amino group reactivity. *Polymer Degradation and Stability* 86 (1), 33–41.

National Fire Protection Association (NFPA). 1999. NFPA IIA Foam System.

Nekrasov, A.G., Tatiev, S.S., Todes, O.M. and Shubin, I.F. 1989. Thermal characteristics of water foams. *Journal of Engineering Physics (Inzhenerno-Fizicheskii Zhurnal)* 55 (2), 897–902.

Olson, K. 2008. Foam fire protection. *Hydrocarbon Engineering (Suppl.)* 83–86.

Place, B.J. and Field, J.A. 2012. Identification of novel fluorochemicals in aqueous film-forming foams used by the US military. *Environmental Science and Technology* 46 (13), 7120–7127.

Sabadra, R. and Shroff, U.K. 2011. Coal mine fires: Evaluation of class A foam. *Journal of Mines, Metals and Fuels* 59 (3–4), 111–115.

Tsai, T.P., Yang, H.C. and Liao, P.H. 2011. The application of concurrent engineering in the installation of foam fire extinguishing piping system. *Procedia Engineering* 14, 1920–1928.

Wang, D., Zhang, X., Yang, D. and Li, S. 2012. Study of preparation and properties of fire-retardant melamine formaldehyde resin foam. *Advanced Materials Research* 510, 634–638.

Xia, J., Fu, X., Zhang, X., Hu, Y. and Wang, R. 2013. Design and application of endoscope with aided lens for observation of fire-fighting foam. *Applied Mechanics and Materials* 271 (Part 1), 823–828.

Yu, S., Wang, Y., Duan, B., Zhou, J., Yang, F., Wang, X. and Liang, D. 2011. Fireproof performance of foam concrete insulation board. *Advanced Materials Research* 250–253, 474–479.

Zatorski, W., Brzozowski, Z.K. and Kolbrecki, A. 2008. New developments in chemical modification of fire-safe rigid polyurethane foams. *Polymer Degradation and Stability* 93 (11), 2071–2076.

9

Fire-Fighting and Fire Protection Facilities

9.1 Introduction

Water has always been the primary extinguishing agent for firefighting for its

1. Availability
2. Economy
3. Effectiveness
4. Ease of storage and transfer

Firefighters must have knowledge of water supply systems and how to use these systems to supply water for fire-fighting efforts in both offshore and onshore facilities.

The following sections are general guidelines for providing fire control equipment on a platform. Some fires can be effectively controlled and extinguished by isolating the fuel source and thereby preventing escalation of the fire. Fuel sources should be isolated by manually or automatically closing block valves, with nonreturn valves, and with depressurization.

Fire control systems on offshore platforms should use water only, chemicals only, or combinations of water and chemicals.

Many considerations influence the design of the fire control system to provide the desired level of protection. These include the size and complexity of the platform, the nature of operations, number and skill of operators for the fire extinguishing equipment, the areas protected by the system, availability of additional equipment not located on the platform, and the consequences of a major fire. The equipment described by these guidelines should be adequate to allow one or two persons to control and/or extinguish most fires or allow a platform fixed water spray system to protect the operator selected portions of a structure.

The proper training of personnel and maintenance of this fire control equipment in operable condition is of paramount importance.

The best protection against the occurrence of fires will be realized through the provision of well-designed facilities and the training of personnel to employ safe operating practices. The facility should be designed and operated to account for all phases of the producing operations, including temporary situations such as drilling, work over, construction, and so forth. Facilities and operating practices should be capable of isolating fuel sources should a fire occur.

This chapter lays down the minimum requirements of fire protection facilities at petroleum depots, terminals, and pipeline installations with or without storages, central tank farms (CTFs), lube oil installations, and offshore facilities.

9.2 Fire Prevention Practices in Offshore Facilities

The facility should be designed to prevent fires and minimize damage if a fire occurs. Some specific items that should be considered are detailed in the following sections.

9.2.1 Well and Process Safety Systems

An important role in preventing fires or minimizing their effect can be performed by the platform surface and subsurface safety systems. The purpose of a surface safety system is to detect abnormal conditions and initiate appropriate action to prevent creation of situations that could result in an accidental fire. The action normally initiated by the surface safety system is to shut off process flow, thus eliminating the major fuel source on a platform.

The safety system may also shut down potential ignition sources such as engines, compressors, and heaters. A subsurface safety system can be provided that is designed to shut off flow from the wells.

9.2.2 Equipment Arrangement

Within the limits of practicality, equipment should be arranged on a platform to provide maximum separation of fuel sources and ignition sources. Particular consideration should be given to the location of fired process vessels.

9.2.3 Ignition Prevention Devices

Fired vessels should be placed in a safe location. Natural draft components should be equipped with spark and flame arrestors to prevent spark emission.

9.2.4 Hot Surface Protection

Surfaces with a temperature in excess of 204°C should be protected from liquid hydrocarbon spillage or mist, and surfaces in excess of 482°C should be protected from combustible gases.

9.2.5 Fire Barriers

Barriers constructed from fire-resistant materials will be helpful in special situations to prevent the spreading of flames and to provide a heat shield.

9.2.6 Electrical Protection

Protection from ignition by electrical sources should be provided by designing and installing electrical equipment using the area classifications.

9.2.7 Combustible Gas Detectors

The concentration of a combustible gas can be determined by detection devices that should initiate alarms or shutdowns. The usual practice is to activate an audible and/or visual alarm at a low gas concentration and to initiate action to shut off the gas source and/or ignition source if the concentration reaches a preset limit below the LEL.

9.2.8 Diesel Fuel Storage

Standards for diesel fuel storage facilities should include the following:

- Diesel fuel storage tanks should be isolated as far as practicable from ignition sources. Diesel fuel tanks should be enclosed by curbs, drip pans, or catchments, and so forth that drain to a sump with provisions to prevent vapor return.
- Diesel fuel tanks should be adequately vented or equipped with a pressure/vacuum relief valve and should be electrically grounded.
- Fire detection device(s) such as fusible plugs should be installed in the diesel fuel storage area.

9.3 Fire-Fighting Equipment

9.3.1 Scale of Equipment Provision

The scale of fire-fighting equipment provided for any particular installation or vessel should be governed by the nature and extent of operations to be carried out. Figure 9.1 shows a schematic of piping arrangement for an off-shore platform fire-fighting installation. When drilling is involved the possibility of a fire from a well blowout should be provided for with hydrant equipment directable to the wellhead and bell nipple areas. In the case of production operations the processing areas should have priority attention.

When deciding the range and scale of equipment to be provided, it must always be kept in mind that when fire has to be dealt with onboard an installation, the unit and the personnel onboard will be on their own and external assistance will not be readily available. Therefore the scale, range, and quality of the fire-fighting equipment and facilities should be generous with adequate backup provided.

Consideration should also be given to the emergency support that is available from other operating oil companies with installations already in the area. When the use of outside support for emergencies is planned it is essential that the compatibility of fire hose connections on the support vessels with those on the installation concerned is checked and provision made for any special fittings that would be required.

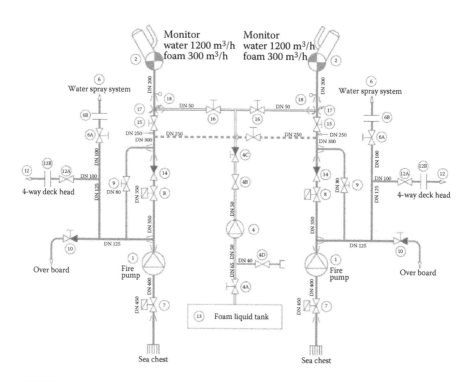

FIGURE 9.1
Piping arrangement for an onshore fire-fighting system.

9.3.2 Fire Pumps, Fire Mains, Hydrants and Hoses (Mobile and Fixed Offshore Installations)

9.3.2.1 Fire Pumps

At least two independently driven power pumps are to be provided, each arranged to draw directly from the sea and discharge into a fixed fire main. Figure 9.2 shows a fire pump designed for offshore installation. However, in units with high suction lifts, booster pumps and storage tanks should be installed, provided such arrangements will satisfy all the requirements of these standards.

At least one of the required pumps is to be dedicated for fire-fighting duties and be available for such duties at all times.

The arrangements of the pumps, sea suctions, and sources of power are to be such as to ensure that a fire in any one space would not put both the required pumps out of action.

The capacity of the required pumps is to be appropriate to the fire-fighting services supplied from the main. However, the total capacity of the pumps should not be less than 180 m³/h.

Where more pumps than required are installed, their capacity is to be not less than 80% of the total required capacity divided by the number of fire pumps.

Each pump is to be capable of delivering at least one jet simultaneously from each of any two fire hydrants, hoses, and 19-mm nozzles while maintaining a

FIGURE 9.2
A fire pump for offshore installation. (Reproduced with permission from Peerless Engineered Systems.)

minimum pressure of 3.5 bar (3.5 kgf/cm²) at any hydrant. In addition, where a foam system is provided for protection of the helicopter deck, the pump is to be capable of maintaining a pressure of 7 bar (7 kgf/cm²) at the foam installation.

The maximum pressure at any hydrant should not exceed that at which the effective control of a fire hose can be demonstrated.

Where either of the required pumps is located in a space not normally manned and is relatively far removed from working areas, suitable provision is to be made for remote start-up of that pump and remote operation of associated suction and discharge valves from a normally manned space or fire-control station.

Every centrifugal pump that is connected to the fire main is to be fitted with a nonreturn valve.

Relief valves are to be provided in conjunction with all pumps connected to the fire main if the pumps are capable of developing a pressure exceeding the design pressure of the fire main, hydrants, and hoses.

Such valves are to be placed and adjusted so as to prevent excessive pressure in any part of the fire main system.

Where intermediate water storage tanks are fitted as permitted, they are to be of such size and be operated so that the lowest water level permitted will ensure that the supply of water is adequate for two hoses at a minimum nozzle pressure of 3.5 bar (3.5 kgf/cm²) at the uppermost hydrant for at least 15 minutes (minimum tank capacity of 10 m³) in order to allow sufficient time for bringing a replenishment pump into operation. Valves and pumps serving the intermediate tank that are not readily accessible are to be provided with means for remote operation.

The system is to be kept charged at the necessary pressure, and the pump supplying the water for the system is to be put automatically into action by a pressure drop in the system.

The pump should be driven by independent internal combustion type machinery but if it is dependent on power being supplied from the emergency generator, that generator is to be arranged to start automatically in case of a main power failure. When the pump is driven by independent internal combustion type machinery it is to be situated so that a fire in the protected space will not affect the air supply to the machinery.

The following features are to be incorporated in a system that includes an intermediate tank:

a. A low-water-level alarm.

b. Two reliable and adequate means to replenish water in the intermediate tank. Such means are to be pumps that are arranged in accordance with the requirements of this section. At least one of the replenishment pumps is to be arranged for automatic operation.

c. If the unit is intended to operate in cold weather, the entire firefighting system is to be protected from freezing. This would include tanks used as water reservoirs.

9.3.2.2 Fire Main

A fixed fire main is to be provided and be so equipped and arranged that water for fire-fighting purposes can be supplied to any part of the installation. The fire main is to be

a. Connected to at least two pumps situated in widely separated parts of the installation
b. With any one pump out of action, capable of delivering at least one jet simultaneously from each of any two fire hydrants, hoses, and 19-mm nozzles while maintaining a minimum pressure of 3.5 bar (3.5 kgf/cm²) at any hydrant

In addition, where a foam system is provided for protection of the helicopter deck and is served by the fire main, a pressure of 7 bar (7 kgf/cm²) at the foam installation is to be capable of being simultaneously maintained.

With any one pump out of action, the aggregate capacity of the remaining pumps is to be not less than 180 m³/h when supplying the fire hydrants only.

The diameter of the fire main and water service pipes is to be sufficient for the effective distribution of the maximum required discharge from the required fire pumps operating simultaneously.

The water velocity in the mains should not exceed 2.13 m/s (7 ft/s).

Where practicable the fire main is to be routed clear of hazardous areas and be arranged in such a manner as to make maximum use of any thermal shielding or physical protection afforded by the structure of the unit.

The fire main is to be provided with isolating valves located so as to permit optimum utilization in the event of physical damage to any part of the main.

The fire main is not to have connections other than those necessary for fire-fighting purposes.

All practical precautions consistent with having water readily available are to be taken to protect the fire main against freezing.

Materials readily rendered ineffective by heat are not to be used for fire mains and hydrants unless adequately protected. The pipes and hydrants are to be placed so that the fire hoses may be easily coupled to them.

A cock or valve is to be fitted to serve each fire hose so that any fire hose may be removed while the fire pumps are at work.

9.3.3 Hydrants (Mobile and Fixed Offshore Installations)

The number and position of the hydrants are to be such that at least two jets of water not emanating from the same hydrant, one of which is to be from a single length of fire hose, should reach any part of the unit normally accessible to those on board while the unit is being navigated or is engaged in drilling operations. A hose is to be provided for every hydrant.

Fire hoses are to be of an approved standard and sufficient in length to project a jet of water to any of the spaces in which they are required to be used. Their length in general is not to exceed 25 m. Every fire hose is to be provided with a dual-purpose jet/spray nozzle and the necessary couplings. Fire hoses together with any necessary fittings and tools are to be kept ready for use in conspicuous positions near the water service hydrants or connections.

9.3.4 Nozzles (Mobile and Fixed Offshore Installations)

Standard nozzle sizes are to be 12, 16, and 19 mm or as near thereto as possible. Larger-diameter nozzles should be permitted if required as a result of special considerations.

For accommodation and service spaces, a nozzle size greater than 12 mm need not be used.

For machinery spaces and exterior locations, the nozzle size is to be such as to obtain the maximum discharge possible from two jets at the pressure specified from the smallest pump, provided that a nozzle size greater than 19 mm need not be used. The jet throw at any nozzle is to be about 12 m.

All nozzles are to be of an approved dual-purpose type (i.e., spray/jet type) incorporating a shutoff.

The number and arrangement of the nozzles are to be such as to ensure an effective average distribution of water over the areas to be protected. Nozzles are to be fitted above areas over which oil fuel is liable to spread and also above other specific fire hazards. Typical water application rates are as shown in Table 9.1.

TABLE 9.1

Typical Water Application Rates

Fire Risk	Application Rate in Fire Risk (L/m²/min)
Boiler fronts or roof firing areas	20
Oil fuel units	20
Centrifugal separators (not oily water separators)	20
Oil fuel purifiers and clarifiers	20
Oil fuel pressure pumps	20
Hot oil fuel pipes near exhaust pipes or similar heater surfaces on main or auxiliary diesel engines	10
Machinery module floors	5
Tank top area	5
Oil tanks not forming part of the structure of the installation	5

9.4 Water Monitors (Mobile and Fixed Offshore Installations)

The minimum number of monitors and their discharge rate, range, and height of trajectory above sea level are to comply with the requirements of Table 9.2.

The monitors are to be so arranged that the required direction, range, and height of trajectory can be achieved with the required number of monitors when they are operating simultaneously.

The monitors are to be capable of adequate adjustment in the vertical and horizontal direction and are to be positioned so that the jets will be unimpeded within the required range of operation.

Means are to be provided for preventing the monitor jets from impinging on the unit's structure and equipment.

The monitors are to be capable of being activated and maneuvered by remote control from a protected position providing a good view of the monitors and the operating area of the water jets.

Water monitors should be operated either remotely or locally. Each monitor arranged solely for local operation is to be

a. Provided with an access route that is remote from the part requiring protection

b. Sited so as to afford the operator maximum protection from the effects of radiant heat

TABLE 9.2

Water Monitor System Capacities

Class Notation	I	II		III
Number of monitors	2	3	4	4
Capacity of each monitor in m³/h	1200	2400	1800	2400
Number of pumps	1–2	2–4		2–4
Total pump capacity in m³	2400	7200		9600
Length of throw in m (measured horizontally from the mean impact area to the nearest part of the vessel when all monitors are in satisfactory operation simultaneously)	120	150		150
Height of throw in m (measured vertically from sea level to mean impact area at a horizontal distance at least 70 m from the nearest part of the vessel)	45	70		70
Fuel oil capacity in hours (capacity for continuous operation of all monitors, to be included in the total capacity of the vessel's fuel oil tanks)	24	96		96

Each monitor is to be capable of discharging under jet and spray conditions.

The notation "Firefighting unit 1" or "Firefighting unit 2" or "Firefighting unit 3" signifies that a unit complies with these standards and is provided with the appropriate fire-fighting equipment described in Table 9.2 with the total discharge capacity of monitors in m^3/h.

9.5 Water Deluge Pumps and Water Deluge Main

Each offshore installation is to be provided with a water deluge system and/ or water monitor system that any part of the installation containing equipment used for storing, conveying, or processing hydrocarbon resources (other than fuels for use on the installation) can be protected in the event of fire. The following list identifies areas containing equipment requiring water protection and is intended to be typical rather than restrictive:

- Well-heads
- Crude/gas separation equipment
- Gas compressors
- Liquefaction plants
- Gas pressure vessels
- Crude oil pumps
- Crude oil and gas manifolds/piping (not fuel gas) including piping routed over bridges between platforms
- Crude oil storage vessels or tanks
- Gas liquids/concentrate storage vessels
- Glycol regeneration plants
- Flare knockout drums
- Pig launchers/receivers
- Deaeration/filtration equipment (if using gas)
- Drill floor/workover areas
- Areas containing equipment (including piping) through which petroleum will flow during well test operations

Water deluge systems and water monitors are to be connected to a continuously pressurized water main supplies by at least two pumps capable, with any one pump out of action, of maintaining a supply of water at a pressure sufficient to enable the system or monitors to operate at the required discharge rates to meet the water demand of the largest single area requiring protection.

The pumps supplying the water main are to be remote from any part of the installation requiring water deluge protection.

Once activated, each pump is to be capable of continuous unattended operation for at least 18 hours.

9.6 Water Deluge Systems

The quantity of water supplied to any part requiring protection is to be at least sufficient to provide exposure protection to the relevant equipment within that part, and where appropriate, local principal load-bearing structural members. "Exposure protection" means the application of water spray to equipment or structural members to limit absorption of heat to a level that will reduce the possibility of failure.

Generally the minimum water application rate is to be not less than 10 liters/minute over each square meter of exposed surface area requiring protection within the appropriate reference area.

Other water application rates in accordance with a suitable standard or code that meets the specification will be considered. A reference area is a horizontal area bounded completely by

a. Vertical A or H class divisions
b. The seaward extremities of the offshore installation
c. A combination of a and b.

Each part requiring water protection is to be provided with a primary means of application, which may be

a. A fixed system of piping fitted with suitable spray nozzles
b. Water monitors
c. A combination of a and b

Water monitors may only be used for the protection of equipment sited in essentially open areas.

9.7 Automatic Sprinkler, Fire Detection, and Fire Alarm Systems for Fixed Offshore Installations

Each installation is to be provided with an automatic sprinkler, fire detection, and fire alarm system complying with the following requirements.

The system required by the above clause is to be capable of indicating the presence of a fire in all accommodation spaces and service spaces, except spaces that afford no substantial fire risk such as void spaces, sanitary spaces, and so forth.

Any required automatic water sprinkler, fire alarm and fire detection system is to be designed for immediate use at any time, so that no action on the part of the installation's personnel is necessary to set it in operation. Where such a system is fitted, it is to be of the wet pipe type but small exposed sections may be of the dry pipe type where this is shown to be a necessary precaution. Any part of the system that is subjected to freezing temperatures in service is to be suitably protected against freezing. It is to be kept charged at the necessary pressure and have provision for a continuous supply of water.

Each section of sprinklers is to include means for giving a visual and audible alarm signal automatically at one or more indicating units whenever any sprinkler comes into operation. Such alarm systems are to be constructed so as to indicate if any fault occurs in the system.

Such units are to give an indication of any fire and its location in any space served by the system and are to be centralized in the main fire-control station, which is to be manned or equipped so as to ensure that any alarm from the system is immediately received by a responsible person.

Sprinklers are to be grouped into separate sections, each of which is to contain no more than 200 sprinklers. Any section of sprinklers is not to serve more than two floors, except where it is satisfactorily shown that the protection of the installation against fire will not thereby be reduced.

Each section of sprinklers is to be capable of being isolated by one stop valve only. The stop valve in each section is to be readily accessible and its location is to be clearly and permanently indicated. Means are to be provided to prevent the operation of the stop valves by any unauthorized person.

A gage indicating the pressure in the system is to be provided at each section's stop valve and at a central station.

The sprinklers are to be resistant to corrosion by marine atmosphere. In accommodation and service spaces the sprinklers are to come into operation within the temperature ranges from 68°C to 79°C, except that in locations such as drying rooms, where high ambient temperatures are expected, the operating temperature should be increased to no more than 30°C above the maximum underside of floor temperature.

A list or plan is to be displayed each unit showing the spaces covered and the location of the zone in respect of each section. Suitable instructions for testing and maintenance are to be available.

Sprinklers are to be placed in an overhead position and spaced in a suitable pattern to maintain an average application rate of not less than 5 L/m²/min over the nominal area covered by the sprinklers. The use of sprinklers providing other amounts of water suitably distributed will be considered provided they are shown to be not less effective.

The tank is to contain a standing charge of freshwater equivalent to the amount of water that would be discharged in 1 minute by the pump and the arrangements are to provide for maintaining such air pressure in the tank to ensure that where the standing charge of freshwater in the tank has been used, the pressure will be not less than the working pressure of the sprinkler plus the pressure exerted by a head of water measured from the bottom of the tank to the highest sprinkler in the system.

Suitable means of replenishing the air under pressure and of replenishing the freshwater charge in the tank are to be provided. A glass gage suitably protected is to be provided to indicate the correct level of the water in the tank.

Means are to be provided to prevent the passage of seawater into the tank.

An independent power pump is to be provided solely for the purpose of continuing automatically the discharge of water from the sprinklers. The pump is to be brought into action automatically by the pressure drop in the system before the standing freshwater charge in the pressure tank is completely exhausted. Consideration should be given to alternative water-supplying arrangements.

The pump and the piping system are to be capable of maintaining the necessary pressure at the level of the highest sprinkler to ensure a continuous output of water sufficient for the simultaneous coverage of a minimum area of 280 m² at various application rates.

The pump is to have fitted on the delivery side a test valve with a short open-ended discharge pipe. The effective area through the valve and pipe is to be adequate to permit the release of the required pump output while maintaining the pressure in the system.

The sea inlet to the pump is to be, wherever possible, in the space containing the pump and is to be arranged so that it will not be necessary to shut off the supply of seawater to the pump for any purpose other than the inspection or repair of the pump.

The sprinkler pump and tank are to be situated in a position reasonably remote from any main machinery space and not in any space required to be protected by the sprinkler system.

There are to be not less than two sources of power supply for the seawater pump and automatic alarm and detection system. Where one of the sources of power for the pump is an internal combustion engine it is to be situated so that a fire in any protected space will not affect the air supply to the machinery.

The sprinkler system is to have a connection from the installation's fire main by way of a lockable screwdown nonreturn valve at the connection that will prevent a backflow from the sprinkler system to the fire main.

A test valve is to be provided for testing the automatic alarm for each section of sprinklers by a discharge of water equivalent to the operation of one sprinkler. The test valve for each section is to be situated near the stop valve for that section.

Means are to be provided for testing the automatic operation of the pump on reduction of pressure in the system.

9.8 Pressure Water-Spraying Systems

9.8.1 Pipes and Nozzles

Any required fixed-pressure water-spraying fire extinguishing system in machinery spaces is to be provided with spraying nozzles of an approved type suitable for extinguishing burning oil.

Precautions are to be taken to prevent the nozzles from becoming clogged by impurities in the water or corrosion of the piping, nozzles, valves, and pump.

The system should be divided into sections, the distribution valves of which are to be operated from easily accessible positions outside the spaces to be protected and which will not be readily cut off by fire in the protected space.

Where applicable the system is to include mobile sprayers ready for immediate use in the firing area of a boiler or in the vicinity of an oil fuel unit.

9.8.2 Hose Stations

Hose stations are to be provided on each side of the unit and should be located considering accessibility from other decks (near a stairway), the possibility of damage from a fire, coordination with other stations, and interference from other platform activities.

Each hose station is to be provided with a hydrant, a hose, and a nozzle capable of producing a jet or a spray and simultaneously a jet and a spray. The hoses are to be 25 m in length and not less than 38 mm or more than 70 mm in diameter. Where hose stations are connected to the monitor supply lines, provision is to be made to reduce the water pressure at the hydrants to an amount at which each fire hose nozzle can be safely handled by one person. The water pressure should be sufficient to produce a water jet throw of at least 12 m.

Fire hoses should be stored on reels or other suitable devices designed for rapid deployment and for protection of the hose. These storage devices should be corrosion-resistant.

9.9 Fire-Extinguishing Systems

9.9.1 Fixed-Gas Fire-Extinguishing Systems

a. The use of a fire-extinguishing medium which, either by itself or under expected conditions of use, gives off toxic gases in such quantities as to endanger persons is not permitted.

b. The necessary pipes for conveying a fire-extinguishing medium into protected spaces are to be provided with control valves that are to be placed so that they will be easily accessible and not readily cut off from use by an outbreak of fire. The control valves are to be marked so as to indicate clearly the spaces to which the pipes are led. Suitable provision is to be made to prevent inadvertent admission of the medium to any space.

c. The piping for the distribution of the fire-extinguishing medium is to be of adequate size and so arranged, and discharge nozzles positioned so that a uniform distribution of medium is obtained. All pipes are to be arranged to be self-draining and where led into refrigerated spaces, the arrangement will be especially considered.

d. Means whereby the individual pipes to all protected spaces can be tested using compressed air are to be provided.

e. Steel pipes fitted in spaces where corrosion is likely to occur are to be galvanized, at least internally.

f. Distribution pipes are not to be smaller than 20-mm bore for carbon dioxide.

g. Means are to be provided to close all openings that may admit air to or allow gas to escape from a protected space.

h. Where the volume of free air contained in air receivers in any space is such that, if released in such a space in the event of fire, such release of air within that space would seriously affect the efficiency of the fixed fire-extinguishing system, an additional quantity of fire-extinguishing medium is to be provided.

i. Means are to be provided for automatically giving audible warning of the release of fire-extinguishing medium into any space in which personnel normally work or to which they have access. The alarm is to operate for a suitable period (depending on time required for personnel to evacuate. before the medium is released).

j. Where pneumatically operated alarms are fitted that require periodic testing, carbon dioxide is not to be used as an operating medium. Air-operated alarms may be used provided that the air supply is clean and dry.

k. Where electrically operated alarms are used, the arrangements are to be such that the electric operating mechanism is located outside any hazardous area; alternatively the equipment is to be suitable for use in hazardous atmospheres.

l. The means of control of any fixed-gas fire-extinguishing system are to be readily accessible and simple to operate and should be grouped together in as few locations as possible at positions not likely to be cut off by a fire in a protected space. At each location there are to be clear instructions relating to the operation of the system with regard to the safety of personnel.

m. Automatic release of fire-extinguishing medium will not be permitted except as allowed by authorities.

n. Where the quantity of extinguishing medium is required to protect more than one space, the quantity of medium available need not be more than the largest quantity required for any one space so protected.

o. Except as otherwise permitted, pressure containers required for the storage of fire-extinguishing media are to be located outside protected spaces.

p. Means are to be provided for personnel to safely check the quantity of medium in the containers.

q. Containers for the storage of fire-extinguishing media and associated pressure components are to be designed and tested according to local standards with regard to their locations and the maximum ambient temperatures expected in service.

r. When the fire-extinguishing medium is stored outside a protected space, it is to be stored in a room that is situated in a safe and readily accessible position and effectively ventilated. Any entrance to such a storage room is preferably to be from the open deck and in any case be independent of the protected space. Access doors are to open outward, and walls and decks including doors and other means of closing any opening therein that form the boundaries between such rooms and adjoining enclosed spaces are to be gastight.

9.9.2 Carbon Dioxide Systems

The quantity of carbon dioxide carried is to be sufficient to give a minimum volume of free gas equal to 35% of the gross volume of the largest machinery space protected, provided that if two or more spaces are not entirely separate they are to be considered as forming one space.

For the purpose of this standard the volume of free carbon dioxide is to be calculated at $0.56 \ m^3/kg$.

The fixed piping system is to be such that 85% of the gas can be discharged into the space within 2 minutes.

Systems in which refrigerated liquid carbon dioxide is stored in bulk will be especially considered.

Duplicate refrigerating units are to be provided; one is to be operable from two sources of power, while the other is to be the emergency source of electrical power.

9.9.3 High-Expansion Foam Systems

a. Any required fixed high-expansion foam system is to be capable of discharging rapidly through fixed discharge outlets a quantity of foam sufficient to fill the greatest space to be protected within

10 minutes and at a rate of at least 1 m in depth per minute. The quantity of foam-forming liquid available is to be sufficient to produce a volume of foam equal to five times the volume of the largest space to be protected. Figure 9.3 shows a typical water pump and foam supply system.

b. The expansion ratio of the foam is not to exceed 1000 to 1.

c. When the gross horizontal area of the protected space exceeds 400 m², at least two foam generators are to be provided.

d. Alternative arrangements and discharge rates will be permitted provided that equivalent protection is achieved.

e. Supply ducts for delivering foam, air intakes to the foam generator, and the number of foam-producing units are to be such as will provide effective foam production and distribution.

f. Where major fire hazards exist in high positions in the protected space, foam is to be separately led to them by suitable ducts and permanent means are to be provided to contain the foam around and above the hazard.

g. The arrangement of the foam-generator delivery ducting is to be such that a fire in the protected space will not affect the foam-generating equipment.

h. The foam generator, its sources of power supply, foam-forming liquid, and means of controlling the system are to be readily accessible and simple to operate and are to be grouped in as few locations as possible at positions not likely to be cut off by fire in the protected space.

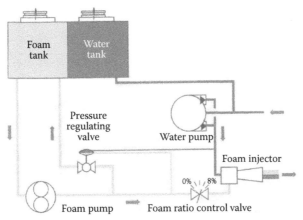

Balanced pressure foam proportioning system

FIGURE 9.3
A water pump and foam supply system.

9.10 Storage of Gas Cylinders

Where more than one cylinder of oxygen and more than one cylinder of acetylene are carried simultaneously, such cylinders are to be arranged in accordance with the following.

Permanent piping systems for oxy-acetylene are acceptable provided that

a. All fixed piping is of steel or other approved material and suitable joints are fitted

b. Material containing more than 70% copper is not used in the system except for welding or cutting tips

c. Allowance is made for expansion of the piping

d. The piping system is suitable for the intended pressures

Where two or more cylinders of each gas are intended to be carried in enclosed spaces, separate dedicated storage rooms are to be provided for each gas.

Storage rooms are to be constructed of steel and be well ventilated and accessible from the open deck.

Provision is to be made for the expeditious removal of cylinders in the event of fire.

Where cylinders are stored in open locations, means are to be provided to

a. Protect cylinders and associated piping from physical damage

b. Minimize exposure to hydrocarbons

c. Ensure suitable drainage

d. Protect cylinders against solar radiation

9.11 Fire Protection for Onshore Installations

Water is the most commonly used agent for controlling and fighting a fire, by cooling adjacent equipment and for controlling and/or extinguishing the fire either by itself or combined as a foam. It can also provide protection for firefighters and other personnel in the event of fire. Water should therefore be readily available at all the appropriate locations, at the proper pressure, and in the required quantity.

Depending on the nature of risk, the following fire protection facilities should be provided in the installation:

- Fire water system
- Foam system

- Clean agent protection system
- First aid fire-fighting equipment
- Mobile fire-fighting equipment
- Fire detection, alarm, communication system

Fire water should not be used for any other purpose. Unless otherwise specified or agreed, the company requirements that are given for major installations such as refineries, petrochemical works, crude oil production areas where large facilities are provided, and for major storage areas should be applied.

In determining the quantity of fire water (i.e., the required fire water rate), protection of the following areas should also be considered:

a. General process
b. Storage (low pressure, including pump stations, manifolds, in-line blenders, etc.)
c. Pressure storage (LPG, etc.)
d. Refrigerated storage (LNG, etc.)
e. Jetties
f. Loading
g. Buildings
h. Warehouses

Basically, the requirements consist of an independent fire grid main or ring main fed by permanently installed fire pumps taking suction from a suitable large-capacity source of water such as a storage tank, cooling tower basin, river, sea, and so forth. The actual source will depend on local conditions and is to be agreed on with the company.

The water will be used for direct application to fires and for the cooling of equipment. It will also be used for the production of foam.

9.12 Water Supplies

Water is used for fire extinguishments, fire control, cooling of equipment, and exposure protection of equipment and personnel from heat radiation.

9.12.1 Nature of Supply

The choice of water supplies should be made in cooperation with the relevant authorities.

9.12.2 Public Water Systems

One or more connections from a reliable public water system of proper pressure and adequate capacity furnishes a satisfactory supply. A high static water pressure should not, however, be the criterion by which the efficiency of the supply is determined. If this cannot be done, the post indicator valves should be placed where they will be readily accessible in case of fire and not liable to injury. Where post indicator valves cannot readily be used, as in a city block, underground valves should conform to these provisions and their locations and direction of turning to open should be clearly marked.

Adequacy of water supply should be determined by flow tests or other reliable means. Where flow tests are made, the flow in (L/min) together with the static and residual pressures should be indicated on the plan.

Public mains should be of ample size, in no case smaller than 15 cm (6 inches).

No pressure-regulating valve should be used in water supply except by special permission of the authority concerned. Where meters are used they should be of an approved type.

Where connections are made from public waterworks systems, it may be necessary to guard against possible contamination of the public supply. The requirements of the public health authority should be determined and followed.

Connections larger than 50.8 mm to public water systems shall be controlled by post indicator valves of a standard type and located not less than 12.2 m from the buildings and units protected.

9.13 Bases for a Fire-Fighting Water System

A ring main system should be laid around processing areas or parts thereof, utility areas, loading and filling facilities, tank farms, and buildings while one single line should be provided for jetties and a fire-fighting training ground, complete with block valves and hydrants.

The water supply should be obtained from at least two centrifugal pumps of which one is electric motor-driven and one driven by a fully independent power source (e.g., a diesel engine, the latter serving as a spare pump).

The water quantities required are based on the following considerations:

a. There will be only one major fire at a time.

b. As a recommendation in processing units the minimum water quantity is 200 dm³/s of air-foam-making and exposure protection. It is assumed that approximately 30% of this quantity is blown away and evaporates; the balance of this quantity, which is 140 dm³/s per processing unit, should be drained via a drainage system. Note: specification is based on one major fire only.

The quantity of fire water required for a particular installation should be assessed in relation to fire incidents that could occur on that particular site, taking into account the fire hazard, the size, duties, and location of towers, vessels, and so forth. The fire water quantity for installations having a high potential fire hazard should normally be not less than 820 m^3/h and no more than 1360 m^3/h.

c. For storage areas, the quantity needed for making air foam for extinguishing the largest cone-roof tank on fire and for exposure protection of adjacent tanks.

d. For pressure storage areas, the quantity needed for exposure protection of spheres by means of sprinklers.

e. For jetties, the quantity needed for fighting fires on jetty decks and ship manifolds with air foam as well as for exposure protection in these areas.

f. The policy for a single major fire or more to occur simultaneously should be decided on by the authorities concerned.

For new installations the quantities required for items a to f mentioned above should be compared and the largest figure should be adhered to for the design of the fire-fighting system.

The system pressure should be such that at the most remote location a pressure of 10 bar can be maintained during a water takeoff required at that location.

Fire-fighting water lines should be provided with permanent hydrants.

Hydrants with four outlets should be located around processing units, loading facilities, storage facilities for flammable liquids, and on jetty heads and berths.

Hydrants with two outlets should be located around other areas, including jetty approaches.

Fire hose reels should be located in each process unit, normally 31–47 m apart at certain strategic points.

The water will be applied by means of hose and branch pipes using jet, spray, or fog nozzles, or by fixed or portable monitors preferably with interchangeable nozzles for water or foam jets.

The fire water ring main should be provided all around the perimeter of the installation with hydrants/monitors spaced at intervals not exceeding 30 m when measured aerially.

9.13.1 Components of the Fire Water System

The main components of the system are fire water storage, fire water pumps, and distribution piping network.

9.13.2 Basis

The fire water system in an installation should be designed to meet the fire water flow requirement to fight a single largest risk at a time.

9.13.3 Design Flow Rate

- Fire water flow rate for a tank farm should be the aggregate of the following:

  Water flow calculated for cooling a tank on fire at a rate of 3 lpm/ m^2 of tank shell area.

  Water flow calculated for exposure protection for all other tanks falling within a radius of $R + 30$ m from the center of the tank on fire (R = radius of tank on fire) and situated in the same dike at a rate of 3 lpm/m^2 of the tank shell area.

  Water flow calculated for exposure protection for all other tanks falling outside a radius of $R + 30$ m from the center of the tank on fire and situated in the same dike at a rate of 1 lpm/m^2 of the tank shell area.

  For water flow calculations, all tanks farms having class A or B petroleum storage should be considered irrespective of diameter of tanks and whether a fixed water spray system is provided.

  Water flow required for applying foam on a single largest tank by way of a fixed foam system, where provided, or by use of water/foam monitors.

  Various combinations should be considered in the tank farm for arriving at different fire water flow rates and the largest rate to be considered for design.
- Fire water flow for pump house shed at cross-country pipeline installations should be at a rate of 10.2 lpm/m^2.
- Fire water flow rate for supplementary streams should be based on using four single hydrant outlets and one monitor simultaneously. The capacity of each hydrant outlet as 36 m^3/h and of each monitor as 144 m^3/h minimum may be considered at a pressure of 7 kg/cm^2 g.

9.13.4 Header Pressure

A fire water system should be designed for a minimum residual pressure of 7 kg/cm^2(g) at the hydraulically remotest point in the installation considering the single largest risk scenario.

9.13.5 Storage

Water for the firefighting should be stored in easily accessible surface or underground or above-ground tanks of steel, concrete, or masonry.

The effective capacity of the reservoir/tank above the level of the suction point should be minimum 4 hours aggregate rated capacity of pumps. However, where reliable makeup water supply is 50% or more of design flow rate, the storage capacity may be reduced to 3 hours aggregate rated capacity of pumps.

Freshwater should be used for fire-fighting purposes. If seawater or treated effluent water is used for fire-fighting purposes, the material of the pipe selected should be suitable for the service.

The installation should have facilities for receiving and diverting all the water coming to the installation to fire water storage tanks in case of an emergency.

The storage reservoir should be in two equal interconnected compartments to facilitate cleaning and repairs. In the case of steel tanks there should be a minimum of two tanks, each having 50% of required capacity.

Large natural reservoirs having water capacity exceeding 10 times the aggregate water requirement of fire pumps may be left unlined.

9.13.6 Fire Water Ring Main System, General

Fire water ring mains of the required capacity should be laid to surround all processing units, storage facilities for flammable liquids, loading facilities for road vehicles and rail cars, bottle filling plants, warehouses, workshops, utilities, training centers, laboratories, and offices. Normally, these units will also be bounded by service roads. Large areas should be subdivided into smaller sections, each enclosed by fire water mains equipped with hydrants and block valves.

A single fire water pipeline is only acceptable for a fire-fighting training ground. Fire water to jetties should be supplied by a single pipeline provided that it is interconnected with a separate pipeline for water spray systems. The fire water pipelines from the fire pumps to the jetty should be provided with isolating valves for closing in the event of serious damage to the jetty. These valves should close without causing high surge pressures.

The fire water mains should be provided with full bore-valved flushing connections so that all sections and dead ends can be properly flushed out. The flushing connections should be sized for a fluid velocity in the relevant piping of not less than 80% of the velocity under normal design conditions but for not less than 2 m/s.

Fire water mains should normally be laid underground in order to provide a safe and secure system, and which will also provide protection against freezing for areas where the ambient temperature can drop below 0°C. When in exceptional circumstances, fire water mains are installed above ground, they should be laid alongside roads and not in pipe tracks where they could be at risk from spill fires.

The basic requirements consist of an independent fire grid main or ring main fed by permanently installed fire pumps. The size of ring main and fire pumps should be such as to provide a quantity of water sufficient for the largest single risk identified within the overall installation.

Suction will be from a suitable large-capacity source of water such as a storage tank, cooling tower, basin, river, sea, and so forth. The actual source will depend on local conditions and is to be investigated. Pump suction lines should be positioned in a safe and protected location and incorporate permanent but easily cleanable strainers or screening equipment for the protection of fire pumps.

Advantage should be taken where available in obtaining additional emergency water supplies through a mutual aid scheme or by recycling, but mandatory national or local authority requirements may modify these to a considerable extent.

9.13.7 Fire Water Ring Main/Network Design

The fire water mains network pipe sizes should be calculated and based on design rates at a pressure of 10 bar gage at the takeoff points of each appropriate section, and a check calculation should be made to prove that pressure drop is acceptable with a blocked section of piping in the network. The maximum allowable flow/velocity in the system should be 3.5 m/s.

Fire water rates should, however, be realistic quantities since they determine the size of fire water pumps, the fire water ring main system, and the drainage systems that have to cope with the discharged fire water. If the drainage system is too small or becomes blocked, major hazards such as burning hydrocarbons floating in flooded areas may occur to escalate the fire. Facilities for cleaning should therefore be provided. For large areas such as pump floors, and in pipe tracks, fire stops should be provided to minimize the spillage area. It is assumed that 30% of fire water evaporates or is blown away while extinguishing a fire. This figure should be taken into account for the design of drainage systems.

Under nonfire conditions, the system should be kept full of water and at a pressure of 2–3 bar gage by means of a jockey pump, by a connection to the cooling water supply system, or by static head from a water storage tank. If a jockey pump is used, it should be "spared" and both pumps should have a capacity of 15 m^3/h to compensate for leakages.

A single water line connected to the ring main system should run along the jetty approach to the jetty deck. This line should be fitted with a block valve located at a distance of about 50 m from the jetty deck.

For small chemical plants, depots, minor production and treatment areas, and so forth for which precise commensurate with the size of risk involved, requirements should be as specified or agreed on with authorities.

9.13.8 Design Criteria for Fire Protection System

Facilities should be designed on the basis that a city fire water supply is not available close to the installation. One single largest risk should be considered for providing facilities.

The hazardous areas should be protected by a well-laid combination of hydrants and monitors. The following installations are exempted from this provision:

i. The installation having aggregate above-ground storage capacity of less than 1000 kl (class A+B+C) other than AFS
ii. Pipeline installation having only scrapper stations or sectionalizing valve stations

Tank wagon (TW)/tank truck (TT) loading/unloading facilities and the manifold area of the product pump house and exchange pit should be fully covered with a well laid-out combination of hydrants and water-cum-foam monitors.

The installations storing class A petroleum in above-ground tanks should have a fixed water spray system.

However, installations above 1000 kl storage fulfilling the following both conditions are exempted from the provision of fixed water spray system as follows:

• Aggregate above-ground storage of class A and B petroleum up to 5000 kl
• Floating-roof tank storing class A petroleum having a diameter up to 9 m

Class B above-ground petroleum storage tanks (fixed roof or floating roof) of a diameter larger than 30 m should be provided with a fixed water spray system.

When class A and B above-ground storage tanks are placed in a common dike, the fixed water spray system should be provided on all tanks except for small installations.

TW loading gantries should be provided with manually operated fixed water spray system. If an automatic fixed water spray system is provided, the gantry may be divided into a suitable number of segments (each segment having a minimum length of 15 m and width of 12 m) and three segments operating at a time should be considered as single risk for calculating the water requirement. Accordingly, a provision should be made to actuate the water spray system from a safe approachable central location (i.e., the affected zone and adjoining zones).

A fixed foam system or semifixed foam system should be provided on tanks (floating roof or fixed roof) exceeding 18 m diameter storing class A or class B petroleum.

Portable foam and/or water-cum-foam monitors should be provided for suppression of pool fire in a tank farm area. An automatic actuated rim seal fire-extinguishing system may be provided based on foam or clean agent flooding mechanism on floating-roof tanks having a diameter larger than 60 m.

This is in addition to the fixed water spray system and fixed foam system or semifixed foam system on all floating-roof tanks storing class A and B petroleum.

Clean agent flooding system: The selection and design of clean agent based flooding system should be in line with the Standard on Clean Agent Fire Extinguishing Systems, NFPA 2001 (latest edition).

Listed clean agents like Trifluroiodide and others can be used as a fire suppressant for a floating-roof rim seal fire protection system. Listed clean agents like Fluroketone and others can be used as a fire suppressant in control rooms and computer rooms.

The clean agent based protection system consists of an in-built fire detection, control, and actuation mechanism. If a rim seal fire occurs, its heat causes one or more spray nozzles to open and the extinguishing gas (clean agent) is applied on the surface of the fire and a simultaneous alarm is also sounded.

A clean agent (halon substitute) based flooding system may be considered for control rooms, computer rooms, and pressurized rooms in major locations having automated pipeline receipt/dispatch and/or TW/TT loading facilities.

Selection of a clean agent and the design of a fire protection system for control rooms, computer rooms, and pressurized rooms should follow the Standard on Clean Agent Extinguishing Systems, NFPA 2001 (latest edition), including its safety guidelines with respect to hazards to personnel, electrical clearance, and environmental factors in line with the environmental considerations of the Kyoto Protocol. A clean agent like inert gas, or Fluroketone can be used as a fire suppressant in control rooms, computer rooms, and pressurized rooms.

9.13.9 Fire Water Pumps

Fire water should be provided by at least two identical pumps, and each pump should be able to supply the maximum required capacity for a fire water ring main system. Fire water pumps should be of the submerged vertical type when taking suction from open water and of the horizontal type when suction is taken from a storage tank.

The fire water pumps should be installed in a location that is considered to be safe from the effects of fire and clouds of combustible vapor and from collision damage by vehicles and shipping. For example, they should be at least 100 m away from jetty loading points and from moored tankers or barges handling liquid hydrocarbons. They should be accessible to facilitate maintenance and be provided with hoisting facilities.

The main fire water pump should be driven by an electric motor and the second pump, of 100% standby capacity, by some other power source, preferably a diesel engine. Alternatively, three pumps, each capable of supplying 60% of the required capacity may be installed, with one pump driven by an electric motor and the other two by diesel engines.

Note that a refinery with over 100,000 barrels a day capacity should have two electric and two diesel pumps.

When the required pump capacity should exceed 1000 m³, two or more smaller pumps should be installed together with an adequate number of spare pumps. The power of the drives for both the main and standby units should be rated so that it will be possible to start the pumps against an open discharge with pressure in the fire water ring main system under nonfire conditions, normally at 2–3 bar gage unless otherwise agreed by the relevant authorities. The main fire water pump should be provided with automatic starting facilities that will function immediately the fire alarm system becomes operational due to one of the following actions:

- When a fire call point is operated
- When an automatic fire detection system is operated
- When the pressure in the fire water ring main system drops below the minimum required static pressure, which is normally 2–3 bar (ga)

The standby fire water pumps should be provided with automatic starting facilities that will function if the main fire water pumps do not start, or having started, fail to build up the required pressure in the firewater ring main system within 20 seconds.

Manual starting of each pump unit (without the fire alarms coming into operation) should be possible at the pump from the control center, and when necessary, from the gate house. Manual stopping of each pump unit should only be possible at the pump.

Fuel tank capacity: Fuel supply tank(s) should have a capacity at least equal to 1 gal per horsepower (5.07 L/kW), plus 5% volume for expansion and 5% volume for sump. Larger-capacity tanks may be required and should be determined by prevailing conditions such as refill cycle and fuel heating due to recirculation, and be subject to special conditions in each case. The fuel supply tank and fuel should be reserved exclusively for the fire pump diesel engine.

The pumps should have stable characteristic curves exhibiting a decrease in head with increasing capacity from zero flow to maximum flow; a relatively flat curve is preferred with a shutoff pressure not exceeding the design pressure by more than 15%.

The total water supply within the refinery should be capable of supplying the maximum flow for a period of not less than 4–6 hours, consistent with projected fire scenario needs. Where the water system is supplied from a tank or reservoirs, the quantity of water required for fire protection is sufficient. However, where the tank or reservoir is automatically filled by a line from a reliable, separate supply, such as from a public water system or wells, the total quantity in storage may be reduced by the incoming fill rate.

i. Fire water pumps having flooded suction should be installed to meet the design fire water flow rate and head. If fire water is stored in underground tanks, an overhead water tank of sufficient capacity should be provided for flooded suction and accounting for leakages in the network, if any.

ii. The pumps should be capable of discharging 150% of its rated discharge at a minimum of 65% of the rated head. The shutoff head should not exceed 120% of the rated head for horizontal centrifugal pumps and 140% for vertical turbine pumps.

iii. At least one standby fire water pump should be provided up to two numbers of main pumps. For main pumps three numbers and above, minimum two numbers standby pumps of the same type, capacity and head as the main pumps should be provided.

iv. The fire water pump(s) including the standby pump(s) should be of the diesel-engine-driven type. Where electric supply is reliable, 50% of the pumps may be electric-driven. The diesel engines should be a quick-starting type with the help of push buttons located on or near the pumps or located at a remote location. Each engine should have an independent fuel tank adequately sized for 6 hours of continuous running of the pump.

v. Fire water pumps and storage should be located at 30 m (minimum) away from equipment or where hydrocarbons are handled or stored.

vi. Fire water pumps should be exclusively used for fire-fighting purposes only.

vii. Suction and discharge valves of fire water pumps should be kept fully open all the time.

viii. The fire water network should be kept pressurized by a static water tank or jockey pumps.

ix. If a jockey pump is used for pressurization, a standby jockey pump of similar type, capacity, and head should be provided.

9.13.10 Water Network

i. Looping

The fire water network should be laid in closed loops as much as possible to ensure multidirectional flow in the system. Isolation valves should be provided in the network to enable isolation of any section of the network without affecting the flow in the rest. The isolation valves should be located normally near the loop junctions. Additional valves should be provided in the segments where the length of the segment exceeds 300 m.

ii. Above/Underground Network

The fire water network steel piping should normally be laid above ground at a height of at least 300 mm above the finished ground level. Pipes made of composite material should be laid underground. However, the ring main should be laid underground at the following places:

- Road crossings.
- Places where above-ground piping is likely to cause obstruction to operation and vehicle movement.
- Places where above-ground piping is likely to get damaged mechanically.
- Where frost conditions warrants and ambient temperature is likely to fall to subzero, above-ground piping should be laid at least 1 m below the finished grade level to avoid freezing of water. Alternatively, water circulation may be carried out in the above-ground pipelines or any other suitable means.

iii. Protection of Underground Pipeline

If fire water ring mains are laid underground, the following should be ensured:

A. The ring main should have at least a 1-m earth cushion in open ground, a 1.5-m cushion under the road crossings, and in the case of crane movement, area pipeline may be protected with concrete/steel encasement as per design requirement.

B. The underground ring main should be protected against soil corrosion by suitable coating/wrapping with or without cathodic protection.

C. Pipe supports under the pipe line should be suitable for the soil conditions.

iv. Support and Protection of Above-Ground Pipelines

The mains should be supported at regular intervals not exceeding 6 m. For pipeline size less than 150 mm, support interval should not exceed 3 m.

The pipe support should have only point contact.

The system for an above-ground portion should be analyzed for flexibility against thermal expansion and necessary expansion loops, guides/cross guides, and supports provided.

v. Sizing of Pipeline

The fire water ring main should be sized for 120% of the design water flow rate. Design flow rates should be distributed at nodal points to give the most realistic way of water requirements in an emergency. It may be necessary to assume several combinations of flow requirement for the design of the network.

The stand post for hydrants and monitors should be sized to meet the respective design water flow rates.

vi. General

Connections for fixed water monitors on the network should be provided with independent isolation valves.

Fire water mains should not pass through buildings or dike areas.

In the case of underground mains the isolation valves should be located in a brick masonry chamber of suitable size to facilitate operation during emergency and maintenance.

9.13.11 Hydrants and Monitors

i. Hydrants should be located bearing in mind the fire hazards at different sections of the premises to be protected and to give the most effective service. At least one hydrant post should be provided for every 30 m of external wall measurement or perimeter of battery limit in the case of high-hazard areas. For nonhazardous area, they should be spaced at 45-m intervals. The horizontal range and coverage of hydrants with hose connections should not be considered beyond 45 m.

ii. Hydrants should be located at a minimum distance of 15 m from the periphery of a storage tank or equipment under protection. In the case of buildings this distance should not be less than 2 m and not more than 15 m from the face of building. Provision of hydrants within the building should be provided in accordance with IS 3844.

iii. Hydrant/monitors should be located along roadside berms for easy accessibility.

iv. Double-headed hydrants with two separate landing valves or monitor on suitably sized stand posts should be used. All hydrant outlets/monitor isolation valves should be situated at workable height above ground or hydrant/monitor operating platform level.

v. Monitors should be located to direct water on the object as well as to provide a water shield to firefighters approaching a fire. The requirement of monitors should be established based on hazards involved and layout considerations. Monitors should not be installed within

15 m of hazardous equipment. The location of the monitors should not exceed 45 m from the hazard to be protected. However, high-volume, long-range monitors, if provided, should be located more than 45 m from the hazardous equipment and their water header/s sized accordingly to meet the rated water flow rate of the monitor as well as the design water flow requirement of the single largest risk scenario.

vi. TW/TT loading and unloading facilities should be provided with alternate hydrant and water-cum-foam monitors having multipurpose combination nozzles for jet spray and fog arrangement and fire hydrants located at a spacing of 30 m on both sides of the gantry. The hydrants and monitors should be located at a minimum distance of 15 m from the hazard (e.g., TW and TT loading/unloading facilities) to be protected.

vii. Hydrants/monitors should be located preferably with branch connections.

9.13.12 Material Specifications

The materials used in fire water system should be of an approved type as indicated below:

i. Pipes: Carbon steel as per IS:3589/IS:1239/IS:1978 or composite material or its equivalent for freshwater service. In the case saline, brackish or treated effluent water is used, the fire water ring main of steel pipes, internally cement mortar lines or glass reinforced epoxy coated or pipes made of material suitable for the quality of water should be used. Alternately, pipes made of composite materials should be used. The composite material to be used should be as per API 15LR/API 15HR.

ii. Isolation valves: Gate or butterfly type isolation valves made of cast steel having an open/close indication should be used. Other materials such as cupro-nickel for saline/brackish water may be used.

iii. Hydrants: Stand post: carbon steel; outlet valves: gunmetal/aluminum/stainless/steel/aluminum-zinc (Al-Zn) alloy.

iv. Fire hoses: Reinforced rubber lined hose as per IS 636 (Type A)/Non-percolating Synthetic Hose (Type B)/UL or Equivalent Standard.

v. Fire water mains, hydrant, and monitor stand posts and risers of the water spray system should be painted with fire red paint.

vi. Hose boxes, water monitors and hydrant outlets should be painted with luminous yellow paint.

vii. Corrosion-resistant paint should be used in corrosion-prone areas.

9.14 Water Tanks for Fire Protection

This section covers elevated tanks on towers or building structures and grade or below-grade water storage tanks and pressure tanks.

9.14.1 Capacity and Elevation

The size and elevation of the tank should be determined by conditions at each individual property after due consideration of all factors involved.

The capacity of the tank is the number of cubic meters available above the outlet opening. The net capacity between the outlet opening of the discharge pipe and the inlet of the overflow should be at least equal to the rated capacity. For gravity tanks with large plate risers, the net capacity should be the number of cubic meters between the inlet of the overflow and the designated low-water level line. For suction tanks, the net capacity should be the number of cubic meters between the inlet of the overflow and the level of the vortex plate.

The standard sizes of steel tanks are 18.93, 37.85, 56.78, 75.70, 94.63, 113.55, 151.40, 189.25, 227.10, 283.88, 378.50, 567.75, 757.00, 1135.50, and 1892.50 cubic meters net capacity.

The capacity of pressure tanks should be as approved by the authority concerned.

The standard sizes of wooden tanks are 18.93, 37.85, 56.78, 75.70, 94.63, 113.55, 151.40, 189.25, 227.10, 283.88, and 378.50 cubic meters net capacity. Tanks of other sizes may be built.

The standard capacities of coated-fabric tanks are in increments of 378.5 to 3785 m^3 (according to NFPA 22).

9.14.2 Location of Tanks

The location chosen should be such that the tank and structure will not be subject to fire exposure from adjacent units. If lack of yard room makes this impracticable, the exposed steel work should be suitably fireproofed or protected by open sprinklers.

Fireproofing where necessary should include steel work within 6.1 m of combustible buildings, windows, doors, and flammable liquid and gas from which fire might issue.

When steel or iron is used for supports inside the building near combustible construction or occupancy, it should be fireproofed inside the building, 1.8 m above combustible roof coverings and within 6.1 m of windows and doors from which fire might issue. Steel beams or braces joining two building columns that support a tank structure should also be suitably fireproofed when near combustible construction or occupancy. Interior timber should not be used to support or brace tank structures.

Fireproofing, where required, should have a fire resistance rating of not less than two hours.

Foundations or footings should furnish adequate support and anchorage for the tower.

If the tank or supporting trestle is to be placed on a building, the building should be designed and built to carry the maximum loads.

9.14.3 Storage Facilities

Fire water taken from open water is preferred, but if water of acceptable quality for firefighting in the required quantity cannot be supplied from open water, or if it is not economically justified because of distance to install firewater pumps at an open source, water storage facilities should be provided.

Storage facilities may consist of an open tank of steel or concrete or a basin of sufficient capacity. The tank or basin should have two compartments to facilitate maintenance, each containing 60% of the total required capacity and there should be adequate replenishment facilities. A single compartment of 100% capacity is acceptable providing that an alternative source of water (e.g., from temporary storage) will be available during maintenance periods. The replenishment rate should normally not be less than 60% of the total required fire water pumping capacity.

If a 100% replenishment rate is available, the stored fire water capacity may be reduced if agreed by the expert authorities that may be considered for replenishment are plant cooling water, open water, or below-ground water, provided that it is available at an acceptable distance and in sufficient quantity for a minimum of 6 hours uninterrupted firefighting at the maximum required rate. Figure 9.4 shows a typical fire water storage tank.

9.14.4 Buried Tanks

Where lack of space or other conditions require it, pressure tanks may be buried if the following requirements are satisfied.

For protection against freezing the tank should be below the frost line.

The end of the tank and at least 457 mm of its shell should project into the building basement or a pit in the ground, with protection against freezing. There should be adequate space for inspection and maintenance and use of the tank manhole for interior inspection.

The exterior surface of the tank should be fully coated as follows for protection against corrosion conditions indicated by a soil analysis:

a. An approved cathodic system of corrosion protection should be provided

b. At least 305 mm of sand should be backfilled around the tank

FIGURE 9.4
A typical fire water storage tank.

The tank should be above the maximum groundwater level so that buoyancy of the tank when empty will not force it upward. An alternative would be to provide a concrete base and anchor the tank to it.

The tank should be designed with strength to resist the pressure of earth against it.

A manhole should be located preferably on the vertical centerline of the tank end to clear the knuckle while remaining as close as possible to it.

Types of materials should be limited to steel, wood, concrete, and coated fabric.

The elevated wood and steel tanks should be supported on steel or reinforced concrete towers.

9.15 Sample Calculation of Fire Water Flow Rate for Storage Tanks

9.15.1 Design Basis

The fire water system in an installation should be designed to meet the fire water flow requirement of fighting a single largest fire scenario.

9.15.2 Fire Water Demand for a Single Largest Fire

Consider various areas under fire and calculate fire water demand for each area based on the design basis.

9.15.3 Fire Water Flow Rate for Floating Roof Tank Protection

Data:

Total storage capacity in one dike area = 32,000 m³
Number of tanks = 2
Capacity of each tank = 16,000 m³
Diameter of each tank = 40 m
Height of each tank = 14.4 m

a. Cooling water flow rate/cooling water required for tank on fire:
 Cooling water rate = 3 lpm/m² of tank area for tank on fire
 Cooling water required = 3.142 × 40 m × 14.4 m × 3 lpm/m²
$$= 5426 \text{ lpm}$$
$$= (5426 × 60 \text{ m}^3/\text{h})/1000 = 326 \text{ m}^3/\text{h}$$
 assuming that a second tank is located within the tank dike at a distance more than 30 m from the tank's shell. Therefore, in such a case the cooling required is at the rate of 1 lpm/m² of tank shell area.
  Cooling water required for a tank falling beyond $R + 30$ from the center of the tank on fire:
 Cooling water rate = 1 lpm/m² of tank area
 Cooling water required = 3.142 × 40 m × 14.4 m × 1 lpm/m²
$$= 1809 \text{ lpm}$$
$$= (1809 × 60 \text{ m}^3/\text{h})/1000 = 109 \text{ m}^3/\text{h}$$

b. Foam water flow rate:
 Foam solution application rate = 12 lpm/m² of rim seal area of tank
 Foam solution required = 3.142 × 40 m × 0.8 m × 12 lpm/m²
 Foam water required (3%) = 1207 lpm
$$= 0.97 × 1207 \text{ lpm} = 1171 \text{ lpm}$$
$$= (1171 × 60 \text{ m}^3/\text{h})/1000 = 71 \text{ m}^3/\text{h}$$

c. Total water flow rate:
 Tank cooling = 326 + 109 = 435 m³/h
 Foam solution application = 71 m³/h
 Total = 506 m³/h
 Say = 510 m³/h

9.15.4 Fire Water Flow Rate for Cone Roof Tank Protection

Data:

Total storage capacity in one dike area = 50,000 m³
Number of tanks = 4
Capacity of each tank = 12,500 m³
Diameter of each tank = 37.5 m
Height of each tank = 12 m

a. Cooling water flow rate/cooling water required for tank on fire:
   Cooling water rate = 3 lpm/m² of tank area for tank on fire
   Cooling water required = 3.142 × 37.5 m × 12 m × 3 lpm/m²
   $$= 4242 \text{ lpm}$$
   $$= (4242 \times 60 \text{ m}^3/\text{h})/1000 = 255 \text{ m}^3/\text{h}$$
   Cooling water required for tanks falling within $(R + 30)$ from center of tank on fire:
   Cooling water rate = 3 lpm/m² of tank area
   Cooling water required = 3.142 × 37.5 m × 12 m × 3 lpm/m² × 3
   $$= 12,726 \text{ lpm}$$
   $$= (12,726 \times 60 \text{ m}^3/\text{h})/1000 = 764 \text{ m}^3/\text{h}$$
   Total cooling water required = 254 + 762 = 1019 m³/h

b. Foam water flow rate:
   Foam solution application rate = 5 lpm/m² of liquid surface area
   Foam solution required = 3.142 × (18.75 m)² × 5 lpm/m²
   $$= 5523 \text{ lpm}$$
   Foam water required = 0.97 × 5523 lpm = 5357 lpm
   $$= (5357 \times 60 \text{ m}^3/\text{h})/1000 = 321 \text{ m}^3/\text{h}$$
   Total foam water required = 1019 m³/h

c. Total water flow rate:
   Tank cooling = 1019 m³/h
   Foam solution application = 321 m³/h
   Total = 1340 m³/h

9.16 Fire Protection System, Inspection, and Testing

The fire protection equipment should be kept in good working condition all the time. It should be periodically tested for proper functioning and logged for record and corrective actions.

9.16.1 Fire Water Pumps

Every pump should be test run for at least half an hour or as per available guidelines, whichever is higher, twice a week at the rated head and flow. Every pump should be checked, tested, and its shutoff pressure observed once a month.

Pumps should be checked and tested for its performance once every six months by opening the required numbers of hydrants/monitors depending on the capacity of the pump to verify that the discharge pressure, flow, and motor load are in conformity with the design parameters. They should be test run continuously for four hours at their rated head and flow using a circulation line of fire water storage tanks and observations logged once a year. The testing of a standby jockey pump, if provided, shall be checked weekly. Frequent starts and stops of the pump indicate that there are water leaks in the system that should be attended to promptly.

9.16.2 Fire Water Ring Mains

The ring main should be checked for leaks once in a year by operating one or more pumps and keeping the hydrant points closed to get the maximum pressure.

The ring mains, hydrant, monitor, and water spray header valves should be visually inspected for any missing accessories, defects, damage, and corrosion every month and records maintained. All valves on the ring mains, hydrants, monitors, and water spray headers should be checked for leaks, smooth operation, and lubricated once a month.

9.16.3 Fire Water Spray System

The water spray system should be tested for performance (i.e., its effectiveness and coverage) once every six months. Spray nozzles should be inspected for proper orientation, corrosion, and cleaned if necessary at least once a year. The strainers provided in the water spray system should be cleaned once every quarter and records maintained.

9.16.4 Fixed/Semifixed Foam System

The fixed/semifixed foam system on storage tanks should be tested once every six months. This should include the testing of the foam maker/

chamber. The foam maker/chamber should be designed suitably to facilitate discharge of foam outside the cone-roof tank. After testing the foam system, the piping should be flushed with water.

9.16.5 Clean Agent System

The clean agent fire extinguishing system should be checked as follows:

Agent quantity and pressure of refillable containers should be checked once every 6 months. The complete system should be inspected for proper operation once a year (refer to the latest NFPA 2001 (2004 edition) for details of inspection of various systems).

Further Readings

American Petroleum Institute (API). 2001. *Fire Protection in Refineries,* 7th Edition.

Ang, W.K. and Jowitt, P.W. 2005. Some new insights on informational entropy for water distribution networks. *Engineering Optimization* 37 (3), 277–289.

Anon. 1997. Fire engineering 120-year retrospective. *Fire Engineering* 150 (3), 138–143.

Brennan, T. 2005. Nozzles schmozzles! *Fire Engineering* 158 (6), 140.

British Standards Institution (BSI). 2009. BS 5306 Part 1 System Design Water Supply.

Dubay, C. 1996. Effects of water mist on interior firefighting. *Fire Engineering* 149 (11), 78–79.

Fleming, R. 2002. Safety and fire protection: Water-based fire suppression systems. *International Hydrocarbon* 2002, 77.

Hadjisophocleous, G.V. and Richardson, J.K. 2005. Water flow demands for firefighting. *Fire Technology* 41 (3), 173–191.

Hansen, R. 2012. Estimating the amount of water required to extinguish wildfires under different conditions and in various fuel types. *International Journal of Wildland Fire* 21 (5), 525–536.

Ibaraki, H. and Hashimoto, H. 2009. Improvement of water nozzle shape for fire protection based on optimum design. *Transactions of the Japan Society of Mechanical Engineers, Part C* 75 (757), 2596–2603.

Klein, R. 2002. Water power. *Fire Prevention* (362), 26–27.

Liao, J.-H. and Shaw, D. 2013. Cooling the hydro-powered synergistic nozzle light by water. *Applied Mechanics and Materials* 284–287, 657–661.

National Fire Codes (NFC) (NFPA). 1996. Section 1231 Water Supply; NFC Section 15 Water Spray System; NFC Section 22 Water Tanks; NFC Section 24 Mains Water Supplies; NFPA-20 Standard for the Installation Pumps for Fire Protection; NFPA-15 Standard for Water Spray Fixed Systems for Fire Protection.

Sardqvist, S. and Holmstedt, G. 2000. Correlation between firefighting operation and fire area: Analysis of statistics. *Fire Technology* 36 (2), 109–130.

Shupe, J. 2005. Fighting fires in "monster houses." *Fire Engineering* 158 (2), 77–86.

Torvi, D., Hadjisophocleous, G., Guenther, M.B. and Thomas, G. 2001. Estimating water requirements for fire fighting operations using FIERA system. *Fire Technology* 37 (3), 235–262.

Xinmin, Y., Feixue, H. and Xinbin, Y. 1999. Deploying fire trucks and water sources. *Fire Technology* 35 (2), 179–183.

Glossary

ACB: Accepted Certification Bodies

accelerator: A device that reduces the delay in operation of a dry alarm valve or composite alarm valve in dry mode by early detection of the drop in air pressure when a sprinkler operates.

alarm set point: The concentration of gas to which an alarm is set.

alarm test valve: A valve through which water will be drawn to test the operation of the water motor fire alarm and/or of any associated electric fire alarm.

alarm valve: A check valve, of the wet, dry, or composite type, that also initiates the water motor fire alarm when the sprinkler installation operates.

alarm valve, composite: An alarm valve suitable for a wet, dry, or alternate installation.

alarm valve, dry: An alarm valve suitable for a dry installation, and/or in association with a wet alarm valve for an alternate installation.

alarm valve, preaction: An alarm valve suitable for a preaction installation.

alarm valve, recycling: An alarm valve suitable for a recycling installation.

alarm valve, wet: An alarm valve suitable for a wet installation.

alcohol or polar solvent: There are two types of foams that are resistant to destruction by water-miscible polar compounds. One type is based on protein foam, which is called alcohol. A second type of material, usually termed polar-solvent resist, contains a water-soluble polymer.

ambient air: The normal atmosphere surrounding the instrument.

aqueous film-forming foam (AFFF): A mixture of fluorocarbon carbon and hydrocarbon surfactants.

aqueous foam: A mixture of water and a foaming agent.

arm pipe: A pipe, other than the last section of a range pipe, feeding a single sprinkler.

aspirated apparatus: Apparatus in which the transfer of gas from the atmosphere to the gas-sensing element is by means of hand or powered pump.

aspirated instrument: Combustible gas detecting instrument that obtains the gas by drawing it to the gas sensor by means of a hand-operated or electric pump.

assembly: A combination of two or more pieces of equipment, with components if necessary, placed on the market and/or put into service as a single functional unit.

assemblies with a fully specified configuration of parts: These are put together and placed on the market as a single functional unit by the manufacturer of the assembly. The manufacturer assumes responsibility for compliance of the integral assembly with the directive and must therefore provide clear instructions for assembly/installation/operation/maintenance, and so forth. The EC declaration of conformity as well as the instructions for use must refer to the assembly as a whole. It must be clear which is/are the combination(s) that form(s) the assemblies.

assemblies forming a modular system: In this case the parts are not necessarily put together by the manufacturer of the assembly and placed on the market as a single functional unit. The manufacturer is responsible for the compliance of the assembly with the directive as long as the parts are chosen from the defined range and selected and combined according to his instructions.

associated apparatus: Electrical apparatus that contains both intrinsically safe and nonintrinsically safe circuits and is constructed so that the nonintrinsically safe circuits cannot adversely affect the intrinsically safe circuits.

assumed maximum area of operation (AMAO): The maximum area over which it is assumed, for design purposes, that sprinklers will operate in a fire.

assumed maximum area of operation, hydraulically most favorable location: The location in a sprinkler array of an AMAO of specified shape at which the water flow is the maximum for a specific pressure.

assumed maximum area of operation, hydraulically most unfavorable location: The location in a sprinkler array of an AMAO of specified shape at which the water supply pressure is the maximum needed to give the specified design density.

ATEX: Atmosphères Explosibles

ATEX 137: Directive 1999/92/EC—Minimum requirements for improving the safety and health protection of workers potentially at risk from explosive atmospheres.

ATEX 95: Directive 94/9/EC—Equipment and protective systems for use in potentially explosive atmospheres. Formerly known as ATEX 100a, it is aimed at manufacturers. It applies to equipment and protective devices intended for use in potentially explosive atmospheres. Safety and controlling devices for use outside the hazardous area but essential for the safe operating of equipment inside it are also covered. The directive applies to electrical as well as mechanical equipment and applies to gases, vapors, mists, and dust atmospheres.

ATR: Assessment and Test Report.

Attestation of Conformity for Components: Declaration by the manufacturer that the components conform with the provisions of Directive 94/9/EC and includes details on how to be incorporated into equipment or protective systems.

automatic/manual or manual-only changeover device: A device that can be operated before a person enters a space protected by a fire-extinguishing system preventing the fire detection system from activating the automatic release of carbon dioxide.

auxiliary equipment: Listed equipment used in conjunction with the dry chemical systems (i.e., to shut down powder, fuel, or ventilation to the hazard being protected or to initiate signaling devices).

balanced system: A powder fire extinguishing system, with more than one discharge nozzle, in which the powder flow divides equally at each junction in the pipework.

BASEEFA: British Approval Service for Electrical Equipment in Flammable Atmospheres—UK Notified Body

basic method of firefighting: Three factors should be considered as follows:
1. Smothering: Reducing the feed of oxygen to the fire in order to prevent combustion
2. Cooling: Keeping the surrounding areas of fire cooled by water to prevent spread of fire
3. Starving: Cutting off the fuel supplying the fire

bell-nipple: A short piece of pipe at the entry to a well that is belled at the top to guide tools into the hole. Usually has side connections for the fill-up and mud return lines.

blowout: An uncontrolled and often violent escape of reservoir fluids from a drilling well when a high-pressure reservoir has been encountered and efforts to prevent or control the escape have failed. Production wells can also blow out due to surface equipment failure or if well servicing operations get out of control.

boiling point: The temperature of a liquid at which the vapor pressure of the liquid equals the atmospheric pressure.

booster pump: An automatic pump supplying water to a sprinkler system from an elevated private reservoir or a town main.

boundary: Boundary of the equipment is the term used in a processing facility by an imaginary line that completely encompassed the defined site. The term distinguishes areas of responsibility and defines the processing facility for the required scope of work.

bounding area: The area of the real or notional surface (sides, bottom, and top) of an enclosure around a hazard protected by a total flooding system.

CAD: Chemical Agents Directive 98/24/EC—Protection of the health and safety of workers from risks related to chemical agents at work.

caking: A phenomenon that occurs when moisture chemically reacts with a dry chemical fire-extinguishing agent. This reaction results in materials that, being hydrated by moisture, stick together to form a large agglomerate, or what is more commonly referred to as lumps. For the purpose of this book, lumps are defined as those that do not crumble into particles when dropped from a height of 101 mm onto a hard surface.

calculation and design: The process of computing with the use of equations, graphs, or tables, the system characteristics such as flow rate, pipe size, area, or volume protected by each nozzle, nozzle pressure, and pressure drop. This information is not required for listed pre-engineered systems since these systems must be installed in accordance with their pre-tested limitations described in the manufacturer's installation manual.

catalytic sensor: A sensor that the operation of which depends on the oxidation of gases on an electrically heated catalytic element.

CEC: Canadian Electrical Code

CEN: European Committee for Standardisation (nonelectrical)

CENELEC: European Committee for Electrotechnical Standardisation

classification of construction: Classifying types of construction by surveying and allocating proper construction classification numbers to different types of structures.

classification of fires: Standardized system of classifying fires in terms of the nature of the fuel. The four classes of fires are A, B, C, and D.

classification of hazard: Selection of extinguishing agents for the specific class(es) of occupancy hazards to be protected and is classified as follows:

    a. Light (low) hazard

    b. Ordinary (moderate) hazard

    c. Extra (high) hazard

classification of occupancy hazard: Classifying occupancies hazard in relation to their quantity and combustibility of contents by surveying and allocating proper occupancy hazard classification numbers to different occupancies.

clearance: Shortest distance in air between two conductive parts

CoC: Certificate of Conformity (e.g., IECEx-Scheme)

combustible dust: Finely divided solid particles, 500 μm or less in nominal size, which may be suspended in air, settle out of the atmosphere under their own weight, burn or glow in air, and form explosive mixtures with air at atmospheric pressure and normal temperatures.

components: Any item essential to the safe functioning of equipment and protective systems but with no autonomous function.

concentration: The percent of foam concentrate contained in a foam solution.

conductive dust: Dust with electrical resistivity equal to or less than 103 Ωm.

construction classification numbers: A series of numbers from 0.5 to 1.50 that are mathematical factors used in a formula to determine the total water supply requirement of this book.

continuous grade source of release: A source that will release continuously or is expected to release for long periods or for short periods that occur frequently.

coupling: A device for connecting lengths of hose so as to secure continuity from the source of a water supply to the delivery point.

creepage distance: Shortest distance along the surface of an insulating medium between two conductive parts.

cutoff sprinkler: A sprinkler protecting a door or window between two areas, only one of which is protected by sprinklers.

Da, Db, Dc: Equipment protection levels for dusts.

deep-seated fire: A fire involving solids subject to smoldering.

degrees of flammability hazards: The degrees of hazards are ranked according to the susceptibility of materials to burning and are numbered from 4 to 0. Number 4 has a severe hazard that materials will burn readily and number 0 indicates that materials will not burn.

degrees of health hazards: Degrees of hazards are ranked according to the probable severity of hazard to personnel and are numbered from 4 to 0. Number 4 has a severe hazard degree and number 0 offers no hazard degree.

degrees of reactivity hazards: The degrees of hazards are ranked according to ease, rate, and quantity of energy released, and are numbered from 4 to 0. Number 4 is for materials that readily detonate or explode at normal temperature and pressure and number 0 is for materials that are normally stable and not reactive with water.

deluge installation: An installation or tail-end extension fitted with open sprayers and either a deluge valve or a multiple control arrangement so that an entire area is sprayed with water on operation of the installation.

deluge valve: A valve suitable for use in a deluge installation. Note that the valve is operated manually and usually also automatically by a fire detection system.

design density: The minimum density of discharge, in mm/min of water, for which a sprinkler installation is designed, determined from the discharge of specified group of sprinklers, in L/min, divided by the area covered, in m².

design point: A point on a distribution pipe of a precalculated installation downstream of which pipework is sized from table and upstream of which pipework is sized by hydraulic calculation.

detector sprinkler: A sealed sprinkler mounted on a pressurized pipeline used to control a deluge valve. Operation of the detector sprinkler causes loss of air pressure to open the valve.

diffusion apparatus: An apparatus in which the transfer of gas from the atmosphere to the gas sensing element takes place by diffusion (i.e., there is no aspirated flow).

diffusion instrument: An instrument in which the transfer of gas from the atmosphere to gas sensor take place by diffusion. There is no aspirated flow.

dike: An earth or concrete wall providing a specified liquid retention capacity.

distribution pipe: A pipe feeding either a range pipe directly or a single sprinkler on a nonterminal range pipe more than 300 mm long.

distribution pipe spur: A distribution pipe from a main distribution pipe to a terminal branched pipe array.

diversion wall: An earth or concrete wall that directs spills to a safe disposal area.

drencher: A sprayer used to distribute water over a surface to provide protection against fire exposure.

drop: A vertical pipe feeding a distribution or range pipe.

dry type sprinkler: A sprinkler intended for installation in the upright or pendent position, designed to distribute water so that approximately 40% of the discharge is directed upward and 60% is directed downward. When installed in the upright position, this discharge will cover a 3.05 m diameter circle, 3 m below the sprinkler, when the sprinkler is discharging water at the rate of 0.95 L/s.

DSEAR: Dangerous Substances and Explosive Atmospheres Regulations

dust-ignition-proof: Enclosed in a manner that will exclude dusts and will not permit arcs, sparks, or heat otherwise generated inside of the enclosure to cause ignition of exterior layers or clouds of a specified dust on or in the vicinity of the enclosure.

dust-protected: Enclosure in which the ingress of dust is not totally prevented but dust does not enter in sufficient quantities to interfere with the safe operation of the equipment; dust must not accumulate in a position within the enclosure where it is liable to cause an ignition hazard.

Apparatus enclosed in a case that is capable of withstanding an explosion of a specified gas or vapor that may occur within it and of preventing the ignition of a specified gas or vapor surrounding the enclosure by sparks, flashes, or explosion of the gas or vapor within and that operates at such an external temperature that a surrounding flammable atmosphere will not be ignited thereby.

dust-tight: Constructed so that dust particles will not enter the enclosure.

EC: European Community

EC Declaration of Conformity: Declaration by the manufacturer that the equipment complies with the EHSRs of Directive 94/9/EC and any other relevant directives that apply.

educator (inductor): A device that uses the venture principle to introduce a proportionate quantity of foam concentrate into a water stream. The pressure at the throat is below atmospheric pressure and will draw in liquid from atmospheric storage.

EHSR: Essential Health and Safety Requirements

electrical area classification, extent: Areas are classified into three zones:

1. Zone 0 is an area in which an explosive gas atmosphere is present continuously or is present for a long period

2. Zone 1 is an area in which an explosive gas atmosphere is likely to occur in normal operation

3. Zone 2 is an area in which an explosive atmosphere is not likely to occur in normal operation and if it does occur it will exist for a short period only

end-center array: A pipe array with range pipes on both sides of distribution pipe.

end-side array: A pipe array with range pipes on one side only of a distribution pipe.

engineered system: Those requiring individual calculation and design to determine the flow rates, nozzle pressures pipe size, area, or volume protected by each nozzle, quantities of dry chemicals, and the number and types of nozzles and their placement in a specific system.

equipment: Machines, apparatus, fixed or mobile devices, control components and instrumentation thereof and detection or prevention systems which, separately or jointly, are intended for the generation, transfer, storage, measurement, control and conversion of energy for the processing of material and which are capable of causing an explosion through their own potential sources of ignition.

equipment protection level (EPL): Level of protection assigned to equipment based on its likelihood of becoming a source of ignition and distinguishing the differences between explosive gas atmospheres, explosive dust atmospheres, and the explosive atmospheres in mines susceptible to firedamp.

EU: European Union

Ex component: Part of electrical equipment or a module, marked with the symbol "U", which is not intended to be used alone and requires additional consideration when incorporated into electrical equipment or systems for use in explosive atmospheres.

exhauster: A device to exhaust the air from a dry or alternate installation to atmosphere on sprinkler operation to give more rapid operation of the alarm valve.

ExNB: Ex notified bodies

expansion: The ratio of air to water in foam. A measure of the volume of foam produced for each volume of foam solution used.

expellant gas: The medium used to discharge dry chemical from its container (i.e., CO_2 and nitrogen).

explosion-proofing: Any electrical equipment used in hazardous area including gas detection equipment must be tested and approved to ensure that even under fault condition it cannot initiate an explosion.

explosion-protected apparatus: Any form of apparatus with a recognized type of protection.

explosive gas atmosphere: A mixture with air under normal atmospheric conditions of flammable materials in the form of gas, vapor, or mist, in which, after ignition, combustion spreads throughout the unconsumed mixture.

explosive limit:
- **Lower explosive limit (LEL):** The concentration of flammable gas, vapor, or mist in air, below which an explosive gas atmosphere will not be formed
- **Upper explosive limit (UEL):** The concentration of flammable gas, vapor, or mist in air, above which an explosive gas atmosphere will not be formed

explosive range: The range of gas or vapor mixture with air between the explosive (flammable) limits over which the gas mixture is explosive.

ExTL: Ex testing laboratory

fastener: A device for attaching pipe hanger components to a building structure or racking.

fault signal: An audible, visible, or other indication that instrument is not working satisfactorily.

filling density: The ratio of mass of carbon dioxide charged in a container to the container volume.

fire
1. Process of combustion characterized by the emission of heat accompanied by smoke or flame or both
2. Combustion spreading uncontrolled in time and space

fire and gas detection system (FGDS): The combination of a fire and gas detection system connected to emergency shutdown system and also activating automatic extinguishing systems.

firedamp: A combustible gas formed in coal mines.

fire detection system: Fire detectors and associated control panel to detect and alarm to personnel for evacuation of the plant area and building as well as to indicate the location of the incident to fire brigade to proceed to the scene of the incident (if available).

fire hazard: Any situation process, material, or condition that on the basis of applicable data may cause a fire or explosion or provide a ready fuel supply to augment the spread or intensity of the fire or explosion and that poses a threat to life or property.

fire hazard properties: Properties measured under laboratory conditions may be used as elements of fire risk assessment only when such assessment takes into account all of the factors that are pertinent to the evaluation of the fire hazard of a given situation.

fire hydrant (underground fire hydrant): An assembly contained in a pit or box below ground level and comprising a valve and outlet connection from a water supply main.

fire hydrant pillar: A fire hydrant whose outlet connection is fitted to a vertical component projecting above ground level.

fire point: The lowest temperature at which a liquid in an open container will give off sufficient vapors to burn when once ignited. It generally is slightly above the flash point.

fire-resistive: Fire resistance rating, as the time in minutes or hours, that materials or assemblies have to withstand a fire exposure as established in accordance with the test of NFPA 251.

FISCO: fieldbus intrinsically safe concept

flammable (explosive) limits: All combustible gases and vapors are characterized by flammable limits between which the gas or vapor mixed with air is capable of sustaining the propagation of flame. Lower limit and upper limit are usually expressed as percentage of the material mixed with air by volume.

flammable (explosive) range: The range of flammable vapor or gas air mixture between the upper and lower flammable limits is known as the flammable range or explosive range.

flammable gas or vapor: Gas or vapor which, when mixed with air in certain proportions, will form an explosive gas atmosphere.

flammable liquid: A liquid capable of producing a flammable vapor or mist under any foreseeable operating conditions.

flammable material: Material consisting of flammable gas, vapor, liquid, and/or mist (see also Appendix C).

flammable mist: Droplets of flammable liquid dispersed in air so as to form an explosive atmosphere.

flash point: Minimum temperature of liquid that it gives off sufficient vapor to form an ignitable mixture with the air near the surface of the liquid or within the vessel used.

fluoroprotein: Conventional protein foam modified by the addition of fluorocarbon surfactants.

foam: A mass of bubbles formed by the mechanical agitation of foam and water solution.

foam concentrate: A concentrated liquid foaming agent as received from the manufacturer.

foam inlet: Fixed equipment consisting of an inlet connection, fixed piping, and a discharge assembly, enabling firemen to introduce foam into an enclosed compartment.

foam solution: A homogeneous mixture of water and foam concentrate in the proper proportions.

foam-water spray system: A foam-water spray system is a special system pipe-connected to a source of foam concentrate and to a water supply, and equipped with foam-water spray nozzles for extinguishing-agent discharge (foam or water sequentially in that order or in reverse order) and distribution over the area to be protected. System-operation arrangements parallel those for foam-water sprinkler systems as described in the foregoing paragraph.

foam-water sprinkler system: A foam-water sprinkler system is a special system pipe-connected to a source of foam concentrates and to a water supply, and equipped with appropriate discharge devices for extinguishing agent discharge and for distribution over the area to be protected. The piping system is connected to the water supply through a control valve that is usually actuated by operation of automatic detection equipment installed in the same areas as the sprinklers. When this valve opens, water flows into the piping system, foam concentrate is injected into the water, and the resulting foam solution discharging through the discharge devices generates and distributes foam.

Upon exhaustion of the foam concentrate supply, water discharge will follow the foam and continue until shut off manually. Systems may be used for discharge of water first, followed by discharge of foam for a definite period, and this is followed by water until manually shut off. Existing deluge sprinkler systems that have been converted to the use of aqueous film forming foam are classed as foam-water sprinkler systems.

(fully) hydraulically calculated: A term applied to pipework sized as specified, or an installation in which all the pipework downstream of the main installation control valve set is sized as specified in BS 5306: Part 2.

Ga, Gb, Gc: Equipment protection levels for gas.

gas container system: A system in which the propellant gas is separately contained in a gas container, not in the powder container.

general artificial ventilation: Movement of air and its replacement with fresh air by artificial means (e.g., fans) and applied to a general area.

grades of sources of release: There are three basic grades of sources of release as listed below in order of decreasing likelihood of the release occurring:

1. Continuous grade
2. Primary grade
3. Secondary grade

A source of release may be one of the above three grades or may be a combination of two or three, in which case it is regarded as a multigrade source of release.

gridded configuration pipe array: A pipe array in which water flows to each sprinkler by more than one route.

group i instrument: Portable, transportable, and fixed instrument for sensing the presence of combustible gas concentration with air. The instrument or part thereof may be used or installed in mines susceptible to firedamp.

group ii instrument: Apparatus for use in potentially explosive atmosphere other than mines susceptible to firedamp.

hanger: An assembly for suspending pipework from elements of building structure.

harmonized standards: Standards developed specifically to allow a presumption of conformity with the EHSR of ATEX 95.

hazardous area: An area in which an explosive gas atmosphere is present, or may be expected to be present, in quantities such as to require special precautions for the construction, installation, and use of electrical apparatus.

health hazard: Any property of material that either directly or indirectly can cause injury or incapacitation, either temporary or permanent from exposure by contact, inhalation, or ingestion.

hermetically sealed component: A component that is sealed against entrance of an external atmosphere and in which the seal is made by fusion, such as soldering, brazing, welding, or the fusion of glass to metal.

high-flash stock: Those having a closed-up flash point of 55°C or over (such as heavy fuel oil, lubricating oils, transformer oils, etc.). This category does not include any stock that may be stored at temperatures above or within 8°C of its flash point.

high-pressure storage: Storage of carbon dioxide at ambient temperature.

high-rise system: A sprinkler system in which the highest sprinkler is more than 45 m above the lowest sprinkler or the sprinkler pumps, whichever is lower.

high expansion: Foam having an expansion ratio higher than 200 (generally about 500).

hose reel: Fire-fighting equipment, consisting of a length of tubing fitted with a shutoff nozzle and attached to a reel, with a permanent connection to a pressurized water supply.

hose reel system: A system including a hose, stowed on a reel or a rack, with a discharge nozzle that is manually directed and operated.

hydrant outlet: The component of fire hydrant to which the standpipe is connected.

ignition temperature: Minimum temperature under prescribed test conditions at which the material will ignite and sustain combustion when mixed with air at normal pressure, without initiation of ignition by spark or flame.

increased safety: A type of protection in which additional measures are applied to give increased security against the possibility of excessive temperatures and the occurrence of arcs and sparks inside and on external parts of electrical apparatus that does not produce arcs or sparks in normal service.

infrared sensor: A sensor that the operation of which depends on the absorption of infrared radiation by the gas being detected.

installation: A combination of two or more pieces of equipment that were already placed on the market independently by one or more manufacturers.

3.41 installation, wet (pipe): An installation in which the pipework is always charged with water.

installation alternate: An installation in which the pipework is selectively charged with either water or air according to ambient temperature conditions.

installation, dry (pipe): An installation in which the pipework is charged with air under pressure.

installation, preaction: One of two types of dry, or alternate in dry mode, installation in which the alarm valve can be opened by an independent fire detection system in the protected area.

installation, recycling: A preaction installation in which the alarm valve can be opened and closed repeatedly by a heat-detection system.

intrinsically safe: Any electrical equipment that can be used in zones 0, 1, or 2 being designed as such that even if two faults develop, an explosion will not happen.

intrinsically safe apparatus: Apparatus in which all the circuits are intrinsically safe

intrinsically safe circuit: A circuit in which any spark or thermal effect produced is not capable of causing ignition of a mixture of flammable or combustible material in air under prescribed test conditions.

IP: ingress protection

jockey pump: A small pump used to replenish minor water loss to avoid starting an automatic suction or booster pump unnecessarily.

landing valve: An assembly comprising a valve and outlet connection from a wet or dry riser.

life safety: A term applied to sprinkler systems forming an integral part of measures required for the protection of life.

local application system: An automatic or manual fire extinguishing system in which a fixed supply of carbon dioxide is permanently connected to fixed piping with nozzles arranged to discharge the carbon dioxide directly to a fire occurring in a defined area that has no enclosure surrounding it, or is only partially enclosed and that does not produce an extinguishing concentration throughout the entire volume containing the protected hazard.

local artificial ventilation: Movement of air and its replacement with fresh air by artificial means (usually extraction) applied to a particular source of release or local area.

looped configuration: A pipe array in which there is more than one distribution pipe route along which water may flow to a range pipe.

low expansion: Foam having an expansion ratio up to 20 (generally about 10).

lower explosive limit (LEL): The concentration of combustible (flammable) gas, vapor, or mist in air below that an explosive gas/atmosphere will not be formed.

low-flash stocks: Those having a closed-up flash point under 55°C such as gasoline, kerosene, jet fuels, some heating oils, diesel fuels, and any other stock that may be stored at temperatures above or within 8°C of its flash point.

low-pressure storage: Storage of carbon dioxide in pressure containers at a controlled low temperature of –18°C. Note that the pressure in this type of storage is approximately 21 bar.

low-rise system: A sprinkler system in which the highest sprinkler is not more than 45 m above ground level or the sprinkler pumps.

main distribution pipe: A pipe feeding a distribution pipe

manual: Pertaining to a fire extinguishing system that under specified conditions functions by means of intervention of a human operator.

manual hose reel system: A manual fire extinguishing system consisting of a hose, stowed on a reel or a rack, with a manually operated discharge nozzle assembly, all connected by a fixed pipe to a supply of carbon dioxide.

material conversion factor (MCF): A numerical factor that should be used when the minimum design concentration of carbon dioxide for the material at risk exceeds 34% to increase the basic quantity of carbon dioxide as obtained by application of the volume factor required for protection against surface fires.

mechanical pipe joint: A component part of pipework other than threaded tubulars, screwed fittings, lead or compound sealed spigots, and socket and flanged joints used to connect pipes and to produce a seal both against pressure and vacuum.

medium expansion: Foam having an expansion ratio between 20 and 200 (generally about 100).

melting point: The temperature at which a solid of a pure substance changes to a liquid.

MESG: maximum experimental safety gap

MIC: minimum ignition current

monitor system: A system of fixed piping with nozzles that can be manually directed and operated locally and/or remotely.

multifunction: A detecting instrument that detects 0%–100% LEL, 0%–25% oxygen, 0–25 ppm hydrogen sulfide, and 0–50 ppm carbon monoxide.

Multigrade source of release: A source of release that is a combination of two or three of the above-mentioned grades and

   a. Is basically graded continuous or primary

   b. Gives rise to a release under different conditions that create a larger zone but less frequently and/or for a shorter duration than as determined for the basic grade

Note that different conditions mean, for example, different release rate of flammable material but under the same ventilation conditions.

A source of release that is basically graded continuous may in addition be graded primary if the rate of release of flammable material, for the primary grade frequency and/or duration, exceeds that for the continuous grade.

It may, additionally or alternatively to the primary grade, also be graded secondary if the rate of release of flammable material, for the secondary grade frequency and/or duration, exceeds that for continuous and, if applicable, the primary grade.

Similarly, a source of release which that is basically graded primary may in addition be graded secondary if the rate of release of flammable material for the secondary grade frequency and/or duration exceeds that for the primary grade.

multiple control: A valve, normally held closed by a temperature-sensitive element, suitable for use in a deluge system or for the operation of a pressure switch.

natural ventilation: Movement of air and its replacement with fresh air due to the effects of wind and/or temperature gradients.

NEC: National Electrical Code

no ventilation: No ventilation exists where no arrangements have been made to cause air replacement with fresh air.

node: A point in pipework at which pressure and flow(s) are calculated; each node is a data point for the purpose of hydraulic calculations in the installation.

Noncombustible: Material incapable of igniting or supporting combustion.

nonconductive dust: Combustible dust with electrical resistivity greater than 103 Ωm.

nonhazardous area: An area in which an explosive gas atmosphere is not expected to be present in quantities such as to require special precautions for the construction, installation, and use of electrical apparatus.

nonincendive circuit: A circuit in which any arc or thermal effect produced under intended operating conditions of the equipment is not capable, under the test conditions specified, of igniting the specified flammable gas- or vapor-air mixture.

nonincendive component: A component having contacts for making or breaking an incendive circuit and the contacting mechanism should be constructed so that the component is incapable of igniting the specified flammable gas- or air-air mixture. The housing of a nonincendive component is not intended to exclude the flammable atmosphere or contain an explosion.

nonincendive field circuit: A circuit that enters or leaves the equipment enclosure and that under intended operating conditions is not capable, under the test conditions specified, of igniting the specified flammable gas- or air-air mixture or combustible dust.

nonsparking apparatus: Apparatus that has no normally arcing parts or thermal effects capable of ignition. Normal use excludes the removal or insertion of components with the circuit energized.

normal operation: The situation when the plant equipment is operating within its design parameters.

Minor releases of flammable material may be part of normal operation. For example, releases from seals that rely on wetting by the fluid being pumped are considered to be minor releases.

Failures (such as the breakdown of pump seals, flange gaskets, or spillages caused by accidents) that involve repair or shutdown are not considered to be part of normal operation.

Notified Body: A Notified Body, in the European Union, is an organization that has been accredited by a Member State to assess whether a product meets certain preordained standards. Assessment can include inspection and examination of a product, its design, and manufacture.

occupancy hazard classification number: A series of numbers from 3 through 7 that are mathematical factors to be used in calculating total water supplies for firefighting and fire protection.

NFC Standards has allocated number 3, which is the lowest occupancy hazard number, as the highest hazard grouping and number 7, which is the highest occupancy hazard number, as the lowest hazard grouping.

offshore installation: The term used to describe any offshore unit for the drilling or producing oil or gas.

open-ended pipework: Pipework between a valve (including a relief valve) and open nozzle that cannot be under a continuous pressure.

open-path infrared sensor: A sensor that is capable of detecting gas at any location along an open path traversed by an infrared beam.

ordinary: General masonry walls with wood roof and/or wood floors; also all frame construction.

pipe array: The pipes feeding a group of sprinklers. Note that pipe arrays may be looped, gridded, or branched.

pipe rack: The pipe rack is the elevated supporting structure used to convey piping between equipment. This structure is also utilized for cable trays associated with electric-power distribution and for an instrument tray.

plot plan: The plot plan is the scaled plan drawing of the processing facility.

portable apparatus: Apparatus that is designed to be readily carried by the user from place to place as required.

precalculated: A term applied to pipework sized as specified in 18.1(b) of BS 5306: Part 2 or an installation in which pipes downstream of the design point are sized as specified in 18.1 (b) of BS 5306: Part 2.

premixed foam solution: Produced by introducing a measured amount of foam concentrate into a given amount of water in a storage tank.

pressurization: The process of supplying an enclosure with a protective gas with or without continuous flow at sufficient pressure to prevent the entrance of a flammable gas or vapor, a combustible dust, or an ignitable fiber.

primary grade source of release: A source that can be expected to release periodically or occasionally during normal operation.

proportioning: Is the continuous introduction of foam concentrate at the recommended ratio into the water stream to form foam solution.

protective systems: Design units, that are intended to halt incipient explosions immediately and/or to limit the effective range of explosion flames and explosion pressures. Protective systems may be integrated into equipment or separately placed on the market for use as autonomous systems.

protein: A mixture of hydrolyzed animal protein with various stabilizing materials.

PTB: German Notified Body

purging: The process of supplying an enclosure with a protective gas at a sufficient flow and positive pressure to reduce the concentration of any flammable gas or vapor initially present to an acceptable level.

QAR: IECEx-Scheme Quality Assessment Report

range pipe: A pipe feeding sprinklers directly or via arm pipes of restricted length.

reactivity hazard: Susceptibility of materials to release energy either by themselves or in combination with water.

relative density of a gas or a vapor: The density of a gas or a vapor relative to the density of air at the same pressure and at the same temperature (air is equal to 1.0).

relative vapor density: The mass of a given volume of the material in its gaseous or vapor form compared with the mass of an equal volume of dry air at the same temperature and pressure.

riser: A vertical pipe feeding a distribution or range pipe above.

rising main, dry (dry riser): A vertical pipe installed in a building for firefighting purposes, fitted with inlet connections at fire brigade access level and landing valves at a specified point, which is normally dry but is capable of being charged with water usually by pumping from fire service appliances.

rising main, wet (wet riser): A vertical pipe installed in a building for firefighting purposes and permanently charged with water from a pressurized supply, and fitted with landing valves at specified points.

rosette (sprinkler rosette): A plate covering the gap between the shank or body of a sprinkler projecting through a suspended ceiling and the ceiling.

safety devices, controlling devices and regulating devices: Devices intended for use outside potentially explosive atmospheres but required for or contributing to the safe functioning of equipment and protective systems with respect to the risks of explosion.

sampling probe: A separate sample line that is attached to the instrument as required. It is usually short (1 m) and rigid but may be connected by a flexible tube to the instrument.

sealed device: A device that is constructed so that it cannot be opened, has no external operating mechanisms, and is sealed to restrict entry of an external atmosphere without relying on gaskets. The device may contain arcing parts or internal hot surfaces.

secondary grade source of release: A source that is not expected to release in normal operation and if it releases is likely to do so only infrequently and for short periods.

section: That part (which may be one or more zones) of an installation on a particular floor fed by a particular riser.

semiconductor sensor: A sensor, the operation of which depends on changes of electrical conductance of a semiconductor due to chemical absorption of the gas being detected at its surface.

sensing element: That part of a sensor that reacts in the presence of a flammable gas mixture to produce some physical change that can be used to activate a measuring or alarm function or both.

service spaces: Those are used for galleys, pantries containing cooking appliances, lockers, and store rooms, workshops other than those forming part of the machinery spaces and similar spaces and trunks to such spaces.

sheathed incombustible or incombustible: Wood frame, incombustible sheathing.

short-term detector tubes: Tubes and associating aspirating pumps used for evaluating atmospheric contaminants at concentration in the range occupational exposure limit (OEL). It covers color tubes that are designed to give indication of concentration over a short period of time.

shutoff nozzle: A device that is coupled to the outlet end of hose reel tubing and by means of which the jet of water or spray is controlled.

simple apparatus: An electrical component or combination of components of simple construction with well defined electrical parameters that is compatible with the intrinsic safety of the circuit in which it is used.

sleepers: The sleepers comprise the grade-level supporting structure for piping between equipment for facilities (e.g., tank farm or other remote areas).

sling rod: A rod with a sling eye or screwed ends for supporting pipe clips, rings, band hangers, and so forth.

source of release: A point or location from which a gas, vapor, mist, or liquid may be released into the atmosphere so that an explosive gas atmosphere could be formed.

specific gravity: The ratio of the weight of the substance to the weight of the same volume of water or air, whichever is applicable.

spot-reading apparatus: Apparatus that is intended to be used for a short period of time as required.

sprayer: A sprinkler that gives a downward conical pattern discharge.

sprayer, high velocity: An open nozzle used to extinguish fires of high flash-point liquids.

sprayer, medium velocity: A sprayer of a sealed or open type used to control fires of lower flash-point liquids and gases or to cool surfaces.

sprinkler, automatic: A temperature-sensitive sealing device that opens to discharge water for fire extinguishing. Note that the term "automatic sprinkler" is now rarely used. The term "sprinkler" does not include "open sprinkler."

sprinkler, ceiling of flush pattern: A pendent sprinkler for fitting partly above but with the temperature-sensitive element below the lower plane of the ceiling.

sprinkler, concealed: A recessed sprinkler with a cover plate that disengages when heat is applied.

sprinkler, conventional pattern: A sprinkler that gives a spherical pattern of water discharge. (See also: cutoff sprinkler, detector sprinkler.)

sprinkler, dry pendent pattern: A unit comprising a sprinkler and a dry drop pipe unit with a valve at the head of the pipe held closed by a device maintained in position by the sprinkler head valve.

sprinkler, dry upright pattern: A unit comprising a sprinkler and dry rise pipe unit with a valve at the base of the pipe held closed by a device maintained in position by the sprinkler head valve.

sprinkler, foam water type: An air-aspirating open type sprinkler constructed to discharge water or foam water solutions.

sprinkler, fusible link: A sprinkler that opens when a component provided for the purpose melts.

sprinkler, glass bulb: A sprinkler that opens when a liquid-filled glass bulb bursts.

sprinkler, horizontal: A sprinkler in which the nozzle directs water horizontally.

sprinkler installation: Part of a sprinkler system comprising a set of installation main control valves, the associated downstream pipes, and sprinklers.

sprinkler, intermediate: A sprinkler installed below and additional to the roof or ceiling sprinklers.

sprinkler, open: A device, otherwise like a sprinkler (automatic sprinkler), not sealed by a temperature-sensitive element.

sprinkler, pendent: A sprinkler in which the nozzle directs water downward.

sprinkler, recessed: A sprinkler in which all or part of the heat-sensing element is above the plane of the ceiling.

sprinkler, roof or ceiling: A sprinkler protecting the roof or ceiling.

sprinkler, sidewall pattern: A sprinkler that gives an outward half-paraboloid discharge.

sprinkler, spray pattern: A sprinkler that gives a downward paraboloid pattern discharge.

sprinkler system: The entire means of providing sprinkler protection in the premises comprising one or more sprinkler installations, the pipework to the installations, and the water supply/supplies except town mains and bodies of water such as lakes or canals.

sprinkler, upright: A sprinkler in which the nozzle directs water upward.

sprinkler yoke (arms): The part of a sprinkler that retains the heat-sensitive element in load-bearing contact with the sprinkler head valve.

staggered (sprinkler) layout: An offset layout with the sprinklers displaced one-half pitch along the range pipe relative to the next range or ranges.

standard (sprinkler) layout: A rectilinear layout with the sprinklers aligned perpendicular to the run of the ranges.

stored pressure system: A system in which the propellant gas is stored within and permanently pressurizes the powder container(s).

subsurface foam injection: Discharge of foam into a storage tank below the liquid surface near the tank bottom.

suction pump: An automatic pump supplying water to a sprinkler system from a suction tank, river, lake, or canal.

suitable for sprinkler use: A term applied to equipment or components accepted by the authorities as for a particular application in a sprinkler system, either by particular test or by compliance with specified general criteria.

supply pipe: A pipe connecting a water supply to a trunk main or the installation main control valve set(s), or a pipe supplying water to a private reservoir, suction tank, or gravity tank.

surface fire: A fire involving flammable liquids, gases, or solids not subject to smoldering.

surfactant: Also known as syndet or detergent foam.

suspended open cell ceiling: A ceiling of regular open cell construction through which water from sprinklers can be discharged freely.

tail-end alternate (wet and dry pipe) extension: A part of a wet installation that is selectively charged with water or air according to ambient temperature conditions and which is controlled by a subsidiary dry or alternate alarm valve.

tail-end dry extension: A part of a wet or alternate installation that is charged permanently with air under pressure.

tank diameter: Where tank spacing is expressed in terms of tank diameter, the following criteria governs:
   a. If tanks are in different services, or different types of tanks are used, the diameter of the tank that requires the greater spacing is used
   b. If tanks are in similar services, the diameter of the largest tank is used

tank spacing: The unobstructed distance between tank shells or between tank shells and the nearest edge of adjacent equipment, property lines, or buildings.

terminal main configuration: A pipe array with only one water supply route to each range pipe.

terminal range configuration: A pipe array with only one water supply route from a distribution pipe.

thermal conductivity sensor: A sensor that the operation of which depends on the changes of heat loss by conduction of an electrically heated element located in the gas to be measured compared with that of a similar element located in a reference gas cell.

thermal semiconductivity sensor: A sensor that the operation of which depends on the condition of gases on an electrically heated catalytic element.

toe wall: A low earth, concrete, or masonry unit curb without capacity requirements for the retention of small leaks or spills.

toggle support: A swivel device for securing hangers to hollow section ceilings or roofs.

total flooding system: An automatic or manual fire extinguishing system in which a fixed supply of carbon dioxide is permanently connected to fixed piping with nozzles arranged to discharge the carbon dioxide into an enclosed space in order to produce a concentration sufficient to extinguish fire throughout the entire volume of the enclosed space.

transportable apparatus: Apparatus that is not intended to be portable, but which can be readily moved from one place to another.

trunk main: A pipe connecting two or more water supply pipes to the installation main control valve set(s).

"U" symbol: The symbol used as a suffix to a certificate reference to denote special conditions for safe use.

UEL: upper explosive limit

unbalanced system: A dry chemical powder fire extinguishing system with more than one discharge nozzle in which the powder flow divides unequally at one or more junctions in the pipework.

upper explosive limit (UEL): The concentration of combustible (flammable) gas, vapor, or mist in air above which an explosive gas/atmosphere will not be formed.

user: The person responsible for or having effective control over the fire safety provision adopted in or appropriate to the premises or the building.

vessel diameter: Where vessel spacing is expressed in terms of vessel diameter, the diameter of the largest vessel is used. For spheroids, the diameter at the maximum equator is used.

vessel spacing: The unobstructed distance between vessel shells or between vessel shells and nearest edge of adjacent equipment, property lines, or buildings.

volume factor: A numerical factor that, when applied to the volume of an enclosure, indicates the basic quantity of carbon dioxide (subject to a minimum appropriate to the volume of the enclosure) required for protection against surface fires.

water solubility: The extent to which a substance mixes with pure water to form a molecular homogeneous system at a given temperature.

wellhead: An assembly on top of the well casing strings with outlets and valves for controlling flow of production.

wet water: Water to which a compatible wetting agent has been added.

wet water foam: An admixture of wet water with air to form a cellular structure foam that breaks down rapidly into its original liquid state at temperatures below the boiling point of water at a rate directly related to the heat to which it is exposed in order to cool the combustible on which it is applied.

wetting agent: A chemical compound that when added to water in proper quantities, materially reduces its surface tension, increases its penetrating and spreading abilities, and can also provide emulsification and foaming characteristics.

"X" symbol: The symbol used as a suffix to a certificate reference to denote special conditions for safe use.

zones: Hazardous areas are classified in zones based on the frequency of the appearance and the duration of an explosive gas atmosphere as follows:

  zone 0: An area in which an explosive gas atmosphere is present continuously or is present for long periods

  zone 1: An area in which an explosive gas atmosphere is likely to occur in normal operation

  zone 2: An area in which an explosive gas atmosphere is not likely to occur in normal operation and if does occur it will exist for a short period only

Appendix A

Comparing IEC, NEC, and CEC Zone Standards with NEC/CEC Class/Division Standards

TABLE A.1

Class I Area Classification Comparison[a]

Zone 0	Zone 1	Zone 2
Where ignitable concentrations of flammable gases, vapors, or liquids are present continuously or for long periods of time under normal operating conditions	Where ignitable concentrations of flammable gases, vapors, or liquids: • Are likely to exist under normal operating conditions • May exist frequently because of repair, maintenance operations, or leakage	Where ignitable concentrations of flammable gases, vapors, or liquids: • Are not likely to exist under normal operating conditions • Occur for only a short period of time • Become hazardous only in case of an accident or some unusual operating condition
	Division 1	**Division 2**
	Where ignitable concentrations of flammable gases, vapours, or liquids: • Are likely to exist under normal operating conditions • Exist frequently because of maintenance/repair work or frequent equipment failure	Where ignitable concentrations of flammable gases, vapours, or liquids: • Are not likely to exist under normal operating conditions • Are normally in closed containers where the hazard can only escape through accidental rupture or breakdown of such containers or in case of abnormal operation of equipment

[a] Note that per NEC Article 505-10(b)(1), a division classified product may be installed in a zone-classified location but the reverse is not true. Typically, a zone-classified product provides protection utilizing a protection method not available in the class/division scheme.

TABLE A2

Typically Zone-Classified Products

Zone	Class/Division
IIC, acetylene and hydrogen	A/acetylene
	B/hydrogen
IIB, ethylene	C/ethylene
IIA, propane	D/propane

TABLE A3

Class 1 Protection Method Comparison

Zone 0	Zone 1	Zone 2
• Intrinsically safe (2 fault) • Intrinsically safe, ia (2 fault) class I, division 1 (U.S. only)	• Encapsulation, m • Flame-proof, d • Increased safety, e • Intrinsically safe, ib (1 fault) • Oil Immersion, o • Powder-filled, q • Purged/Pressurized, p • Any class I, zone 0 method • Any class I, division I method (U.S. only)	• Energy limited, nC • Hermetically sealed, nC • Nonincendive, nC • Nonsparking, nA • Restricted breathing, nR • Sealed device, nC • Any class I, zone 0 or 1 method • Any class I, division 1 or 2 method (U.S. only)

Division 1	Division 2
• Explosion-proof • Intrinsically safe (2 fault) • Purged/pressurized (type X or Y)	• Hermetically sealed • Nonincendive • Nonsparking • Oil immersion • Sealed device • Purged/pressurized (type Z) • Any class I, zone 1 or 2 method (U.S. only) • Any class I, division 1 method

TABLE A4

Class 1 Temperature Class Comparison

Zone 0, 1, and 2	Division 1 and 2	Maximum Temperature
T1	T1	450°C (842°F)
T2	T2	300°C (572°F)
	T2A	280°C (536°F)
	T2B	260°C (500°F)
	T2C	230°C (446°F)
	T2D	215°C (419°F)
T3	T3	200°C (392°F)
	T3A	180°C (356°F)
	T3B	165°C (329°F)
	T3C	160°C (320°F)
T4	T4	135°C (275°F)
	T4A	120°C (248°F)
T5	T5	100°C (212°F)
T6	T6	85°C (185°F)

Appendix B: Inspection Schedules for Gas Atmospheres

TABLE B1

Inspection Schedule for Ex d, Ex e, and Ex n Installations

	Ex d			Ex e			Ex n		
	Grade of Inspection								
Check	D	C	V	D	C	V	D	C	V
A: Equipment									
Equipment is appropriate to area classification (EPL or zone)	✓	✓	✓	✓	✓	✓	✓	✓	✓
Equipment group is correct	✓	✓		✓	✓		✓	✓	
Equipment temperature class is correct	✓	✓		✓	✓		✓	✓	
Equipment circuit identification is correct	✓			✓			✓		
Equipment circuit identification is available	✓	✓	✓	✓	✓	✓	✓	✓	✓
Enclosure, glasses, and glass-to-metal sealing gaskets and/or compounds are satisfactory	✓	✓	✓	✓	✓	✓	✓	✓	✓
There are no unauthorized modifications	✓			✓			✓		
There are no visible unauthorized modifications		✓	✓		✓	✓		✓	✓
Bolts, cable entry devices (direct and indirect), and blanking elements are of the correct type and are complete and tight									
• Physical check	✓	✓		✓	✓		✓	✓	
• Visual check			✓			✓			✓
Flange faces are clean and undamaged and gaskets, if any, are satisfactory	✓								
Flange gap dimensions are within maximal values permitted	✓	✓							
Lamp rating, type, and position are correct	✓			✓			✓		
Electrical connections are tight				✓			✓		
Condition of enclosure gaskets is satisfactory				✓			✓		
Enclosed-break and hermetically sealed devices are undamaged							✓		
Restricted breathing enclosure is satisfactory							✓		
Motor fans have sufficient clearance to enclosure and/or covers	✓			✓			✓		
Breathing and draining devices are satisfactory	✓	✓		✓	✓		✓	✓	

(continued)

TABLE B1 (Continued)

Inspection Schedule for Ex d, Ex e, and Ex n Installations

	Ex d			Ex e			Ex n		
	Grade of Inspection								
Check	D	C	V	D	C	V	D	C	V
B: Installation									
Type of cable is appropriate	✓			✓			✓		
There is no obvious damage to cables	✓	✓		✓	✓		✓	✓	✓
Sealing of trunking, ducts, pipes, and/or conduits is satisfactory	✓	✓		✓	✓		✓	✓	✓
Stopper boxes and cable boxes are correctly fitted	✓								
Integrity of conduit system and interface with mixed system is maintained	✓			✓			✓		
Earthing connections, including any supplementary earthing bonding connections are satisfactory (e.g., connections are tight and conductors are of sufficient cross section)	✓			✓			✓		
• Physical check		✓	✓		✓	✓		✓	✓
• Visual check									
Fault loop impedance (TN system) or earthing resistance (IT system) is satisfactory	✓			✓			✓		✓
Insulation resistance is satisfactory	✓			✓			✓		✓
Automatic electrical protective devices operate within permitted limits	✓			✓			✓		✓
Automatic electrical protective devices are set correctly (auto-reset not possible)	✓			✓			✓		✓
Special conditions of use (if applicable) are complied with	✓			✓			✓		✓
Cables not in use are correctly terminated	✓			✓			✓		✓
Obstructions adjacent to flameproof flanged joints are in accordance with EN 60079-14	✓	✓	✓						
Variable voltage/frequency installation in accordance with documentation	✓	✓		✓	✓		✓	✓	
C: Environment									
Apparatus is adequately protected against corrosion, weather, vibration, and other adverse factors	✓	✓	✓	✓	✓	✓	✓	✓	✓
No undue accumulation of dust and dirt	✓	✓	✓	✓	✓	✓	✓	✓	✓
Electrical insulation is clean and dry				✓			✓		

Note: D = detailed; C = close; V = visual.

TABLE B2

Inspection Schedule for Ex i Installations

Check	Grade of Inspection		
	D	C	V
A: Equipment			
Circuit and/or apparatus documentation is appropriate to area classification (EPL and zone)	✓	✓	✓
Equipment installed is that specified in the documentation • Fixed apparatus only	✓	✓	
Circuit and/or apparatus category and group correct	✓	✓	
Equipment temperature class is correct	✓	✓	
Installation is clearly labeled	✓	✓	
Enclosure, glasses, and glass-to-metal sealing gaskets and/or compounds are satisfactory	✓		
There are no unauthorized modifications	✓		
There are no visible unauthorized modifications		✓	✓
Safety barrier units, relays, and other energy-limiting devices are of the approved type, installed in accordance with the certification requirements, and securely earthed where required	✓	✓	✓
Electrical connections are tight	✓		
Printed circuit boards are clean and undamaged	✓		
B: Installation			
Cables are installed in accordance with the documentation	✓		
Cable screens are earthed in accordance with the documentation	✓		
There is no obvious damage to cables	✓	✓	✓
Sealing of trunking, ducts, pipes, and/or conduits is satisfactory	✓	✓	✓
Point-to-point connections are all correct	✓		
Earth continuity is satisfactory for nongalvanic isolated circuits (e.g., connections are tight and conductors are of sufficient cross section)	✓		
Earth connections maintain the integrity of the type of protection	✓	✓	✓
The intrinsically safe circuit is isolated from earth and the earthing is sufficient	✓		
Separation is maintained between intrinsically safe and nonintrinsically safe circuits in common distribution boxes or relay cubicles	✓		
As applicable, short-circuit protection of the power supply is in accordance with the documentation	✓		
Special conditions of use (if applicable) are complied with	✓		
Cables not in use are correctly terminated	✓	✓	✓
C: Environment			
Equipment is adequately protected against corrosion, weather, vibration, and other adverse factors	✓	✓	✓
No undue accumulation of dust and dirt	✓	✓	✓

Note: D = detailed; C = close; V = visual.

TABLE B3

Inspection Schedule for Ex p or pD Installations

Check	D	C	V
	Grade of Inspection		
A: Equipment			
Equipment is appropriate to area classification (EPL and zone)	✓	✓	✓
Equipment group is correct	✓	✓	
Equipment temperature class is correct	✓	*	
Equipment circuit identification is correct	✓		
Equipment circuit identification is available	✓	✓	✓
Enclosure, glasses, and glass-to-metal sealing gaskets and/or compounds are satisfactory	✓	✓	✓
There are no unauthorized modifications	✓		
There are no visible unauthorized modifications		✓	✓
Lamp rating, type, and position are correct	✓		
B: Installation			
Type of cable is appropriate	✓		
There is no obvious damage to cables	✓	✓	✓
Earthing connections, including any supplementary earthing bonding connections, are satisfactory (e.g., connections are tight and conductors are of sufficient cross section) • Physical check • Visual check	✓	✓	✓
Fault loop impedance (TN system) or earthing resistance (IT system) is satisfactory	✓		
Automatic electrical protective devices operate within permitted limits	✓		
Automatic electrical protective devices are set correctly	✓		
Protective gas inlet temperature is below maximum specified	✓		
Ducts, pipes and enclosures are in good condition	✓	✓	✓
Protective gas is substantially free from contaminants	✓	✓	✓
Protective gas pressure and/or flow is adequate	✓	✓	✓
Pressure and/or flow indicators, alarms, and interlocks function correctly	✓		
Conditions of spark and particle barriers of ducts for exhausting the gas in hazardous area are satisfactory	✓		
Special conditions of use (if applicable) are complied with	✓		
C: Environment			
Equipment is adequately protected against corrosion, weather, vibration, and other adverse factors	✓	✓	✓
No undue accumulation of dust and dirt	✓	✓	✓

Note: D = detailed; C = close; V = visual.

TABLE B4

Inspection Schedule for Ex tD Installations

Check	Grade of Inspection		
	D	C	V
A: Equipment			
Equipment is appropriate to area classification (EPL and zone)	✓	✓	✓
IP rating of the equipment is correct for conductive dust	✓	✓	✓
Maximum surface temperature is correct	✓	✓	
Equipment circuit identification is correct	✓		
Equipment circuit identification is available	✓	✓	✓
Enclosure, glasses, and glass-to-metal sealing gaskets and/or compounds are satisfactory	✓	✓	✓
There are no unauthorized modifications	✓		
There are no visible unauthorized modifications		✓	✓
Bolts, cable entry devices (direct and indirect), and blanking elements are of the correct type and are complete and tight			
• Physical check	✓	✓	
• Visual check			✓
Lamp rating, type, and position are correct	✓		
Electrical connections are tight	✓		
Condition of enclosure gaskets is satisfactory	✓		
Motor fans have sufficient clearance to enclosure and/or covers	✓		
B: Installation			
Accumulation of dust and dirt is avoided	✓	✓	✓
Type of cable is appropriate	✓		
There is no obvious damage to cables	✓	✓	✓
Sealing of trunking, ducts, pipes, and/or conduits is satisfactory	✓	✓	✓
Earthing connections, including any supplementary earthing bonding connections are satisfactory (e.g., connections are tight and conductors are of sufficient cross section)			
• Physical check	✓		
• Visual check		✓	✓
Fault loop impedance (TN system) or earthing resistance (IT system) is satisfactory	✓		
Insulation resistance is satisfactory	✓		
Automatic electrical protective devices operate within permitted limits	✓		
Special conditions of use (if applicable) are complied with	✓		
Cables not in use are correctly terminated	✓	✓	
C: Environment			
Equipment is adequately protected against corrosion, weather, vibration, and other adverse factors	✓	✓	✓
No undue accumulation of dust and dirt	✓	✓	✓

Note: D = detailed; C = close; V = visual.

Appendix C: Inspection Schedules for Dust Atmospheres

TABLE C1

Inspection Schedule for Ex i, iD, and nL Installations

Check	Grade of Inspection		
	D	C	V
A: Equipment			
Circuit and/or equipment documentation is appropriate to the EPL/ zone requirements of the location	✓	✓	✓
Equipment installed is that specified in the documentation (fixed equipment only)	✓	✓	
Circuit and/or equipment category and group is correct	✓	✓	
Equipment temperature class is correct	✓	✓	
Installation is clearly labeled	✓	✓	
Enclosure, glass parts, and glass-to-metal sealing gaskets and/or compounds are satisfactory	✓		
There are no unauthorized modifications	✓		
There are no visible unauthorized modifications		✓	✓
Safety barrier units, relays, and other energy-limiting devices are of the approved type, installed in accordance with the certification	✓	✓	✓
Electrical connections are tight	✓		
Printed circuit boards are clean and undamaged	✓		
B: Installation			
Cables are installed in accordance with the documentation	✓		
Cable screens are earthed in accordance with the documentation	✓		
There is no obvious damage to cables	✓	✓	✓
Sealing of trunking, ducts, pipes, and/or conduits is satisfactory	✓	✓	✓
Point-to-point connections are all correct	✓		
Earth continuity is satisfactory (e.g., connections are tight, conductors are of sufficient cross section) for nongalvanically isolated circuits	✓		
Earth connections maintain the integrity of the type of protection	✓	✓	✓
Intrinsically safe circuit earthing and insulation resistance is satisfactory	✓		
Separation is maintained between intrinsically safe and nonintrinsically safe circuits in common distribution boxes or relay cubicles	✓		
As applicable, short-circuit protection of the power supply is in accordance with the documentation	✓		

(continued)

TABLE C1 (Continued)

Inspection Schedule for Ex i, iD, and nL Installations

Check	Grade of Inspection		
	D	**C**	**V**
Specific conditions of use (if applicable) are complied with	✓		
Cables not in use are correctly terminated	✓		
C: Environment			
Equipment is adequately protected against corrosion, weather, vibration, and other adverse factors	✓	✓	✓
No undue external accumulation of dust and dirt	✓	✓	✓

Note: D = detailed; C = close; V = visual.

TABLE C2

Inspection Schedule for Ex p and pD Installations

Check	Grade of Inspection		
	D	**C**	**V**
A: Equipment			
Equipment is appropriate to the EPL/zone requirements of the location	✓	✓	✓
Equipment group is correct	✓	✓	
Equipment temperature class or surface temperature is correct	✓	✓	
Equipment circuit identification is correct	✓		
Equipment circuit identification is available	✓	✓	✓
Enclosure, glasses, and glass-to-metal sealing gaskets and/or compounds are satisfactory	✓	✓	✓
There are no unauthorized modifications	✓		
There are no visible unauthorized modifications		✓	✓
Lamp rating, type, and position are correct	✓		
B: Installation			
Type of cable is appropriate	✓		
There is no obvious damage to cables	✓	✓	✓
Earthing connections, including any supplementary earthing bonding connections, are satisfactory (e.g., connections are tight and conductors are of sufficient cross section)			
• Physical check	✓		
• Visual check		✓	✓
Fault loop impedance (TN systems) or earthing resistance (IT systems) is satisfactory	✓		
Automatic electrical protective devices operate within permitted limits	✓		
Automatic electrical protective devices are set correctly	✓		

(continued)

TABLE C2 (Continued)

Inspection Schedule for Ex p and pD Installations

Check	Grade of Inspection		
	D	C	V
Protective gas inlet temperature is below maximum specified	✓		
Ducts, pipes, and enclosures are in good condition	✓	✓	✓
Protective gas is substantially free from contaminants	✓	✓	✓
Protective gas pressure and/or flow is adequate	✓	✓	✓
Pressure and/or flow indicators, alarms, and interlocks function correctly	✓		
Conditions of spark and particle barriers of ducts for exhausting the gas in hazardous area are satisfactory	✓		
Specific conditions of use (if applicable) are complied with	✓		
C: Environment			
Equipment is adequately protected against corrosion, weather, vibration, and other adverse factors	✓	✓	✓
No undue accumulation of dust and dirt	✓	✓	✓

Note: D = detailed; C = close; V = visual.

TABLE C3

Inspection Schedule for Ex tD Installations

Check	Grade of Inspection		
	D	C	V
A: Equipment			
Equipment is appropriate to the EPL/zone requirements of the location	✓	✓	✓
IP grade of equipment is appropriate to conductivity of dust	✓	✓	✓
Equipment maximum surface temperature is correct	✓	✓	
Equipment circuit identification is available	✓	✓	✓
Equipment circuit identification is correct	✓		
Enclosure, glasses, and glass-to-metal sealing gaskets and/or compounds are satisfactory	✓	✓	✓
There are no unauthorized modifications	✓		
There are no visible unauthorized modifications		✓	✓
Bolts, cable entry devices, and blanking elements are of the correct type and are complete and tight			
• Physical check	✓	✓	
• Visual check			✓
Lamp rating, type, and position are correct	✓		
Electrical connections are tight	✓		
Condition of enclosure gaskets is satisfactory	✓		
Motor fans have sufficient clearance to enclosure and/or covers	✓		

(*continued*)

TABLE C3(Continued)

Inspection Schedule for Ex tD Installations

Check	Grade of Inspection		
	D	**C**	**V**
B: Installation			
The installation is such as to minimize the risk of dust accumulations	✓	✓	✓
Type of cable is appropriate	✓		
There is no obvious damage to cables	✓	✓	✓
Sealing of trunking, ducts, pipes, and/or conduits is satisfactory	✓	✓	✓
Earthing connections, including any supplementary earthing bonding connections are satisfactory			
• Physical check	✓		
• Visual check		✓	✓
Fault loop impedance (TN systems) or earthing resistance (IT systems) is satisfactory	✓		
Insulation resistance is satisfactory	✓		
Automatic electrical protective devices operate within permitted limited	✓		
Specific conditions of use (if applicable) are compiled with	✓		
Cables not in use are correctly terminated	✓	✓	
C: Environment			
Equipment is adequately protected against corrosion, weather, vibration, and other adverse conditions	✓	✓	✓
No undue accumulation of dust and dirt	✓	✓	✓

Note: D = detailed; C = close; V = visual.

Index

Page numbers followed by f and t indicate figures and tables, respectively.